This timely volume provides the first comprehensive and coherent introduction to modern quantum cosmology – the study of the universe as a whole according to the laws of quantum mechanics. In particular, it presents a useful survey of the many profound consequences of supersymmetry (supergravity) in quantum cosmology.

After a general introduction to quantum cosmology, the reader is led through Hamiltonian supergravity and canonical quantization and quantum amplitudes through to models of supersymmetric mini-superspace and quantum wormholes. The book is rounded off with a look at exciting further developments, including the possible finiteness of supergravity.

Ample introductory material is included, ensuring this topical volume is well suited as a graduate text. Researchers in theoretical and mathematical physics, applied maths and cosmology will also find it of immediate interest.

CAMBRIDGE MONOGRAPHS ON MATHEMATICAL PHYSICS

SUPERSYMMETRIC QUANTUM COSMOLOGY

CAMBRIDGE MONOGRAPHS ON MATHEMATICAL PHYSICS

A. M. Anile *Relativistic Fluids and Magneto-Fluids*
J. A. de Azcárraga and J. M. Izquierdo *Lie Groups, Lie Algebras, Cohomology and Some Applications in Physics*
J. Bernstein *Kinetic Theory in the Early Universe*
G. F. Bertsch and R. A. Broglia *Oscillations in Finite Quantum Systems*
N. D. Birrell and P. C. W. Davies *Quantum Fields in Curved Space*†
D. M. Brink *Semiclassical Methods in Nucleus–Nucleus Scattering*
J. C. Collins *Renormalization*†
P. D. B. Collins *An Introduction to Regge Theory and High Energy Physics*
M. Creutz *Quarks, Gluons and Lattices*†
F. de Felice and C. J. S. Clarke *Relativity on Curved Manifolds*†
B. DeWitt *Supermanifolds, 2nd edition*†
P. D. D'Eath *Supersymmetric Quantum Cosmology*
P. G. O. Freund *Introduction to Supersymmetry*†
F. G. Friedlander *The Wave Equation on a Curved Space-Time*
J. Fuchs *Affine Lie Algebras and Quantum Groups*†
J. A. H. Futterman, F. A. Handler and R. A. Matzner *Scattering from Black Holes*
M. Göckeler and T. Schücker *Differential Geometry, Gauge Theories and Gravity*†
C. Gómez, M. Ruiz Altaba and G. Sierra *Quantum Groups in Two-dimensional Physics*
M. B. Green, J. H. Schwarz and E. Witten *Superstring Theory, volume 1: Introduction*†
M. B. Green, J. H. Schwarz and E. Witten *Superstring Theory, volume 2: Loop Amplitudes, Anomalies and Phenomenology*†
S. W. Hawking and G. F. R. Ellis *The Large-Scale Structure of Space-Time*†
F. Iachello and A. Arima *The Interacting Boson Model*
F. Iachello and P. van Isacker *The Interacting Boson–Fermion Model*
C. Itzykson and J.-M. Drouffe *Statistical Field Theory, volume 1: From Brownian Motion to Renormalization and Lattice Gauge Theory*†
C. Itzykson and J.-M. Drouffe *Statistical Field Theory, volume 2: Strong Coupling, Monte Carlo Methods, Conformal Field Theory, and Random Systems*†
J. I. Kapusta *Finite-Temperature Field Theory*†
V. E. Korepin, A. G. Izergin and N. M. Boguliubov *The Quantum Inverse Scattering Method and Correlation Functions*
D. Kramer, H. Stephani, M. A. H. MacCallum and E. Herlt *Exact solutions of Einstein's Field Equations*
N. H. March *Liquid Metals: Concepts and Theory*
I. M. Montvay and G. Münster *Quantum Fields on a Lattice*
L. O'Raifeartaigh *Group Structure of Gauge Theories*†
A. Ozorio de Almeida *Hamiltonian Systems: Chaos and Quantization*†
R. Penrose and W. Rindler *Spinors and Space-time, volume 1: Two-Spinor Calculus and Relativistic Fields*†
R. Penrose and W. Rindler *Spinors and Space-time, volume 2: Spinor and Twistor Methods in Space-Time Geometry*†
S. Pokorski *Gauge Field Theories*†
V. N. Popov *Functional Integrals and Collective Excitations*†
R. Rivers *Path Integral Methods in Quantum Field Theory*†
R. G. Roberts *The Structure of the Proton*†
J. M. Stewart *Advanced General Relativity*
A. Vilenkin and E. P. S. Shellard *Cosmic Strings and Other Topological Defects*
R. S. Ward and R. O. Wells Jr *Twistor Geometry and Field Theories*†

† Issued as a paperback

SUPERSYMMETRIC QUANTUM COSMOLOGY

P. D. D'EATH

Department of Applied Mathematics and Theoretical Physics
University of Cambridge

Published by the Press Syndicate of the University of Cambridge
The Pitt Building, Trumpington Street, Cambridge CB2 1RP
40 West 20th Street, New York, NY 10011-4211, USA
10 Stamford Road, Oakleigh, Melbourne 3166, Australia

First published 1996

Printed in Great Britain at the University Press, Cambridge

A catalogue record for this book is available from the British Library

Library of Congress cataloguing in publication data available

ISBN 0 521 55287 7 hardback

TAG

For Stephen
and
for Kathy

Contents

	Preface	xi
	Acknowledgements	xiv
1	**Introduction**	1
2	**Quantum cosmology**	9
2.1	Introduction	9
2.2	Parametrized particle dynamics	14
2.3	Hamiltonian treatment of general relativity	18
2.4	Classical boundary-value problem	28
2.5	Feynman path integral in non-relativistic quantum mechanics	31
2.6	Path integral in quantum gravity	38
2.7	Mini-superspace models	45
2.8	Bosonic structure in the universe	58
2.9	Spin-1/2 fermions in quantum cosmology	62
	2.9.1 Two-component spinors	63
	2.9.2 Hamiltonian treatment of the Einstein–Dirac theory	65
	2.9.3 Spectral boundary conditions	72
	2.9.4 Quantum state	79
3	**Hamiltonian supergravity and canonical quantization**	86
3.1	Introduction	86
3.2	Hamiltonian formulation of supergravity	89
3.3	Quantum representation	97
3.4	The quantum constraints	103
3.5	The path integral	107
4	**The quantum amplitude**	112
4.1	Semi-classical expansion of the quantum amplitude	112
4.2	Two-loop finiteness of supergravity with boundaries	121
4.3	Auxiliary fields	131
4.4	Time evolution of the quantum amplitude	133

5 Supersymmetric mini-superspace models 141
5.1 Introduction 141
5.2 Reduction of four-dimensional $N = 1$ supergravity to one dimension 144
5.3 Reduced action for $N = 1$ supergravity model 149
5.4 Quantization of supergravity model 152
5.5 Locally supersymmetric model with scalar spin-1/2 matter 156
5.6 Quantization of supersymmetric model with scalar spin-1/2 matter 159
5.7 Supersymmetric Bianchi models 162
5.8 Bianchi-IX model 165
5.9 Supersymmetric $k = +1$ Friedmann model with Λ-term 170

6 Supersymmetric quantum wormhole states 172
6.1 Wormholes 172
6.2 Quantum wormhole states 177
6.3 Quantization of supergravity–supermatter model 181
6.4 Quantum wormhole states with supersymmetry 183

7 Ashtekar variables 190
7.1 General relativity 190
7.1.1 Necessary results 190
7.1.2 Lagrangian form 192
7.1.3 Hamiltonian formulation 193
7.1.4 Reality conditions 195
7.1.5 Algebra of constraints 196
7.1.6 Quantization 197
7.2 Supergravity 197
7.2.1 Lagrangian form 198
7.2.2 Hamiltonian formulation 198
7.2.3 Reality conditions 200
7.2.4 Algebra of constraints 202
7.2.5 Quantization 202
7.2.6 Cosmological constant 203
7.3 Chern–Simons state 204

8 Further developments 210
8.1 Local boundary conditions and possible finiteness 210
8.2 Spectral boundary conditions and possible finiteness 214
8.3 Cosmology 220
8.4 Supergravity with supermatter 223

9 Conclusion 232

References 241

Index 251

Preface

Quantum mechanics, as for example in the case of a non-relativistic particle, can be treated in either of two ways. One can work with the differential-equation form of the theory, by studying the Schrödinger equation. Alternatively, one can study the Feynman path integral, which gives the integral form of the Schrödinger differential approach. The Feynman path integral has the advantage of incorporating the boundary conditions on the particle, for example that the particle is at spatial position $\mathbf{x}_a$ at an initial time t_a, and at position $\mathbf{x}_b$ at final time t_b. The path integral leads naturally to a semi-classical expansion of the quantum amplitude, valid asymptotically as the action of the classical solution of the equations of motion becomes large compared to Planck's constant $\hbar$.

One moves from quantum mechanics to quantum gravity by replacing the spatial argument $\mathbf{x}$ of the wave function by the three-dimensional spatial geometry $h_{ij}(\mathbf{x})$. A typical quantum amplitude is then the amplitude to go from an initial three-geometry h_{ijI} to a final geometry h_{ijF}, specified (say) on identical three-surfaces Σ_I, Σ_F. To complete the description in the asymptotically flat case, one needs to specify asymptotic parameters such as the time T between the two surfaces, measured at spatial infinity. To make the classical boundary-value problem elliptic and (one hopes) well-posed, one rotates to imaginary time $-iT$. The Feynman path integral would again give a semi-classical expansion of the quantum amplitude, were it not for the infinities present in the loop amplitudes. To have any hope of a less divergent field theory, one should instead study supergravity, in which there is an extra local invariance, local supersymmetry, which rotates bosons into fermions and *vice versa*.

In quantum gravity, possibly coupled to matter fields as in supergravity, one again has a choice between the Feynman path-integral approach and the differential-equation approach. Here one encounters a characteristic feature of quantum cosmology, taking here for definiteness the case in which the three-geometries such as (h_{ij}, Σ) now represent spatial sections through cosmological spacetimes: the fullest understanding is only released when one combines the path-integral with the differential-equation or canonical approach. Both methods may be needed in order to solve a problem, e.g. one to provide insight, and the other for concrete calculations. Both methods are indeed useful for concrete calculations, and sometimes (perhaps surprisingly) it is the canonical method which is more powerful, for example in evaluating loop amplitudes in supergravity or in supersymmetric quantum cosmology.

The principal differential equation of quantum gravity or quantum cosmology is the Wheeler–DeWitt equation. This is a second-order functional differential equation, which can be regarded loosely as being of Klein–Gordon type for a wave function of the form $\Psi(h_{ij})$. If one imposes symmetry restrictions on the three-geometries h_{ij} which are being studied, such as isotropy or homogeneity, then the Wheeler–DeWitt equation becomes a partial differential equation in a finite number of variables. In supergravity, there are extra fermionic differential equations or quantum constraints, corresponding to the freedom to make local supersymmetry transformations. These quantum supersymmetry constraints may be regarded as functional Dirac equations; their anticommutator gives back the 'Klein–Gordon' Wheeler–DeWitt equation. When one uses a certain natural representation of the fermionic variables, one of the quantum supersymmetry constraints is of first order in bosonic derivatives, and linear in fermions. This constraint can be solved exactly, with the help of the auxiliary fields of supergravity [Wess & Bagger 1992], with interesting consequences, for example finiteness.

When one combines supergravity with quantum cosmology under the assumption of an isotropic or homogeneous geometry, particularly simple solutions of the quantum constraints appear. Where in quantum gravity, in the anisotropic homogeneous case, one had a large family of (bosonic) solutions to the Wheeler–DeWitt equation, in supergravity one has a unique bosonic solution of the form $\exp(-I/\hbar)$, with I being a certain bosonic action functional. There are, however, more complicated quantum states which are proportional (e.g.) to two powers of fermions [Csordás & Graham 1995]. In quantum cosmology, there are two preferred quantum states defined by path integrals. The Hartle–Hawking state corresponds to filling in the compact region inside the boundary (with three-metric h_{ij}) with an arbitrary four-geometry $g_{\mu\nu}$ and matter variables, agreeing with the boundary data, and carrying out the path integral. The wormhole ground state is given by summing over all four-geometries $g_{\mu\nu}$ and matter variables outside the bounding three-surface, subject to the requirement of asymptotic flatness in all directions. The bosonic state $\exp(-I/\hbar)$ for homogeneous models corresponds to the wormhole ground state. If, however, one modifies the fermionic variables in a way related to the positive- and negative-frequency decomposition on the bounding surface, then one can instead obtain the Hartle–Hawking state.

The wormhole ground quantum state for a model with supergravity coupled to scalar-spin-1/2 matter is treated here. In general [Hawking 1988], the presence of microscopic wormholes in spacetime implies that scalar particles acquire an effective mass of the order of the Planck mass. It is found here that massless particles give a massless effective theory, in the case of local supersymmetry. Similarly, small-mass particles must have a small effective mass. This alone is a very strong argument in favour of local supersymmetry.

P. D. D'Eath
December 1995

Acknowledgements

My profoundest gratitude is to Stephen Hawking, for his continuing help and advice over more than two decades, and for a warm friendship. In particular, he has made suggestions and had crucial ideas over the past two years which have immensely improved the central sections of the book. In addition, I am deeply indebted to Jim Hartle, a valued colleague and warm friend, who has been generous with encouragement, advice, and also hospitality for sabbaticals on more than one occasion. Terms spent at UCSB were both professionally and personally rewarding, and advanced the work decidedly.

The original inspiration for the application of canonical methods to supergravity came out of conversations at St John's College, Cambridge, with Paul Dirac during his summer visits in the early 1980s. Dirac's pithy and insightful remarks and criticisms helped to reveal ways forward through apparently impenetrable difficulties. Without his encouragement and insistence on 'elegance', this research would have been much less enjoyable. How delighted he would have been had he known that a canonical treatment may well render supergravity a finite theory, if the ideas of chapters 4 and 8 turn out to be correct.

Early parts of this research developed in the delightful surroundings of the University of Texas at Austin in the early 1980s, where Bryce DeWitt and John Wheeler offerred a warm welcome. There could hardly be a more appropriate place in which to have taken Dirac's modern approaches further, in parallel with the well-known contributions of Wheeler and DeWitt, to canonical quantum gravity. Over ten years later, while the most recent developments of this research were still evolving, DeWitt yet again expressed interest and enthusiasm, in quite different surroundings, namely, Durham, England.

I am grateful to a number of other colleagues for interest, encouragement, and often extremely helpful criticism during the (at times) difficult stages of work on the finiteness argument. Igor Volovich worked through with me the details of the first version of the revised material, after Don Page's crucial insight – which led to the uncovering of an early error made in 1984, and its solution. Valery Frolov and Karel Kuchař listened patiently, while Hideo Kodama offered much-needed encouragement at a crucial stage. Interest, useful questions, and, indeed, elucidating dissent came from Gary Horowitz, Stanley Deser, and Hermann Nicolai on visits to Cambridge. Grisha Vilkovisky demanded clarifications by the Bay of Naples during his year at Sorrento and my week at Naples, which Roberto Pettorino's generosity made possible.

Gratitude should also be expressed to Octavio Obregón and Robert Graham for stimulating discussion, as well as collaboration on several aspects of this work, and for their hospitality at Mexico City and in Germany. I would also like to thank Anthony Cheng, Fay Dowker, Jonathan Halliwell, Matthias Wulf, and David Hughes for many hours of discussion and collaboration, which led to progress on various related topics. Two anonymous readers for Cambridge University Press offered useful comments on the proposed book, and constructive criticism from one referee for *Physical Review D* showed me a way forward out of a major difficulty. David Nordstrom is owed warm thanks for trying to find suitable referees and obtaining helpful comments during a period of some error and uncertainty, as is Peter Landshoff as editor of *Physics Letters B*.

To Stuart Rankin, our computer officer at DAMTP, is owed many, many thanks for hours of indispensable guidance in the technical side of the preparation of the manuscript. Brenda Surridge and Rita Gaggs also made invaluable contributions, and Brenda is now much missed since her departure. Additionally, Margaret Downing, draftswoman for the department, transformed my sketches into professional drawings.

Appreciation is owed both to DAMTP and to the University of Cambridge for several terms of sabbatical leave, without which this work would have been much delayed. The editors of *Cambridge Monographs*, Rufus Neal and Adam Black at Cambridge University Press, and the series editor, Steven Weinberg, saw this work through to its final form.

1
Introduction

The application of canonical methods to gravity has a long history [DeWitt 1967]. In [Dirac 1950] a general Hamiltonian approach was presented, which allowed for the presence of constraints in a theory, due to the momenta not being independent functions of the velocities. In particular, this occurs in general relativity, because of the underlying coordinate invariance of gravity. The general approach above was applied to general relativity in [Dirac 1958a,b, 1959] and further described in [Dirac 1965]. It was seen that there are four constraints, usually written $\mathscr{H}_i (i = 1, 2, 3)$ and $\mathscr{H}_\perp$, associated with the freedom to make coordinate transformations in the spatial and normal directions relative to a hypersurface t = const. in the Hamiltonian decomposition. Classically, these four constraints must vanish for allowed initial data. In the quantum theory, as will be seen in chapter 2, these constraints become operators on physically allowed states Ψ, which must obey $\mathscr{H}_i \Psi = 0$, $\mathscr{H}_\perp \Psi = 0$. Here, in the simplest representation, Ψ is a functional of the spatial metric $h_{ij}(x)$. It was shown in [Higgs 1958, 1959] that the constraints $\mathscr{H}_i \Psi = 0$ precisely describe the invariance of the wave function under spatial coordinate transformations. The Hamiltonian formulation of gravity was also studied by [Arnowitt *et al.* 1962], who provided the standard definition of the mass or energy M of a spacetime, as measured at spatial infinity.

This work was continued by Wheeler [Wheeler 1968] and by DeWitt, leading to the explicit formulation of the constraint $\mathscr{H}_\perp \Psi = 0$, known as the Wheeler–DeWitt equation [Wheeler 1968, DeWitt 1967]. The application of these ideas to cosmology was taken up by a number of authors, following DeWitt's work on the Friedmann universe. One can study model universes in which the geometry is restricted (say) to be of the homogeneous Bianchi type [Ryan & Shepley 1975]. Then the three-metric $h_{ij}(x)$ is defined by a small number of parameters, which are the arguments of the wave function. Thus one is effectively studying quantum mechanics, rather than field theory; the only condition to be satisfied is the Wheeler–DeWitt equation $\mathscr{H}_{\perp}\Psi = 0$, which always has the form of an equation of Klein–Gordon type (in fact a wave equation in a suitable curved metric) with a potential. Some of this work on quantum cosmology is described in [Misner 1972, Ryan 1972].

After this period of intense interest in canonical methods in quantum gravity, the subject entered a quieter period. A great deal more interest was kindled by the proposal of Hartle and Hawking [Hawking 1982, Hartle & Hawking 1983] for a preferred 'no-boundary' quantum state in the case of a compact three-geometry. This is defined by regarding the three-geometry as the boundary of a compact four-geometry (Fig. 1.1), and carrying out a path integral, where one sums over all infilling four-geometries, weighted by $\exp(-I/\hbar)$, with I being the Euclidean action [Hawking 1979] of the gravitational field together with any matter fields present. This path integral automatically obeys the quantum constraint equations, but also incorporates boundary conditions on the wave function. One theme of this book will be the interplay between the Feynman-integral form of the theory and the differential-equation quantum-constraint formulation. Both can be powerful tools, in the appropriate contexts. Following the appearance of [Hartle & Hawking 1983], the Hartle–Hawking or no-boundary state was studied in many examples. These examples include isotropic Friedmann $k = +1$ or homogeneous Bianchi 'mini-superspace' models, which are parametrized by a finite number of coordinates. They also include scalar-gravity [Halliwell & Hawking 1985] or scalar-spin-1/2-gravity [D'Eath & Halliwell 1987] models in which the perturbation modes of all fields around isotropy are studied. Such models lead to testable results for fluctuations in the cosmic microwave background radiation [Smoot *et al.* 1992, White *et al.* 1994]. Thus, although it might initially seem that quantum cosmology is a very speculative subject, one can remark first that it is built out of well-established foundations such as general relativity and the Feynman path integral, and second that it has become an observational subject. An introduction to quantum cosmology is given in chapter 2.

Since supersymmetry is expected to be important at high energies, it is then natural, by analogy, to study the Hamiltonian formulation of supergravity with a view to quantization [Deser *et al.* 1977a, Fradkin & Vasiliev 1977, Pilati 1978]. This approach has applications both in cosmology and in scattering, as will be seen in this book. In supergravity, the gravitational field is combined with a supersymmetry partner, the gravitino, whose physical degrees of freedom are given by a spin-3/2 field. The supersymmetry is local, in that the spinor supersymmetry parameters $\epsilon^A(x), \tilde{\epsilon}^{A'}(x)$ defining an infinitesimal supersymmetry transformation depend on spacetime position. This is crucial in the quantization. For canonical quantization, it is necessary to use two-component spinors, and

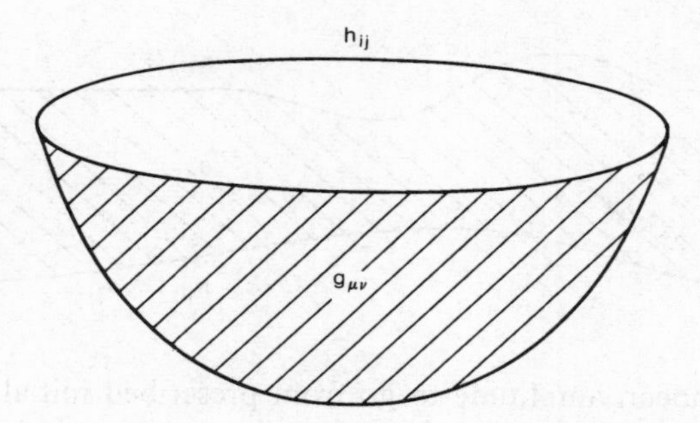

Fig. 1.1. In the Hartle–Hawking path integral, one sums over all four-geometries $g_{\mu\nu}$ on compact manifolds with prescribed three-metric h_{ij} on the boundary.

the corresponding Hamiltonian treatment [D'Eath 1984] and canonical quantization are described in chapter 3. The local supersymmetry generators in the classical Hamiltonian theory are denoted by S_A and $\tilde{S}_{A'}$ $(A = 0, 1; A' = 0', 1')$, and their Dirac bracket [Hanson *et al.* 1976] gives $\mathscr{H}_{AA'}$, which is the spinor version of $\mathscr{H}_i$ and $\mathscr{H}_\perp$ combined into a four-vector. In the quantization, one has fermionic operators S_A and $\bar{S}_{A'}$, whose anticommutator is proportional to $\mathscr{H}_{AA'}$. One also has generators J_{AB} and $\bar{J}_{A'B'}$ of the local Lorentz rotations on spinor indices, needed in a theory with fermions. Then it is sufficient to solve the quantum supersymmetry constraints $S_A\Psi = 0$, $\bar{S}_{A'}\Psi = 0$ and the rotational invariance constraints $J_{AB}\Psi = 0$, $\bar{J}_{A'B'}\Psi = 0$, in order to solve the remaining constraint $\mathscr{H}_{AA'}\Psi = 0$, which includes the Wheeler–DeWitt equation. Since the quantum supersymmetry constraints are of first order in the gravitational momentum, they may be easier to solve than the Wheeler–DeWitt equation. The supersymmetry constraints are analogous to the Dirac equation, while, as already pointed out, their 'square' – the supersymmetric version of the Wheeler–DeWitt equation – is analogous to the Klein–Gordon equation. Much of this book is concerned with solving the supersymmetry constraints, which in the example of quantum cosmology are more restrictive than $\mathscr{H}_{AA'}\Psi = 0$, having fewer solutions.

One can then begin to study the quantum amplitude to go from an initial hypersurface with suitable data prescribed on it to a final hypersurface with prescribed data (chapter 4). As usual in quantum field theory, in the case of an asymptotically flat geometry, it is most natural to evaluate the amplitude assuming that the time interval between the surfaces at spatial infinity is imaginary, corresponding to a Euclidean time separation τ (Fig. 1.2). After the amplitude is found, one can in principle rotate back to a real time separation t. In the case of 'Euclidean' boundary data, one expects to be able to solve the elliptic boundary-value problem for the classical infilling four-metric $g_{\mu\nu}$ obeying the vacuum Einstein field

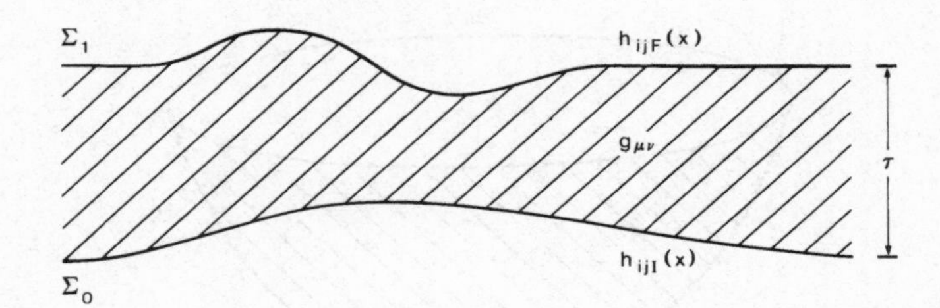

Fig. 1.2. A Euclidean amplitude to go from prescribed initial data h_{ijI} to final data h_{ijF}, in Euclidean time τ at infinity.

equations. Based on the classical gravitational solution, one will then find the infilling classical gravitino field at linear order in fermions, in the case that the fermionic boundary data obey the supersymmetry constraints. By iteration of the coupled classical field equations of supergravity, one will then arrive at a complete classical solution, with Euclidean action I_{class} which is a functional of the boundary data. The quantum amplitude K about a classical solution has the asymptotic form

$$K \sim (A + \hbar A_1 + \hbar^2 A_2 + \cdots) \exp(-I_{\text{class}}/\hbar), \tag{1.1}$$

where the one-, two-,$\cdots$ loop factors $A, A_1, \cdots$ depend on the boundary data. The quantum supersymmetry constraints applied to Eq. (1.1) give restrictive equations obeyed by $A, A_1, \cdots$ which may lead to explicit calculations of loop amplitudes including fermions [D'Eath & Wulf 1995]. In this way canonical methods may be just as effective for explicit computation, if not more so, when compared with Feynman diagrams. One can also use the supersymmetry constraints to show that there are no on-shell one- or two-loop boundary counterterms describing infinite surface contributions to the amplitude in $N = 1$ supergravity [D'Eath 1986a,b]. Since there are no volume counterterms either at one- or two-loop order [Deser *et al.* 1977b], in the case that the spacetime topology is trivial, then amplitudes in supergravity with local boundary conditions for fermions (e.g. the spatial components $\tilde{\psi}^{A'}{}_i (i = 1, 2, 3)$ of the gravitino field fixed initially, and $\psi^A{}_i$ finally) are finite at one- and two-loop order. This is a decided improvement on quantized general relativity, where boundary counterterms already appear (with non-zero coefficient) [Moss & Poletti 1994, Esposito *et al.* 1995] at one loop, and may proliferate at two loops.

The restriction above that the boundary data $\psi^A{}_i$ or $\tilde{\psi}^{A'}{}_i$ should obey the appropriate classical supersymmetry constraint is simplified if one uses the auxiliary fields of $N = 1$ supergravity [Wess & Bagger 1992]. This will be important in the later investigation of finiteness of the quantum theory. Further investigation in this direction can be made by studying the behaviour of the amplitude K in the limit $\tau \to 0_+$. For purely bosonic boundary data, one finds $K \sim \exp(-I_B/\hbar)$ as $\tau \to 0_+$, where I_B is the classical Euclidean gravitational action. Thus the loop factors reduce to 1 in this limit. These results are used later in chapter 8.

One can then move on to supersymmetric quantum cosmology (chapter 5), using mini-superspace models such as Friedmann $k = +1$ and Bianchi types, described by a finite number of parameters. A detailed treatment of the Friedmann $k = +1$ case leads to a wave function $\Psi(a, \psi^A)$, where a is the radius of a three-sphere, and ψ^A is a spin-1/2 object derived from the gravitino. Classically, both a and ψ^A are functions of time only. The general solution of the quantum constraints [D'Eath & Hughes 1988] is

$$\Psi(a, \psi^A) = C \exp(-3a^2/\hbar) + D \exp(3a^2/\hbar)\psi_A\psi^A, \tag{1.2}$$

where C and D are constants. The second state proportional to D is the Hartle–Hawking state, while the first is the ground wormhole state [Hawking 1988] which is studied in more detail in chapter 6. The calculation leading to Eq. (1.2) can be repeated for a more general model with $N = 1$ supergravity coupled to supermatter [D'Eath & Hughes 1992]. For Bianchi models in supergravity, one finds that there is only one purely bosonic state, of the form $\exp(-I_B/\hbar)$, where I_B is a particular Euclidean action; this gives the ground wormhole state. At the top order in fermions one obtains another single state $\psi^6 \exp(I_B/\hbar)$, where ψ^6 denotes a product of all the fermionic variables (there are six independent odd Grassmann fermionic variables, and Lorentz invariants are formed from suitable even combinations of them, up to degree six). In the relevant case of Bianchi type IX [Ryan & Shepley 1975], the top-order state is *not* the Hartle–Hawking state [D'Eath 1994]. These results are an example of the restrictiveness of the supersymmetry constraints; in quantized general relativity there are many solutions of the Wheeler–DeWitt equation, not just one. At the intermediate orders quadratic and quartic in fermions there are, however, more complicated states [Csordás & Graham 1995] which are defined in terms of the solution of a Wheeler–DeWitt equation. The Hartle–Hawking state can be found by working with a different basis of homogeneous fermions [Graham & Luckock 1994]. This will be discussed further in sections 5.8, 8.4 and chapter 9. Quantum wormhole states (chapter 6) [Hawking & Page 1990] are defined by a path integral, where the boundary data are specified on a compact three-surface, and one sums over all Riemannian four-geometries and matter fields on them, subject to the condition that the four-geometry is asymptotically flat in all directions as the distance r from the inner boundary tends to infinity. As described in [Hawking 1988], microscopic wormholes, which are small handles in the spacetime joining possibly distant regions, lead to an effective theory of particle interactions, in which scalar and spin-1/2 particles may acquire a mass of the order of the Planck mass $(\hbar c/G)^{1/2}$, which is characteristic of quantum gravitational effects. In the locally supersymmetric model of the interaction of $N = 1$ supergravity with spin-1/2-scalar matter [D'Eath & Hughes 1992], one can find the wormhole ground state in the massless case [Alty *et al.* 1992]. One then finds that zero mass in the Lagrangian leads to zero effective mass. Similarly, for a small mass in the Lagrangian, one expects a small effective mass. Thus Planckian effective masses are avoided in a locally supersymmetric theory. This alone is a strong argument in favour of local supersymmetry.

Another approach in quantum cosmology has been that using the Ashtekar variables [Ashtekar 1986, 1988, 1991] (chapter 7). These provide a different description of Hamiltonian general relativity and supergravity [Jacobson 1988]. In this approach to general relativity one uses the variables A_{ABi} and $\sigma_{AB}{}^{i}$. Here $A_{ABi} = A_{(AB)i}$ are the spatial components of the four-dimensional connection, and $\sigma_{AB}{}^{i} = \sigma_{(AB)}{}^{i}$ gives the spatial triad, from which the three-metric can be constructed as $h_{ij} = \sigma^{AB}{}_{i}\sigma_{ABj}$. It is simplest to regard A_{ABi} as a coordinate variable and $\sigma_{AB}{}^{i}$ as a momentum. One interesting quantum state in supergravity can then be formed by including fermionic arguments and taking $\Psi = \exp(-I_{CS}/\hbar)$, where I_{CS} is the Chern–Simons action [Sano & Shiraishi 1993] in the case of a positive cosmological constant $\Lambda \geq 0$. This state might conceivably give an exact solution of the quantum constraints, although there are difficulties of factor ordering in testing this. For mini-superspace Friedmann models, this state corresponds to classical solutions for de Sitter spacetime, and to gravitinos propagating in that spacetime.

One can investigate further the question of the possible finiteness of $N = 1$ supergravity (chapter 8). The results of section 8.1 indicate that there may be a bosonic amplitude $\Psi = \exp(-I_B/\hbar)$ corresponding to purely gravitational boundary data (essentially the three-geometry on initial and final three-surfaces, with Euclidean time at infinity τ), with *local* boundary data for the gravitino field set to zero. This further indicates that one may have finite amplitudes in general, with both bosonic and non-zero *local* fermionic boundary data. Scattering calculations, as in [Itzykson & Zuber 1980] for lower-spin fields, require *spectral* boundary conditions, dependent on a splitting of the fermionic boundary data into positive and negative frequencies. At early and late Euclidean times τ, one takes only 'physical gravitinos', corresponding to transverse-traceless fermionic data [D'Eath 1986a,b], and transverse-traceless gravitational perturbations ('physical gravitons'). An argument similar to that in section 8.1 indicates that one may also have finite amplitudes with *spectral* fermionic boundary data (section 8.2). In cosmology (section 8.3), one needs to use spectral boundary conditions on a compact hypersurface [D'Eath & Halliwell 1987] (or alternatively to modify the previous local boundary conditions) when studying the Hartle–Hawking state. In general, one may have zero modes for the gravitino field operator (the Rarita–Schwinger operator). This happens in the Friedmann $k = +1$ example of Eq. (1.2), since ψ^A corresponds to a zero mode. This leads to the $\psi_A \psi^A$ factor in the wave function. If there are no zero modes, then $\Psi = \exp(-I_B/\hbar)$ obeys all the constraint equations, with I_B being the bosonic gravitational action corresponding to the purely gravitational data, for a compact four-geometry bounded by the given three-geometry. But (as yet) one cannot show in any obvious way that $\Psi = \exp(-I_B/\hbar)$ is the Hartle–Hawking state; one can only make this statement plausible. In section 8.4, models with $N = 1$ supergravity coupled in a locally supersymmetric way to supermatter (spins 0, 1/2 and 1) [Wess & Bagger 1992] are studied. These models allow for a wide range of particle interactions. In the massless case, the results indicate that the models may be finite with local boundary conditions, but divergent with spectral boundary conditions. The other main possibility of extending these results to allow for supermatter is to study higher-N supergravity theories [van Nieuwenhuizen 1981]. It is conceivable that results along the lines above may indicate that these theories are also finite, both with local and spectral boundary conditions.

There are a number of possible directions in which one could move forward, based on this work (chapter 9). For example, one could investigate the effects of non-trivial topology, both in the three-dimensional data and in the four-dimensional manifolds summed over in the path integral. Also, as suggested above, one could use the supersymmetry constraints to compute loop amplitudes $A, A_1, A_2, \ldots$ in Eq. (1.1) with fermionic data included; this will provide a check as to whether these expressions are finite. One could also investigate fermionic amplitudes at *high* orders in $\hbar$, by suitable approximation methods – cf. [Brézin *et al.* 1977a,b] for the scalar case. Other examples involve classical bosonic configurations appearing in the bosonic amplitude $\exp(-I_B/\hbar)$. One can study, for example, the amplitude to go from two high-energy gravitons to a final black hole, by considering the Riemannian problem with Euclidean time τ, and then rotating back to Lorentzian time t. The Lorentzian process in which two high-energy objects (e.g. black holes) form a final black hole plus (in this case) gravitational radiation is described in [D'Eath 1978, D'Eath & Payne 1992a,b,c, D'Eath 1996]. Turning the Riemannian geometry upside down, one has a tunnelling process in which a black hole decays into a pair of high(Planckian)-energy gravitons. This might offer a solution to the question of the endpoint of the black-hole evaporation process [Hawking 1975], and connects with a substantial body of theory concerning high-energy processes in quantum field theory, involving quantum gravity, initiated by ['t Hooft 1987]. Further references are given in

[Verlinde & Verlinde 1992]. Yet another application of the bosonic amplitude $\exp(-I_B/\hbar)$ is to thermal equilibrium, where the partition function and hence the entropy can be calculated exactly by considering metrics which are periodic in Euclidean time τ.

2
Quantum cosmology

2.1 Introduction

Before embarking on the full theory of $N = 1$ supergravity in the following chapters, it is necessary to review some of what is known about quantum cosmology based on general relativity, possibly coupled to spin-0 or spin-1/2 (non-supersymmetric) matter. The ideas presented in this chapter, based to a considerable extent but not exclusively on Hamiltonian methods, will recur throughout the book. Perhaps the main underlying idea is that there is an analogy between the classical dynamics of a point particle with position $\mathbf{x}$ and that of a three-geometry $h_{ij}(x)$. The theory of point-particle dynamics, when written in parametrized form [Kuchař 1981] and cast into Hamiltonian form, and the theory of general relativity, again in Hamiltonian form, bear a strong resemblance. In the Hamiltonian form of general relativity, $h_{ij}(x)$ can be taken to be the 'coordinate' variable, corresponding to $\mathbf{x}$ in particle dynamics. In section 2.2, for parametrized particle dynamics, it is shown following [Kuchař 1981] how a constraint arises classically in the Hamiltonian theory, which, when quantized, gives the appropriate Schrödinger or wave equation for the quantum wave function $\psi(\mathbf{x}, t)$. As described in subsequent sections, the quantization of the analogous constraint in general relativity gives the Wheeler–DeWitt equation [DeWitt 1967, Wheeler 1968], a second-order functional differential equation for the wave function $\Psi[h_{ij}(x)]$, which contains all the information in quantum gravity, if only one could solve and interpret it.

The Hamiltonian form of general relativity is derived from the Einstein–Hilbert Lagrangian in section 2.3. The Hamiltonian has the form

$$H = \int d^3x(N\mathscr{H}_\perp + N^i\mathscr{H}_i), \tag{2.1.1}$$

where $\mathscr{H}_\perp$ and $\mathscr{H}_i$ are functions of the coordinates $h_{ij}(x)$ and their conjugate momenta. Here $\mathscr{H}_\perp$ has the interpretation of giving the generator of

Hamiltonian evolution in the normal direction, with $N\delta t$ being the amount of proper time elapsed in infinitesimal coordinate time δt. Similarly, $\mathscr{H}_i$ is the generator of evolution in the spatial direction (i.e. within an initial spacelike hypersurface), with $N^i\delta t$ the infinitesimal spatial displacement. The form (2.1.1) of H, with a Lagrange multiplier for each infinitesimal gauge symmetry, is typical of gauge theories [Hanson *et al.* 1976]. Note that, in deriving Eq. (2.1.1), one finds that an extra boundary term must be included in the Einstein–Hilbert action.

In studying the quantization of general relativity, one would like, as in any quantum theory, to understand the quantum amplitude to go from prescribed initial $h_{ijI}(x)$ describing an initial three-geometry, to a final geometry $h_{ijF}(x)$, say in Lorentzian time T measured at spatial infinity (as in Fig. 1.2). One expects that the quantum amplitude will be approximated by the semi-classical expression $P\exp(iS/\hbar)$. Here S is the Lorentzian action of a classical solution of the Einstein vacuum field equations, joining the initial to final data (if this exists), and P is a quantum prefactor. As pointed out in section 2.4, for weak fields there is not expected to be such a classical solution. Instead one has to complexify the time interval at spatial infinity. For example, one can replace T by the imaginary time interval $-iT$. Then the classical boundary-value problem becomes that for a positive-definite (Riemannian) four-geometry $g_{\mu\nu}$, which is well-posed at least for weak fields [Reula 1987]. The semi-classical approximation to the amplitude becomes $P_E\exp(-I/\hbar)$, where I is the Euclidean (Riemannian) classical action. Having found the action and approximate amplitude, one can then attempt to return to the Lorentzian régime by rotating $-iT$ back towards T. This corresponds to the $+i\epsilon$ prescription in Feynman-diagram theory [Itzykson & Zuber 1980].

As just suggested, one way to evaluate quantum amplitudes is via Feynman diagrams, which are derived from the Feynman path integral [Feynman & Hibbs 1965]. One expects that amplitudes in quantum gravity or cosmology should in principle be derivable from a Feynman path integral; this idea underlies the construction of the Hartle–Hawking or no-boundary state [Hawking 1982, Hartle & Hawking 1983], to be described below. In section 2.5, the Feynman path integral is described for non-relativistic quantum mechanics, and a derivation of the Schrödinger equation is given. This motivates a derivation of the Wheeler–DeWitt equation $\mathscr{H}_\perp\Psi = 0$ of quantum gravity. (The first-order differential equation $\mathscr{H}_i\Psi = 0$ simply describes the invariance of the state $\Psi[h_{ij}(x)]$ under spatial diffeomorphisms.) It is also pointed out that the path integral can be used to obtain the ground state of a quantum theory.

As described above, one can consider the Euclidean amplitude to go

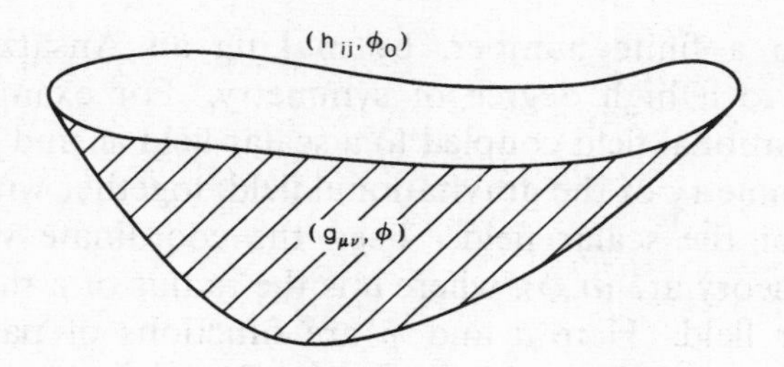

Fig. 2.1. In the Hartle–Hawking state, one sums over all infilling fields $(g_{\mu\nu}, \phi)$.

from an initial three-geometry $h_{ijI}(x)$ to a final three-geometry $h_{ijF}(x)$ in Euclidean time τ. This is given formally by a Euclidean path integral

$$\Psi = \int \mathscr{D}g_{\mu\nu} \exp(-I/\hbar) \tag{2.1.2}$$

over all Riemannian four-metrics $g_{\mu\nu}$ joining the initial and final data (section 2.6). For completeness, one would need to include in I gauge-fixing and Faddeev–Popov terms (see references in [Vilkovisky 1984]). Apart from the semi-classical motivation above for favouring the Euclidean path integral (2.1.2) over the Lorentzian path integral, with $-iS$ replacing I, one might also hope that the reality of the Euclidean path integral would improve the convergence of the integral. This is discussed in section 2.6. Further, it is pointed out that there is a special state defined by a path integral, being somewhat analogous to the ground state described in section 2.5. This is the Hartle–Hawking or no-boundary state, and is naturally defined in the case that the initial three-surface Σ is compact. Suppose, for example, that in addition to gravity there is a scalar field ϕ, which could drive a cosmological expansion or contraction. Then the state is formally

$$\Psi_{HH}(h_{ij}, \phi_0) = \int \mathscr{D}g_{\mu\nu}\mathscr{D}\phi \exp[-I(g_{\mu\nu}, \phi)/\hbar], \tag{2.1.3}$$

where ϕ_0 is the boundary value of ϕ on Σ, and the integral is over all Riemannian geometries $g_{\mu\nu}$ on compact manifolds with boundary Σ and boundary metric h_{ij}, and over matter fields ϕ which agree with ϕ_0 on the boundary (Fig. 2.1). The Hartle–Hawking state plays an important role in quantum cosmology, as an analogue of a ground state. Together with the wormhole ground quantum state of chapter 6, also defined by a path integral, the Hartle–Hawking state describes the most symmetrical quantum state in cosmology or quantum gravity.

The analogy between the motion of a particle and the evolution of a three-geometry becomes still closer in the case of mini-superspace models (section 2.7) [Hartle & Hawking 1983, Hawking 1984]. There one assumes that the degrees of freedom of the gravitational and any matter fields are reduced to a finite number, by making an Ansatz for the fields corresponding to a high degree of symmetry. For example, one might have the gravitational field coupled to a scalar field ϕ, and assume $k = +1$ Friedmann symmetry of the gravitational field, together with homogeneity and isotropy of the scalar field. Then the coordinate variables in the Hamiltonian theory are (a, ϕ), where a is the radius of a three-sphere, and ϕ is the scalar field. Here a and ϕ are functions of parameter time τ only. One makes the above Ansatz in the Euclidean action I, and finds I as a functional constructed from $a, \dot{a}, \phi, \dot{\phi}$ and N, where $N(\tau)$ is as in Eq. (2.1.1) (the lapse function). From this, one can derive the reduced Hamiltonian theory (i.e. the Hamiltonian theory subject to the above symmetries), and the classical generator $\mathscr{H}_{\perp}$. Up to a choice of factor ordering, one then has the Wheeler–DeWitt equation $\mathscr{H}_{\perp}\Psi = 0$, which is now a partial differential equation in a finite number of variables, rather than the second-order functional differential equation of the full theory.

In section 2.7, mini-superspace models in terms of (a, ϕ) are described for the cases of a conformally invariant scalar field with a cosmological constant Λ and for a massive scalar field. In the conformally invariant case, the wave function separates into a product of two functions of one variable. The Hartle–Hawking state can be found, and is proportional to the ground state of the scalar field. The gravitational part of the wave function can be approximated semi-classically in terms of Lorentzian de Sitter spacetime [Hawking & Ellis 1973] or its positive-definite analogue – the four-sphere S^4. One finds that the classical geometry is complex; for example, one can take the geometry to be half of the Riemannian S^4 for $a \leq H^{-1}$, where $H^2 = 2\Lambda/9\pi$, joining on to a Lorentzian de Sitter geometry at the equator $a = H^{-1}$ [Hartle & Hawking 1983, Hartle 1986]. Instead of the geometry 'turning through a right-angle' between real and complex at $a = H^{-1}$, one can instead take a single complex gravitational solution [Lyons 1992]. The massive scalar field model can be treated in a somewhat similar way [Hawking 1984], and turns out to give a de Sitter-like classical evolution at early times, since for large ϕ, there is an effective cosmological constant $\Lambda \simeq m^2\phi^2$. At late time, the scalar field decays and oscillates, corresponding to a matter era. The quantization of this model can again be approximated semi-classically, giving de Sitter-like states at early times.

In the context of $N = 1$ supergravity, mini-superspace models will be studied in chapter 5. The Hamiltonian and diffeomorphism constraints $\mathscr{H}_{\perp}\Psi = 0$, $\mathscr{H}_i\Psi = 0$ are replaced by the supersymmetry constraints $S_A\Psi = 0$, $\bar{S}_{A'}\Psi = 0$, where A, A' are two-component spinor indices (subsection 2.9.1). The supersymmetry constraints imply $\mathscr{H}_{\perp}\Psi = 0$, $\mathscr{H}_i\Psi = 0$. Being of first order in bosonic momenta, the supersymmetry constraints are much more restrictive than $\mathscr{H}_{\perp}\Psi = 0$ and $\mathscr{H}_i\Psi = 0$, and have many fewer solutions. For example, when the symmetrical model is homogeneous, of class A in the Bianchi classification [Ryan & Shepley 1975], the bosonic quantum state is uniquely determined, up to a constant multiple, being of the form const. $\exp(-I/\hbar)$ for a particular action functional I (section 5.7).

Generic spacetimes are not symmetrical; one can begin to understand their classical and quantum behaviour by studying small perturbations about a Friedmann universe (section 2.8). For the massive scalar field model above, this was done in [Halliwell & Hawking 1985], and the Hartle–Hawking state was examined for this bosonic system. For the Hartle–Hawking state, the inhomogeneous or anisotropic modes start out in their ground state, at the time described above at which the classical Riemannian (approximately S^4) geometry joins onto the later-time Lorentzian de Sitter-like classical solution. There is a time-dependent Schrödinger equation describing the evolution of each mode. During the period of exponential expansion, the modes remain in their ground state until their wavelength exceeds the horizon size. During the subsequent non-exponential expansion, the ground state fluctuations are amplified, and the modes re-enter the horizon in the matter- or radiation-dominated era, in a highly excited state. This leads to a scale-free spectrum of density perturbations, which could account for the present structure of the universe. These perturbations would be compatible with observations of the fluctuations in the microwave background [Smoot *et al.* 1992, White *et al.* 1994] if the mass of the scalar field is of order 10^{14} GeV.

Similarly, one can study cosmological perturbations for a model containing massive spin-1/2 fermions [D'Eath & Halliwell 1987] in addition to the massive scalar field, which still drives the cosmological expansion (section 2.9). The Hamiltonian treatment of gravity coupled to a massive spin-1/2 field was described in [Nelson & Teitelboim 1978], and is reviewed here as a prelude to the Hamiltonian treatment of $N = 1$ supergravity to be given in chapter 3. This introduces the notion of Dirac brackets for a constrained theory, and of Poisson and Dirac brackets involving fermions, where the fermions live in a Grassmann algebra. By analogy with the bosonic case above, one can expand the fermionic perturbations in spinor harmonics, here taken to be based on a family of three-spheres. For the quantization to be valid in the massless fermion case, one must use spectral boundary conditions [D'Eath & Halliwell 1987]. The Dirac action is given by an infinite sum of terms, each describing a time-dependent Fermi oscillator. When this system is quantized, one finds that the Wheeler–DeWitt equation decomposes into a set of time-dependent Schrödinger equations, one for each fermionic mode, and a background mini-superspace Wheeler–DeWitt equation, which includes the back–reaction of the fermionic modes on the mini-superspace background. The Hartle–Hawking state is studied; this shows that the fermionic modes start out in their ground state. In the subsequent inflationary evolution, particles are created, and their number is found to be finite, with respect to instantaneous Hamiltonian diagonalization. The back–reaction of the fermions on the background isotropic geometry is found to be negligible. The invariance of the initial state under the group of isometries of de Sitter spacetime [D'Eath & Halliwell 1987] shows that the Green's functions constructed from this state are de Sitter invariant. This invariance is closely connected with the property of being a thermal state (see [Birrell & Davies 1982, Allen 1985] and references therein), at the de Sitter temperature $T = H/2\pi$, where the Riemannian de Sitter metric (S^4) is

$$ds^2 = d\tau^2 + \frac{1}{H^2}\sin^2(H\tau)d\Omega^2, \tag{2.1.4}$$

where $d\Omega^2$ is the metric on a unit three-sphere. Here, for large scalar fields $|\phi|$, $H = M\,|\phi|$, where M is the mass of the scalar field. One can examine the thermal property by introducing a fermionic particle detector. In an exact de Sitter background, a fermionic detector experiences a thermal spectrum at the de Sitter temperature T. Thus, for the Hartle–Hawking state, the universe starts out in a de Sitter-invariant thermal state at temperature T. In the context of this fermionic model (or of the bosonic model of section 2.8), one can examine the back-reaction in more detail. It can be shown [D'Eath & Halliwell 1987] how the semi-classical Einstein field equations arise from the semi-classical limit of the Wheeler–DeWitt equation; this is of importance in interpreting the theory.

2.2 Parametrized particle dynamics

Here we begin by studying an example of parametrized particle motion, which leads to a Hamiltonian constraint and associated quantum constraint equation [Kuchař 1981]. These constraints are directly analogous to those which will be studied in section 2.7 for mini-superspace models in quantum cosmology, where the gravitational and matter fields are taken to be highly symmetrical, described by a finite number of coordinates. Further, the structure of the constraint is closely analogous to the most important constraint $\mathscr{H}_\perp$ of general relativity in the general case without spacetime symmetries. Thus one can use the particle model as a guide in quantum cosmology, the classical motion of a three-geometry $h_{ij}(x)$ being analogous to classical particle motion, with a similar analogy between the quantum versions of the two theories. These comparisons will be made more precise in the following sections.

The canonical action for a single non-relativistic particle in a potential $V(x^i, t)$, with $i = 1, 2, 3$, is

$$S[x^i(t), p_i(t)] = \int dt \left[p_i \frac{dx^i}{dt} - H(x^i, p_i, t) \right], \tag{2.2.1}$$

where p_i is the momentum canonically conjugate to x^i. The Hamiltonian H has the form

$$H = \frac{1}{2m} g^{ij} p_i p_j + V(x^i, t), \tag{2.2.2}$$

where m is the mass of the particle and g^{ij} is the inverse spatial metric. In Newtonian theory, $g^{ij} = \delta^{ij}$.

Now let us parametrize the path $[x^i(t), p_i(t)]$ in phase space by an arbitrary time label τ. Further, adjoin the absolute time t to the configuration variables x^i, giving

$$x^\alpha = (t, x^i),\ x^\alpha = x^\alpha(\tau),\ p_\alpha = p_\alpha(\tau), \tag{2.2.3}$$

where p_α are the momenta canonically conjugate to x^α. The action can be re-expressed as

$$S[x^\alpha(\tau), p_\alpha(\tau)] = \int d\tau (p_i \dot{x}^i - H\dot{t}), \tag{2.2.4}$$

where $(\dot{\ }) = \frac{d}{d\tau}()$. Note that variation of Eq. (2.2.4) with respect to $x^i(\tau)$ and $p_i(\tau)$ yields the usual Hamiltonian equations of motion. Variation with respect to $t(\tau)$ gives

$$\dot{H} = \dot{t} \frac{\partial H}{\partial t}, \tag{2.2.5}$$

a correct equation describing energy balance. Finally, one would like to vary with respect to $p_0(\tau)$. Now Eq. (2.2.4) shows that $p_0 = -H$ is the momentum canonically conjugate to t. But since $H = H(x^\alpha, p_\alpha)$, one can vary p_0 independently in the action: the variables (x^α, p_α) must obey the *constraint*

$$\mathscr{H} \equiv p_0 + H(x^\alpha, p_\alpha) = 0. \tag{2.2.6}$$

In order to be able to vary all the variables (x^α, p_α) freely, one adjoins the constraint to the action, following the general procedure of [Dirac 1965], by means of a Lagrange multiplier N:

$$S[x^\alpha(\tau), p_\alpha(\tau); N(\tau)] = \int d\tau (p_\alpha \dot{x}^\alpha - N\mathscr{H}). \tag{2.2.7}$$

Variation of N gives the constraint $\mathscr{H} = 0$. Variation of p_0 gives the physical meaning of N:

$$N = \dot{t} = \frac{d(\text{absolute time})}{d(\text{label time})}. \tag{2.2.8}$$

The remaining variational equations give back the usual Hamiltonian evolution equations in the τ parametrization, together with an energy balance equation. The form (2.2.7) of the action is invariant under arbitrary re-parametrizations

$$\tau \to \tau'(\tau), \tag{2.2.9}$$

provided N transforms as

$$N \to N' = N\frac{d\tau}{d\tau'}, \tag{2.2.10}$$

and (x^α, p_α) are invariant. The Hamiltonian is then $N\mathscr{H}$, and for a classical solution one has $\mathscr{H} = 0$; this is typical of a theory invariant under re-parametrization.

Again following the general procedure of [Dirac 1965], one can quantize a theory with a constraint. The constraint $\mathscr{H} = 0$ becomes an operator condition on physically allowed wave functions:

$$\mathscr{H}\psi = 0. \tag{2.2.11}$$

This is the Schrödinger equation

$$i\hbar\frac{\partial\psi}{\partial t} + \frac{\hbar^2}{2m}\nabla^2\psi - V\psi = 0, \tag{2.2.12}$$

where (say) we take the representation $\psi = \psi(x^\alpha) = \psi(t, x^i)$ and $p_\alpha \to -i\hbar\partial/\partial x^\alpha$.

One can repeat the calculation for a single free relativistic particle, with the usual Lagrangian

$$L = -m\left(1 - \delta_{ij}\frac{dx^i}{dt}\frac{dx^j}{dt}\right)^{1/2}. \tag{2.2.13}$$

One introduces a parameter λ along the path, and finds the constraint

$$\tilde{\mathscr{H}} \equiv \pi_t + (\delta^{ij}\pi_i\pi_j + m^2)^{1/2} = 0, \tag{2.2.14}$$

where π_i and π_t are the canonical momenta. Equivalently,

$$\begin{aligned} \mathscr{H} &\equiv -\pi_t{}^2 + \delta^{ij}\pi_i\pi_j + m^2 \\ &= \eta^{\alpha\beta}\pi_\alpha\pi_\beta + m^2 = 0, \end{aligned} \tag{2.2.15}$$

where $\eta^{\alpha\beta}$ is the Minkowski metric. One finds that

$$N = \frac{1}{2m}\frac{d(\text{proper time})}{d(\text{label time }\lambda)}. \tag{2.2.16}$$

Quantization leads to the Klein–Gordon equation

$$\mathscr{H}\psi = -\hbar^2\eta^{\alpha\beta}\frac{\partial^2\psi}{\partial x^\alpha\partial x^\beta} + m^2\psi = 0. \tag{2.2.17}$$

The corresponding constraint $\mathscr{H}_\perp \Psi = 0$ of quantum gravity has a structure somewhat analogous to that of Eq. (2.2.17), with a wavelike second-order operator, and the potential term m^2 replaced by a potential which depends on the three-geometry $h_{ij}(x)$.

A closer analogy to the Hamiltonian structure of general relativity may be obtained by studying string theory [Hanson *et al.* 1976, Green *et al.* 1987]. Whereas particle dynamics describes one-dimensional trajectories, the history of a string is a two-dimensional world-sheet in spacetime. The world-sheet is parametrized by two coordinates (τ, σ), which can be regarded as 'time' and 'space' coordinates. A point on the world-sheet is given by position $x^\mu(\tau, \sigma)$ in D dimensions. The Nambu action is of the form

$$S[x^\mu(\tau,\sigma)] = \int\int d\tau d\sigma L\left(\frac{\partial x^\mu}{\partial \tau}, \frac{\partial x^\mu}{\partial \sigma}\right), \qquad (2.2.18)$$

and is invariant under arbitrary two-dimensional reparametrizations

$$\tau, \sigma \to \tau'(\tau,\sigma), \sigma'(\tau,\sigma). \qquad (2.2.19)$$

Corresponding to this invariance, the Hamiltonian can be put in the form

$$H = \int d\sigma (N_1 \mathscr{H}_1 + N_2 \mathscr{H}_2), \qquad (2.2.20)$$

where again τ is being taken as a time coordinate and σ as a space coordinate. Here $N_1(\tau,\sigma)$ and $N_2(\tau,\sigma)$ are Lagrange multipliers, and $\mathscr{H}_1$, $\mathscr{H}_2$ are the generators of infinitesimal two-dimensional coordinate transformations in the directions $\partial/\partial\tau$, $\partial/\partial\sigma$. Classically, the constraints

$$\mathscr{H}_1 = \mathscr{H}_2 = 0 \qquad (2.2.21)$$

hold. Quantum-mechanically these become operator constraints on physical wave-functions $\Psi[x^\mu(\sigma)]$. The position $x^\mu(\sigma)$ of the string at a surface $\tau =$ const. is the analogue of the position $\mathbf{x}$ of a particle at parameter time τ, and hence is used as the argument of the wave function $\Psi[x^\mu(\sigma)]$. The Virasoro constraints [Green *et al.* 1987] are constructed from the Fourier components of $\mathscr{H}_1\Psi$ and $\mathscr{H}_2\Psi$ with respect to σ; these govern the quantum dynamics of first-quantized string theory.

Similarly, in classical gravity, viewed in a Hamiltonian way, one studies the history of three-dimensional hypersurfaces with intrinsic metric h_{ij}, which fill out four-dimensional regions of spacetime. In quantum gravity one then studies the quantum theory of such three-geometries. By analogy with the string example, in gravity there is one scalar constraint $\mathscr{H}_\perp$ corresponding to normal displacements of a spacelike hypersurface with metric $h_{ij}(x)$, and one vector constraint $\mathscr{H}_i$, corresponding to tangential displacements.

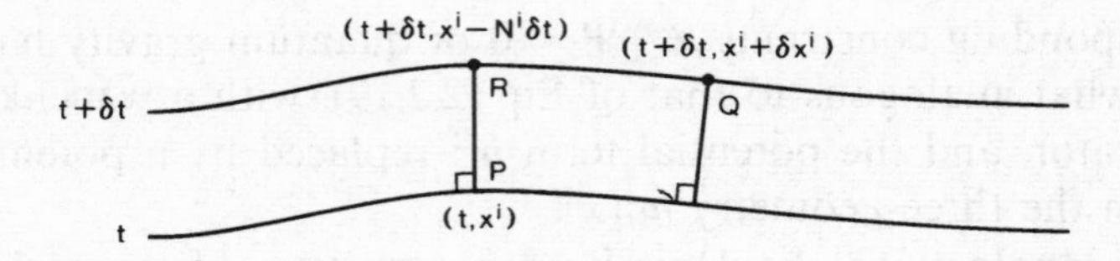

Fig. 2.2. Coordinates for two nearby hypersurfaces labelled by t and $(t + \delta t)$.

2.3 Hamiltonian treatment of general relativity

The Hamiltonian treatment of general relativity is based on a $3+1$ splitting of spacetime. Given a four-geometry, i.e. a spacetime with metric $g_{\mu\nu}$, consider a one-parameter family of three-dimensional spacelike hypersurfaces $t =$ const. , with spatial coordinates x^i $(i = 1, 2, 3)$ [Misner *et al.* 1973]. The *intrinsic metric* of each surface is $h_{ij} = g_{ij}$, with inverse h^{ij}. Three-dimensional indices are raised and lowered using h^{ij} and h_{ij}. The remaining g_{00} and g_{0i} metric components are described geometrically as follows (see Fig. 2.2). Consider a nearby pair of spacelike hypersurfaces labelled by coordinates t and $t + \delta t$. Follow the timelike normal vector from a point P with coordinates (t, x^i) on the initial surface to a point R on the final surface, which will have coordinates $(t + \delta t, x^i - N^i \delta t)$, where $N^i = N^i(t, x^k)$ is called the *shift vector*. The proper time elapsed from P to R is $N\delta t$, where $N = N(t, x^k)$ is the *lapse function*. Let Q be a point on the final surface with coordinates $(t + \delta t, x^i + \delta x^i)$ (see Fig. 2.2). From the Lorentzian geometry in Fig. 2.2, the squared interval $ds^2 = g_{\mu\nu} dx^\mu dx^\nu$ for the infinitesimal vector dx^μ describing PQ is

$$ds^2 = h_{ij}(dx^i + N^i dt)(dx^j + N^j dt) - (N dt)^2. \tag{2.3.1}$$

Hence

$$g_{\mu\nu} = \begin{pmatrix} g_{00} & g_{0k} \\ g_{i0} & g_{ik} \end{pmatrix} = \begin{pmatrix} (N_\ell N^\ell - N^2) & N_k \\ N_i & h_{ik} \end{pmatrix}. \tag{2.3.2}$$

One can verify that the inverse metric is

$$g^{\mu\nu} = \begin{pmatrix} g^{00} & g^{0m} \\ g^{k0} & g^{km} \end{pmatrix} = \begin{pmatrix} -1/N^2 & N^m/N^2 \\ N^k/N^2 & \left(h^{km} - \frac{N^k N^m}{N^2}\right) \end{pmatrix}, \tag{2.3.3}$$

by expanding the relation $g_{\mu\nu} g^{\mu\lambda} = \delta_\nu{}^\lambda$. Note that the volume element has the form

$$(-{}^4g)^{1/2} dx^0 dx^1 dx^2 dx^3 = N h^{1/2} dt dx^1 dx^2 dx^3, \tag{2.3.4}$$

where ${}^4g = \det(g_{\mu\nu})$ and $h = \det(h_{ij})$.

The unit future-directed *normal vector* n^μ to the hypersurface t =const. corresponds to the one-form n_μ given by $-Ndt$, with components

$$n_\mu \equiv (-N, 0, 0, 0). \tag{2.3.5}$$

Then

$$n^\mu = g^{\mu\nu} n_\nu \equiv \left(\frac{1}{N}, \frac{-N^m}{N}\right). \tag{2.3.6}$$

This vector points along PR in Fig. 2.2. Note that this shows that N^m is a vector under spatial coordinate transformations $x^i \to x^{i'}(x^j)$, with no transformation of t.

One can project any tensor orthogonally to n_μ [Hawking & Ellis 1973]. Define the four-dimensional (projection) tensor of rank 3,

$$h_{\mu\nu} = g_{\mu\nu} + n_\mu n_\nu = \begin{pmatrix} N_l N^l & N_k \\ N_i & h_{ik} \end{pmatrix}. \tag{2.3.7}$$

Define $h^{\mu\nu}$ as usual, by raising indices using $g^{\mu\nu}$. Then

$$h^{\mu\nu} = \begin{pmatrix} 0 & 0 \\ 0 & h^{ik} \end{pmatrix}. \tag{2.3.8}$$

Note that $h_\mu{}^\nu$ defines a projection into a three-dimensional subspace, orthogonal to n_μ, since

$$\begin{aligned} h_\mu{}^\nu h_\nu{}^\lambda &= h_\mu{}^\lambda, \\ h_\mu{}^\nu n_\nu &= 0, \\ h_\mu{}^\mu &= 3. \end{aligned} \tag{2.3.9}$$

Then one can (e.g.) project a tensor $T_{\alpha\beta\ldots}$ to

$$^{\|}T_{\lambda\mu\cdots} = h_\lambda{}^\alpha h_\mu{}^\beta \cdots T_{\alpha\beta\cdots}, \tag{2.3.10}$$

or a tensor $X^{\alpha\beta\cdots}$ to

$$^{\|}X^{\lambda\mu\cdots} = h^\lambda{}_\alpha h^\mu{}_\beta \cdots X^{\alpha\beta\cdots}, \tag{2.3.11}$$

where $^{\|}T$ and $^{\|}X$ are orthogonal to n^μ on all indices. Note that $h_i{}^\alpha = \delta_i{}^\alpha$ since $n_i = 0$ in the coordinates that we have been using. Hence

$$^{\|}T_{ij\cdots} = T_{ij\cdots}. \tag{2.3.12}$$

Recall also that there is a natural projection of covariant tensors $T_{\alpha\beta\cdots}$ in four dimensions to covariant tensors $^3T_{ij\cdots}$ on the three-dimensional surface t =const. [Hawking & Ellis 1973], given by

$$^3T_{ij\cdots} = T_{ij\cdots}. \tag{2.3.13}$$

Thus the two projections agree on spatial tensors.

One now needs a geometrical way of denoting the rate of change of the three-metric in the normal direction, in order to be able to discuss gravitational dynamics. This rate of change is given by the *extrinsic curvature* of a three-dimensional hypersurface embedded in a four-geometry. The four-dimensional version of this quantity is given by regarding the unit normal n^μ as a vector field in spacetime, in the case that there is a family of hypersurfaces t =const. One defines

$$v_{\lambda\mu} = -h_\lambda{}^\alpha h_\mu{}^\beta n_{\alpha;\beta}, \tag{2.3.14}$$

where ; denotes a four-dimensional covariant derivative. Since, by the above, $v_{\lambda\mu}$ is orthogonal to n^μ on both indices, one can obtain $v_{\lambda\mu}$ from its spatial components v_{ik}. The corresponding spatial tensor is the *second fundamental form* with components

$$K_{ik} = -n_{i;k} = -N^4\Gamma^0{}_{ik} \tag{2.3.15}$$

in a coordinate basis, where ${}^4\Gamma^0{}_{ik}$ are four-dimensional Christoffel symbols. Hence

$$K_{ik} = K_{ki} \tag{2.3.16}$$

is a symmetric spatial tensor.

If one chooses coordinates on neighbouring surfaces such that $N^i = 0$, then

$$h_\lambda{}^\alpha = \begin{pmatrix} 0 & 0 \\ 0 & \delta_i{}^j \end{pmatrix} \tag{2.3.17}$$

and

$$v_{\lambda\mu} = \begin{pmatrix} 0 & 0 \\ 0 & K_{ik} \end{pmatrix} = v_{\mu\lambda} \tag{2.3.18}$$

is symmetric. But the symmetry of a tensor is preserved under coordinate transformations. Hence

$$v_{\lambda\mu} = v_{(\lambda\mu)} \tag{2.3.19}$$

is symmetric in general.

One can relate K_{ik} to the (normal) time derivative of h_{ik} as follows. Note that

$$\begin{aligned} K_{ik} &= -N^4\Gamma^0{}_{ik} \\ &= -N(g^{00}\,{}^4\Gamma_{0ik} + g^{0m}\,{}^4\Gamma_{mik}), \end{aligned} \tag{2.3.20}$$

where

$$^4\Gamma_{\nu\lambda\mu} = \tfrac{1}{2}(g_{\nu\lambda,\mu} + g_{\nu\mu,\lambda} - g_{\lambda\mu,\nu}). \tag{2.3.21}$$

Hence

$$K_{ik} = \frac{1}{N}({}^4\Gamma_{0ik} - N^m\,{}^3\Gamma_{mik}), \tag{2.3.22}$$

where

$$\begin{aligned} {}^3\Gamma_{mik} &= \tfrac{1}{2}(h_{mi,k} + h_{mk,i} - h_{ik,m}) \\ &= {}^4\Gamma_{mik} \end{aligned} \tag{2.3.23}$$

are the three-dimensional Christoffel symbols. Hence

$$\begin{aligned} K_{ik} &= \frac{1}{2N}\left(\frac{\partial N_i}{\partial x^k} + \frac{\partial N_k}{\partial x^i} - \frac{\partial h_{ik}}{\partial t} - 2\Gamma_{mik}N^m\right) \\ &= \frac{1}{2N}\left(N_{i|k} + N_{k|i} - \frac{\partial h_{ik}}{\partial t}\right), \end{aligned} \tag{2.3.24}$$

where $_|$ denotes a three-dimensional covariant derivative. Another description of K_{ik} is through the Lie derivative [Hawking & Ellis 1973, Misner *et al.* 1973, Wald 1984]

$$v_{\lambda\mu} = -\tfrac{1}{2}(\mathcal{L}_n h)_{\lambda\mu} \tag{2.3.25}$$

of $h_{\lambda\mu}$ with respect to the normal n^μ.

As an example, let us evaluate the second fundamental form in a $k = +1$ Robertson–Walker universe, with metric

$$ds^2 = -N^2 dt^2 + a^2 d\Omega^2, \tag{2.3.26}$$

where

$$d\Omega^2 = d\chi^2 + \sin^2\chi(d\theta^2 + \sin^2\theta d\phi^2) \tag{2.3.27}$$

is the metric of a unit three-sphere, and

$$N = N(t),\ a = a(t). \tag{2.3.28}$$

Note that, for simplicity, the shift has been taken to be $N^i = 0$. Then

$$\begin{aligned} K_{ik} &= -\frac{1}{2N}\frac{\partial h_{ik}}{\partial t} \\ &= -\frac{1}{aN}\frac{da}{dt}h_{ik} \\ &= -\frac{1}{a}\frac{da}{d(\text{proper time}})h_{ik}. \end{aligned} \tag{2.3.29}$$

This will be used in section 2.7 on mini-superspace models.

Note that the second fundamental form K_{ij} of a three-surface only depends on the four-geometry near the three-surface, and on the way in which the three-surface is embedded. It does not depend on the particular choice of a neighbouring family of hypersurfaces $t =$ const. , since the definition $K_{ik} = -n_{i;k}$ makes no mention of other hypersurfaces. The definition (2.3.14) of $v_{\lambda\mu}$ is given for a single hypersurface by taking n^α over the hypersurface, and extending n^α in a smooth way, subject to $n^\alpha n_\alpha = -1$, in a neighbourhood of the hypersurface. Because of the projections in Eq. (2.3.14), only the spatial derivatives of n^α appear. The notion of the second fundamental form for a single hypersurface will be used at the end of this section, in studying boundary corrections to the Einstein–Hilbert action.

For use in the material on curvature which follows, note that there is an alternative description of the three-dimensional covariant derivative. Suppose $X_\mu n^\mu = 0$, i.e. that X_μ is 'spatial'. Define a three-dimensional covariant derivative

$$ {}^3\nabla_\mu X_\nu = h_\mu{}^\alpha h_\nu{}^\beta ({}^4\nabla_\alpha X_\beta), \tag{2.3.30}$$

and similarly for higher-rank tensors. This obeys the linearity and Leibnitz rules of a covariant derivative [Hawking & Ellis 1973]. Note that the connection is metric, in the sense

$$ {}^3\nabla_\mu h_{\lambda\nu} = h_\mu{}^\alpha h_\lambda{}^\beta h_\nu{}^\gamma\, {}^4\nabla_\alpha (g_{\beta\gamma} + n_\beta n_\gamma) = 0. \tag{2.3.31}$$

One can check that the spatial components ${}^3\nabla_i X_j$ agree with the spatial covariant derivative $X_{j|i}$, as follows:

$$\begin{aligned} {}^3\nabla_i X_j &= h_i{}^\alpha h_j{}^\beta ({}^4\nabla_\alpha X_\beta) \\ &= \delta_i{}^\alpha \delta_j{}^\beta ({}^4\nabla_\alpha X_\beta) \\ &= {}^4\nabla_i X_j \\ &= \partial_i X_j - {}^4\Gamma^\gamma{}_{ij} X_\gamma \\ &= \partial_i X_j - g^{\gamma\delta}\, {}^4\Gamma_{\delta ij} X_\gamma \\ &= \partial_i X_j - h^{\gamma\delta}\, {}^4\Gamma_{\delta ij} X_\gamma \text{ (since } X_\gamma n^\gamma = 0) \\ &= \partial_i X_j - h^{kl}\, {}^4\Gamma_{lij} X_k \\ &= \partial_i X_j - h^{kl}\, {}^3\Gamma_{lij} X_k \\ &= X_{j|i}. \end{aligned} \tag{2.3.32}$$

A similar result holds for ${}^3\nabla_i$ acting on higher-rank spatial tensors.

Now one needs to understand the $3+1$ decompostion of curvature, in order to write out the Einstein–Hilbert Lagrangian in $3+1$ form and hence obtain the Hamiltonian form of the theory. In particular, it will be necessary to obtain those components of the Einstein tensor $G_{\mu\nu} = R_{\mu\nu} - \frac{1}{2} g_{\mu\nu} R$ given by contraction once or twice with the normal n^μ.

First, one uses the derivative ${}^3\nabla_i$ to relate the three-dimensional Riemann tensor ${}^3R_{kmij}$ to ${}^4R_{kmij}$. Suppose that $X_\mu n^\mu = 0$. Then

$$\begin{aligned} {}^3\nabla_i {}^3\nabla_j X_k - {}^3\nabla_j {}^3\nabla_i X_k &= X_{k|ji} - X_{k|ij} \\ &= {}^3R_{klij} X^l. \end{aligned} \tag{2.3.33}$$

And

$$^3\nabla_i(^3\nabla_j X_k) = h_i{}^\lambda h_j{}^\mu h_k{}^\nu\, {}^4\nabla_\lambda(h_\mu{}^\rho h_\nu{}^\sigma\, {}^4\nabla_\rho X_\sigma). \tag{2.3.34}$$

One expands out the right-hand side of Eq. (2.3.34), subtracts the term $^3\nabla_j(^3\nabla_i X_k)$, and simplifies using

$$\begin{aligned} h_j{}^\rho n^\sigma(^4\nabla_\rho X_\sigma) &= {}^4\nabla_\rho(n^\sigma X_\sigma)h_j{}^\rho - X_\sigma(^4\nabla_\rho n^\sigma)h_j{}^\rho \\ &= -X_\sigma(^4\nabla_\rho n^\sigma)h_j{}^\rho. \end{aligned} \tag{2.3.35}$$

This is a straightforward but tedious calculation, leading [Hawking & Ellis 1973, Wald 1984] to *Gauss' equation*

$$^3R_{kmij} = {}^4R_{kmij} + K_{mi}K_{jk} - K_{mj}K_{ik}. \tag{2.3.36}$$

Consider the contraction of (2.3.36) with $h^{ki}h^{mj}$. The left-hand side gives ^3R. For the right-hand side, note that

$$\begin{aligned} h^{ki}h^{mj}\, {}^4R_{kmij} &= h^{\lambda\nu}h^{\mu\rho}\, {}^4R_{\lambda\mu\nu\rho} \\ &= (g^{\lambda\nu} + n^\lambda n^\nu)(g^{\mu\rho} + n^\mu n^\rho)\, {}^4R_{\lambda\mu\nu\rho} \\ &= {}^4R + 2n^\lambda n^\nu\, {}^4R_{\lambda\nu} \\ &= 2n^\lambda n^\nu G_{\lambda\nu}, \end{aligned} \tag{2.3.37}$$

where $G_{\mu\nu} = {}^4R_{\mu\nu} - \frac{1}{2}\, {}^4Rg_{\mu\nu}$ is the Einstein tensor. Hence

$$\begin{aligned} 2n^\mu n^\nu G_{\mu\nu} &\equiv G_{\perp\perp} \\ &= {}^3R + K^2 - K_{ik}K^{ik}, \end{aligned} \tag{2.3.38}$$

where

$$K = \text{tr}K = K^i{}_i. \tag{2.3.39}$$

Note that the $G_{\perp\perp} = 0$ vacuum equation does not involve second time derivatives of the three-metric; it is a constraint equation which restricts the form of the initial data for the Einstein equations.

The mixed components $^4G_{\lambda\mu}n^\lambda h^\mu{}_i$ are given by an analogous calculation. Note that $v_{\lambda\mu}$ defined by Eq. (2.3.14) obeys

$$v^\lambda{}_\lambda = h^{\lambda\mu}v_{\lambda\mu} = h^{ij}K_{ij} = K. \tag{2.3.40}$$

Consider

$$\begin{aligned} ^3\nabla_\mu(v^\lambda{}_\lambda) &= h_\mu{}^\alpha\, {}^4\nabla_\alpha(v^\lambda{}_\lambda) \\ &= -h_\mu{}^\alpha\, {}^4\nabla_\alpha(h^{\beta\gamma}n_{\beta;\gamma}) \end{aligned} \tag{2.3.41}$$

and

$$^3\nabla_\lambda(v^\lambda{}_\mu) = -h_\lambda{}^\alpha h_\mu{}^\nu\, {}^4\nabla_\alpha(h^{\lambda\beta}h_\nu{}^\gamma n_{\beta;\gamma}). \tag{2.3.42}$$

Subtracting these, one finds by another tedious calculation that

$$^3\nabla_\lambda(v^\lambda{}_\mu) - {}^3\nabla_\mu(v^\lambda{}_\lambda) = -{}^4R_{\lambda\nu}n^\nu h^\lambda{}_\mu. \quad (2.3.43)$$

Now take the $\mu = i$ components. For $^3\nabla_\lambda(v^\lambda{}_\mu) = h^{\lambda\nu}\,{}^3\nabla_\lambda(v_{\nu\mu})$ this gives $h^{kl}\,{}^3\nabla_k(v_{li}) = K^k{}_{i|k}$. For $^3\nabla_\mu(v^\lambda{}_\lambda)$ this gives $K_{|i}$. Hence one obtains *Codazzi's equation* [Hawking & Ellis 1973, Wald 1984]

$$\begin{aligned} -K^k{}_{i|k} + K_{|i} &= {}^4R_{\lambda\nu}n^\nu h^\lambda{}_i \\ &= G_{\lambda\nu}n^\nu h^\lambda{}_i. \end{aligned} \quad (2.3.44)$$

As with Eq. (2.3.38), the $G_{\lambda\nu}n^\nu h^\lambda{}_i = 0$ vacuum equation is another constraint, restricting the form of the initial data for the Einstein equations.

As can be seen from Eq. (2.3.37), to obtain the $3+1$ version of ^4R, one also needs

$$\begin{aligned} n^\lambda n^\mu\,{}^4R_{\lambda\mu} &= g^{\lambda\nu}n^\mu(n_{\lambda;\mu\nu} - n_{\lambda;\nu\mu}) \\ &= g^{\lambda\nu}\,{}^4\nabla_\nu(n^\mu n_{\lambda;\mu}) - g^{\lambda\nu}n^\mu{}_{;\nu}n_{\lambda;\mu} \\ &\quad - g^{\lambda\nu}\,{}^4\nabla_\mu(n^\mu n_{\lambda;\nu}) + g^{\lambda\nu}n^\mu{}_{;\mu}n_{\lambda;\nu}. \end{aligned} \quad (2.3.45)$$

Hence

$$n^\lambda n^\mu\,{}^4R_{\lambda\mu} = {}^4\nabla_\mu(n^\lambda n^\mu{}_{;\lambda} - n^\mu n^\lambda{}_{;\lambda}) - K_{ik}K^{ik} + K^2. \quad (2.3.46)$$

Note that the last two terms in Eq. (2.3.46) can be evaluated from Eq. (2.3.45) in coordinates in which

$$N^i = 0, g_{\mu\nu} = \begin{pmatrix} -N^2 & 0 \\ 0 & h_{ik} \end{pmatrix}. \quad (2.3.47)$$

Then $n_{\alpha;\beta}n^\alpha = 0$ implies $n_{0;0} = 0$ and $n_{0;i} = 0$. Hence $K = -h^{ij}n_{i;j} = -g^{\mu\nu}n_{\mu;\nu} = -n^\mu{}_{;\mu}$ and $n_{\lambda;\mu}n^{\mu;\lambda} = n_{i;j}n^{j;i} = K_{ik}K^{ik}$. Combining Eq. (2.3.37) and Eq. (2.3.46), one obtains

$$\begin{aligned} {}^4R &= ({}^3R + K^2 - K_{ik}K^{ik}) - 2n^\lambda n^\mu\,{}^4R_{\lambda\mu} \\ &= ({}^3R - K^2 + K_{ik}K^{ik}) - 2\,{}^4\nabla_\mu(n^\lambda n^\mu{}_{;\lambda} - n^\mu n^\lambda{}_{;\lambda}). \end{aligned} \quad (2.3.48)$$

The Einstein–Hilbert action of general relativity is

$$S_{EH} = \int d^4xL, \quad (2.3.49)$$

where

$$L = \frac{1}{2\kappa^2}(-g)^{1/2}\,{}^4R, \quad (2.3.50)$$

with $\kappa^2 = 8\pi$ in units with $c = G = 1$, and $g = {}^4g = \det(g_{\mu\nu})$. From Eqs. (2.3.24) and (2.3.48), one can see that the momentum canonically conjugate to h_{ij} is

$$\frac{\delta L}{\delta \dot{h}_{ij}} \equiv \frac{1}{2\kappa^2}\pi^{ij} = \frac{1}{2\kappa^2}h^{1/2}(K^{ij} - Kh^{ij}), \tag{2.3.51}$$

a tensor density, where a dot denotes a time derivative. The momenta canonically conjugate to N and N^i are zero. Accordingly one is led to take the configuration space to be the set of Riemannian metrics h_{ij} on a given three-manifold Σ. At present, the total divergence in Eq. (2.3.48) is being neglected in L, since

$$\begin{aligned} \int d^4x(-g)^{\frac{1}{2}}\,{}^4\nabla_\mu A^\mu &= \int d^4x \frac{\partial}{\partial x^\mu}[(-g)^{\frac{1}{2}}A^\mu] \\ &= \text{boundary terms}. \end{aligned} \tag{2.3.52}$$

However, these boundary terms are crucial in subsequent work, as we shall see.

When K^{ij} is replaced by π^{ij} in L, following Eq. (2.3.51), one obtains

$$\begin{aligned} 2\kappa^2 L = \pi^{ij}\dot{h}_{ij} + h^{1/2}N\,{}^3R - Nh^{1/2}[\pi^{ij}\pi_{ij} - \tfrac{1}{2}(\text{tr}\pi)^2] \\ - 2\pi^{ij}({}^3\nabla_i N_j) + \text{total divergence}. \end{aligned} \tag{2.3.53}$$

Now it is more convenient to rearrange the term $-2\pi^{ij}({}^3\nabla_i N_j)$, writing

$$-2\pi^{ij}({}^3\nabla_i N_j) = -2\,{}^3\nabla_i(\pi^{ij}N_j) + 2N_j({}^3\nabla_i \pi^{ij}), \tag{2.3.54}$$

where ${}^3\nabla_i(\pi^{ij}N_j)$ is a total divergence $\partial_i(\pi^{ij}N_j)$ since π^{ij} is a spatial tensor density. This spatial divergence will give zero if the spatial hypersurface Σ is compact or if (Σ, h_{ij}) is asymptotically flat with $N_i = 0$ at large distances. There are many cases where these conditions hold. However, in section 4.4 we shall see an example where N_i only falls off slowly at large distances, and the divergence $-2\partial_i(\pi^{ij}N_j)$ gives an important boundary contribution at spatial infinity to the mass and the action.

The Lagrangian can thus be written as

$$2\kappa^2 L = \pi^{ij}\dot{h}_{ij} - \mathscr{H} + \text{total divergence}, \tag{2.3.55}$$

where

$$\mathscr{H} = N\mathscr{H}_\perp + N_i\mathscr{H}^i, \tag{2.3.56}$$

with

$$\mathscr{H}_\perp = h^{-1/2}[\pi^{ij}\pi_{ij} - \tfrac{1}{2}(\text{tr}\pi)^2] - (h^{1/2})\,{}^3R, \tag{2.3.57}$$

$$\mathscr{H}^i = -2\pi^{ik}{}_{|k}. \tag{2.3.58}$$

From the variation of N and N^i, one obtains the *classical constraints*

$$\mathscr{H}_\perp = 0, \tag{2.3.59}$$

$$\mathscr{H}^i = 0, \tag{2.3.60}$$

on the Hamiltonian initial data h_{ij} and π^{ij}.

One can find the corresponding equations for the case of matter coupled to gravity, by noting from Eqs. (2.3.38),(2.3.51) that

$$\begin{aligned}\mathscr{H}_{\perp} &= -2h^{1/2}G_{\perp\perp},\\ &= -2h^{1/2}G_{\mu\nu}n^{\mu}n^{\nu},\end{aligned} \tag{2.3.61}$$

and from Eqs. (2.3.44),(2.3.51) that

$$\mathscr{H}_{i} = -2h^{1/2}G_{\lambda i}n^{\lambda}, \tag{2.3.62}$$

where $G_{\mu\nu}$ is the Einstein tensor. When matter is included, the constraints become

$$\mathscr{H}_{\perp\text{grav}} = -2\kappa^2 h^{1/2}T_{\perp\perp}, \tag{2.3.63}$$

$$\mathscr{H}_{i\text{grav}} = 2\kappa^2 h^{1/2}T_{\lambda i}n^{\lambda}, \tag{2.3.64}$$

when one recalls the Einstein field equations

$$G_{\mu\nu} = \kappa^2 T_{\mu\nu}. \tag{2.3.65}$$

Here $\mathscr{H}_{\perp\text{grav}}$ is the right-hand side of Eq. (2.3.57), and $\mathscr{H}_{i\text{grav}}$ is that of Eq. (2.3.58). Note that the matter contributions $T_{\perp\perp}$ and $T_{\lambda i}n^{\lambda}$ to the total constraints

$$\mathscr{H}_{\perp} = \mathscr{H}_{\perp\text{grav}} + 2\kappa^2 h^{1/2}T_{\perp\perp}$$

and

$$\mathscr{H}_{i} = \mathscr{H}_{i\text{grav}} - 2\kappa^2 h^{1/2}T_{\lambda i}n^{\lambda}$$

never involve second derivatives of the matter field. For example, for a scalar field ϕ, $T_{\perp\perp}$ and $T_{\lambda i}n^{\lambda}$ will involve first derivatives $(\partial_0\phi)^2$ and $(\partial_i\phi)(\partial_0\phi)$. Hence the classical constraints $\mathscr{H}_{\perp} = 0, \mathscr{H}_{i} = 0$ now restrict the Hamiltonian initial data for the coupled gravity–matter system.

Variation of h_{ij} and π^{ij} in the Hamiltonian form (2.3.55–58) of the action leads to the Einstein field equations in first-order form [Misner *et al.* 1973]. Variation of π^{ij} gives back the definition (2.3.51) of π^{ij}, where K_{ij} is written as in Eq. (2.3.24) in terms of $\partial h_{ij}/\partial t$ plus further contributions. Variation of h_{ij} leads to the first-order form $\partial\pi^{ij}/\partial t = \cdots$ of the dynamical Einstein equations, corresponding essentially to the G_{ij} field equations.

To interpret the form of the Hamiltonian

$$H = \frac{1}{2\kappa^2}\int d^3x(N\mathscr{H}_{\perp} + N^i\mathscr{H}_{i}), \tag{2.3.66}$$

one can compare with the discusion of parametrized particle dynamics and string theory in section 2.2. Here the four constraints $\mathscr{H}_{\perp} = 0, \mathscr{H}_{i} = 0$ correspond to the four degrees of gauge freedom, namely diffeomorphisms, in general relativity. N and N^i are the associated *Lagrange multipliers.*

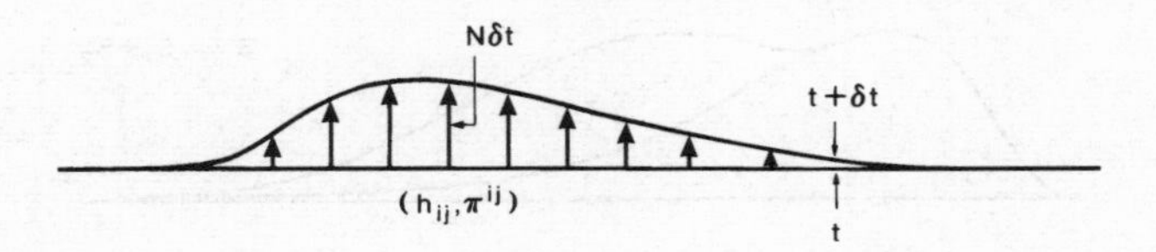

Fig. 2.3. The constraint $\mathscr{H}_\perp$ generates evolution of initial data in the normal direction.

Here $\mathscr{H}_\perp$ generates evolution of initial data (h_{ij}, π^{ij}) in the normal direction (Fig. 2.3), while $\mathscr{H}_i$ generates evolution with $N = 0, N^i \neq 0$, i.e. evolution within the initial surface, corresponding to three-dimensional or spatial diffeomorphisms. As above, the initial data (h_{ij}, π^{ij}) must satisfy the constraints $\mathscr{H}_\perp = 0, \mathscr{H}^i = 0$. The constraints are then preserved under the Hamiltonian evolution. This is because the algebra of Poisson brackets closes on itself [Teitelboim 1980], i.e. the Poisson bracket of any pair of constraints is linear in the constraints. The coefficients of the constraints in the expansion of the right-hand side of such an equation are 'structure functions' rather than 'structure constants', since they may depend on the fields.

So far, the total divergence terms in the Einstein–Hilbert action have not been treated fully. These terms play an important role in the classical action and subsequent quantization. From Eq. (2.3.48), one has

$$\begin{aligned} S_{EH} &= \frac{1}{2\kappa^2}\int d^4x(-g)^{1/2}\,{}^4R \\ &= \frac{1}{2\kappa^2}\int d^4x(-g)^{1/2}({}^3R - K^2 + K_{ik}K^{ik}) \\ &+ \frac{1}{\kappa^2}\int_{\text{boundary}} dS_\mu(n^\mu n^\lambda{}_{;\lambda} - n^\lambda n^\mu{}_{;\lambda}). \end{aligned} \tag{2.3.67}$$

Recall that $n_\mu n^\mu{}_{;\lambda} = 0$, implying $dS_\mu n^\mu{}_{;\lambda} = 0$. Further, $n^\lambda{}_{;\lambda} = \text{tr}K$. Hence in the modified Einstein–Hilbert action [Hawking 1979]

$$S_{\text{mod}} = \frac{1}{2\kappa^2}\int_{\text{volume}} d^4x(-g)^{1/2}\,{}^4R + \frac{1}{\kappa^2}\int_{\text{boundary}} d^3x\, h^{1/2}(\text{tr}K), \tag{2.3.68}$$

with tr$K$ evaluated for the outward normal, the second normal derivatives of the metric have been eliminated. This can be used in either the compact or the non-compact case. Note that, in the 'scattering' case in which the three-surface Σ is asymptotically flat (Fig. 1.2), this modification has already been made at the initial and final boundaries Σ_0, Σ_1 in Eq. (2.3.53) for $2\kappa^2 L = \pi^{ij}\dot{h}_{ij} + \cdots$. The original Einstein–Hilbert action gives instead the term $-\dot{\pi}^{ij}h_{ij}$. An integration by parts has already been carried out, corresponding to the (tr$K$) boundary term in (2.3.68), since $\pi^{ij}h_{ij} \propto$ (tr$K$). Note, however, that one must be careful about boundary contributions at

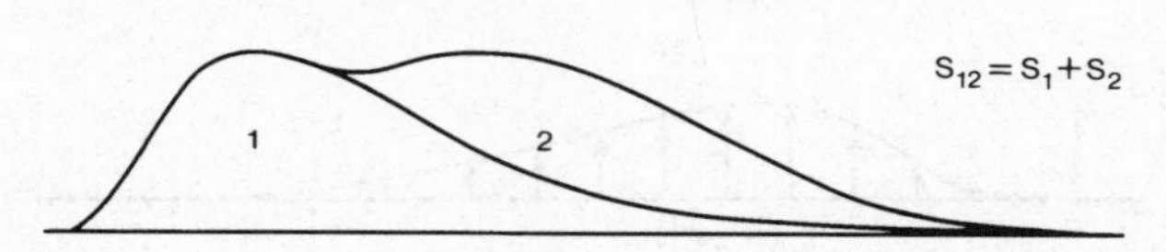

Fig. 2.4. The action $S_{\rm mod}$ including boundary corrections is additive over adjacent spacetime regions.

spatial infinity in the non-compact case, as mentioned above in the case of $\pi^{ij}({}^3\nabla_i N_j)$. In general [Teitelboim 1977b], these give terms in the action proportional to the total mass and momentum of the spacetime.

Including the $(\mathrm{tr}K)$ term in the action is analogous to writing the action in terms of first derivatives of the potential in Maxwell or Yang–Mills theory. The modified action $S_{\rm mod}$ of Eq. (2.3.68) is *additive* over adjacent spacetime regions (Fig. 2.4), when one requires the metric to be continuous at the boundary, but allows for a discontinuity in the first normal derivative, as in the Feynman path integral in quantum theory (section 2.5). This means that the quantum theory describes a Markov process [Simon 1974], in which the quantum amplitude can depend only on boundary data such as that posed naturally (chapters 3,4) on the asymptotically flat surfaces Σ_0, Σ_1 of Fig. 1.2. Finally, note that, at a classical solution, ${}^4R = 0$, so that the classical action is given by the $(\mathrm{tr}K)$ term in $S_{\rm mod}$ of Eq. (2.3.68).

2.4 Classical boundary-value problem

Before moving on to the quantum theory, it is helpful briefly to consider the classical analogue of a quantum amplitude. In the non-relativistic motion of a single particle (section 2.2), one specifies an initial position $\mathbf{x}_1$ on an initial surface at time t_1, and a final position $\mathbf{x}_2$ at a final surface at time t_2; in principle one then can find the quantum amplitude for these boundary data by solving the Schrödinger equation. One can approximate the quantum amplitude by finding classical paths from the initial to the final data (Fig. 2.5). If the action for a classical path is $S_{\rm class}$, then the path contributes to the amplitude an amount $P\exp(iS_{\rm class}/\hbar)$, where P is a quantum prefactor which has an asymptotic expansion in powers of Planck's constant $\hbar$. There may be one, many, or no real classical solutions, depending on the form of the potential V. One should also include complex solutions; such a solution, if it exists, will always contribute to the quantum amplitude.

The corresponding boundary data in the case of an asymptotically flat gravitational field are the initial three-metric h_{ijI} on a three-surface Σ_0, and a final three-metric h_{ijF} on a final three-surface Σ_1 (Fig. 1.2). To complete

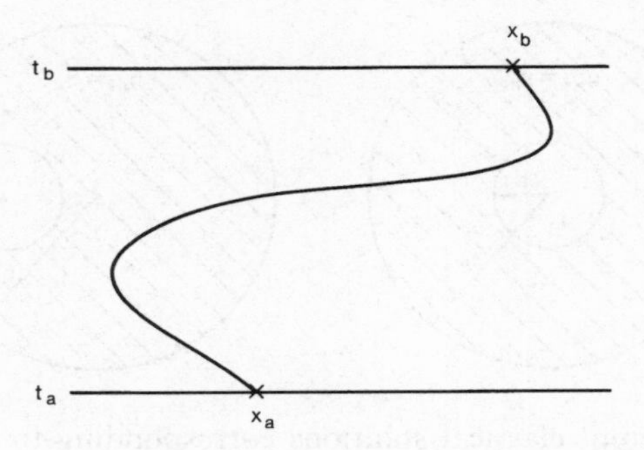

Fig. 2.5. A classical path from initial to final data.

the specification, one gives the proper time interval T between the surfaces, measured at spatial infinity. For completeness [Teitelboim 1977b], one should also include a quantity which specifies the relative spatial translation between the two surfaces. One then asks whether there is an asymptotically flat solution $g_{\mu\nu}$ of the vacuum Einstein equations which agrees with the boundary data, i.e. is such that $g_{ij} = h_{ijI}$ or h_{ijF} at the boundaries, and such that there is a proper time interval T at infinity. Now for weak fields, the transverse-traceless wave modes of the gravitational field obey the flat-space wave equation [Misner *et al.* 1973]. Thus in this case one can model the gravitational boundary-value problem by the boundary-value problem for a scalar field ϕ, obeying the flat-space wave equation $\Box\phi = 0$. As an example, assume that $\phi(t_1 = 0, \mathbf{x}) = 0$ and that $\phi(t_2, \mathbf{x})$ is specified and non-zero. Write $\tilde{\phi}(t_2, \mathbf{k})$ for the spatial Fourier transform of $\phi(t_2, \mathbf{x})$. One obtains the formal solution to the wave equation

$$\phi(t, \mathbf{x}) = \text{const.} \int d^3k \frac{\tilde{\phi}(t_2, \mathbf{k})}{\sin(|\mathbf{k}|\, t_2)} e^{i\mathbf{k}.\mathbf{x}} \sin(|\mathbf{k}|\, t). \qquad (2.4.1)$$

This expression is singular unless $\tilde{\phi}(t_2, \mathbf{k}) = \mathbf{0}$ whenever $|\mathbf{k}| = n\pi/t_2$, for positive integers n. Thus, for generic boundary data, the 'solution' is singular. This is an example of a well-known phenomenon, that the boundary-value problem for a hyperbolic system of equations is not well posed [Garabedian 1964].

The natural arena for a boundary-value problem is, of course, in the context of elliptic equations. Thus, in the gravitational case, one can study the asymptotically flat boundary-value problem of Fig. 1.2, with initial data h_{ijI} and final data h_{ijF}, and an imaginary time interval $-iT$ at infinity. In the case of a compact three-surface Σ, one can study configurations (Fig. 2.1) in which there is a specified three-metric h_{ij} on Σ, which is the only boundary of a compact manifold-with-boundary,

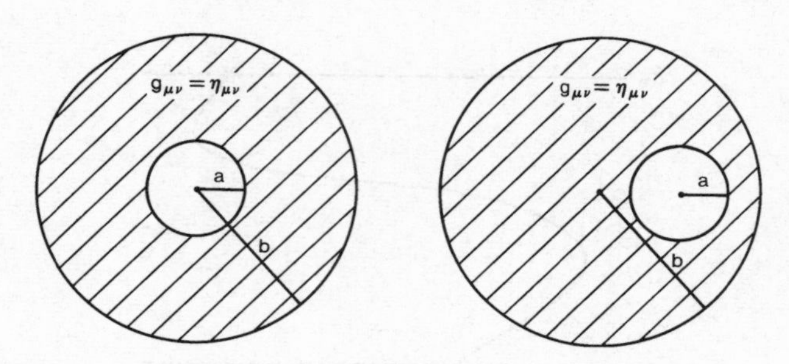

Fig. 2.6. Two different classical solutions corresponding to the same boundary conditions.

with four-metric $g_{\mu\nu}$. One can ask whether there is a (real) Riemannian solution $g_{\mu\nu}$ of the field equations for vacuum or more general matter, subject to these boundary conditions. For weak perturbations of flat space data, under certain conditions, it has been shown in [Reula 1987] that the classical *vacuum* boundary-value problem is well posed; the metric $g_{\mu\nu}$ is real, near Euclidean four-space. The behaviour of the vacuum solution $g_{\mu\nu}$, as the data are deformed further away from flatness, is not known. In some non-vacuum examples (section 2.7) one can only find complex classical solutions in the compact case; the two examples studied there involve gravity with a cosmological constant Λ and a conformal scalar field ϕ, and gravity with a massive scalar field ϕ.

Note that, even if one knows that solutions exist, the solutions may not be unique. This certainly occurs in the case of compact Σ when there is more than one connected component to the boundary (Fig. 2.6). For example, one can take two spherical boundaries, of radii a and b, with $a < b$ (say). As in Fig. 2.6, these can be placed in different relative positions. There is thus a space of classical solutions. One expects that this will still hold for boundary data which are not exactly spherical.

The semi-classical approximation to the gravitational quantum amplitude is $P_E \exp(-I/\hbar)$, where $I = -iS$ is the Euclidean (Riemannian) classical action, and P_E is a quantum prefactor. In the asymptotically flat case, one can rotate $-iT$ back towards T, moving through a family of complex four-metrics $g_{\mu\nu}$, in an attempt to recover the Lorentzian régime. Here one must always work with T which has a (small) negative imaginary part, corresponding to the $+i\epsilon$ prescription in Feynman-diagram theory [Itzykson & Zuber 1980].

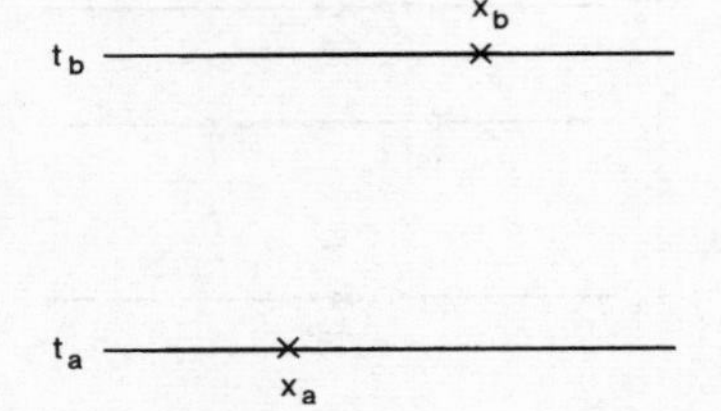

Fig. 2.7. Boundary data for the non-relativistic particle.

2.5 Feynman path integral in non-relativistic quantum mechanics

One would like to understand possible quantum versions of the theories described in this chapter. In the case of non-relativistic motion of a single particle (e.g. in the non-parametrized form), one can obtain the quantum theory, in the form of the quantum amplitude, by means of a Feynman path integral [Feynman & Hibbs 1965]. This will be outlined briefly here. It will be shown that the Schrödinger equation (2.2.12) follows from the path integral, thus linking the differential and integral forms of the theory. Conversely it is possible to derive the path integral from the Schrödinger equation by studying the form of the quantum amplitude for short time intervals [Dirac 1933].

For simplicity, consider the non-relativistic motion of a single particle in one space dimension. The Lagrangian is

$$L(\dot{x}, x, t) = \frac{m}{2}\dot{x}^2 - V(x, t). \tag{2.5.1}$$

Let us write

$$K(b, a) = K(x_b, t_b; x_a, t_a) \tag{2.5.2}$$

for the quantum amplitude to go from position x_a at time t_a to position x_b at time t_b (Fig. 2.7). One can use $K(b, a)$ to evolve physical wave functions ψ:

$$\psi(x_b, t_b) = \int_{-\infty}^{\infty} K(b, a)\psi(x_a, t_a)dx_a. \tag{2.5.3}$$

One can also combine amplitudes, as in (Fig. 2.8)

$$K(b, a) = \int_{-\infty}^{\infty} K(b, c)K(c, a)dx_c, \tag{2.5.4}$$

and this result can be iterated (Fig. 2.9). If neighbouring vertices in Fig. 2.9 are joined by straight lines, the typical path is continuous but not differentiable. Nevertheless, because the action

$$S[b, a] = \int_{t_a}^{t_b} L(\dot{x}, x, t)dt \tag{2.5.5}$$

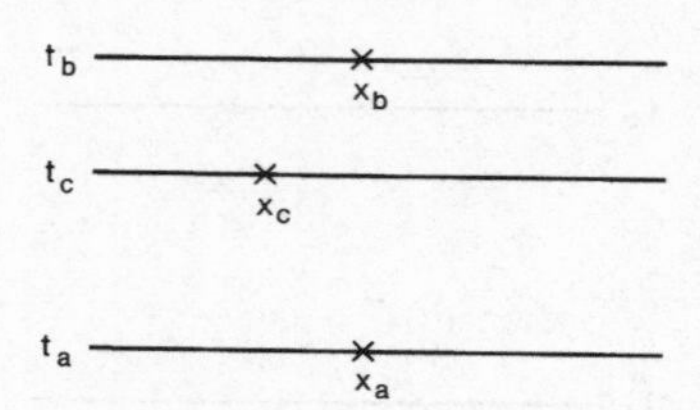

Fig. 2.8. Quantum amplitudes are combined by integrating over x_c.

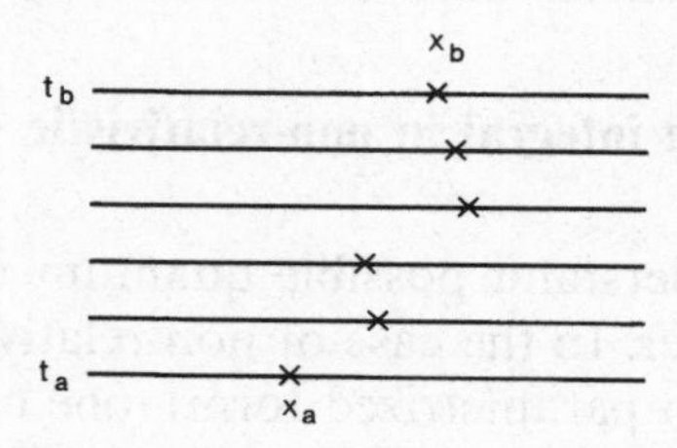

Fig. 2.9. Typical intermediate points to be integrated over in the propagation of a non-relativisitic particle.

with Eq. (2.5.1) only involves first derivatives $\dot{x}$, but not second derivatives, the action of a broken path is additive (Fig. 2.10):

$$S[b,a] = S[b,c] + S[c,a]. \tag{2.5.6}$$

This property is crucial in the path integral. The analogous property in general relativity for the action S_{mod} was described at the end of section 2.3: again, the action S_{mod} is additive over adjacent spacetime regions, where the normal derivative of the three-metric h_{ij} (analogous to the point x in non-relativistic quantum mechanics) may be discontinuous across the boundary. This property for S_{mod} in the general relativistic case occurs analogously because the action only involves first normal derivatives of the metric.

To define the path integral for a single particle, subdivide the interval $[t_a, t_b]$ into M equal parts, with

$$\begin{aligned} M\epsilon &= t_b - t_a, \\ \epsilon &= t_{i+1} - t_i, \ (0 \le i \le M-1), \\ t_0 &= t_a, \ t_M = t_b, \\ x_0 &= x_a, \ x_M = x_b. \end{aligned} \tag{2.5.7}$$

Then [Feynman & Hibbs 1965] one can show that the amplitude $K(b,a)$ is given by

$$K(b,a) = \lim_{\epsilon\to 0}\frac{1}{A}\int\int\cdots\int \exp\left(\frac{i}{\hbar}S[b,a]\right)\frac{dx_1}{A}\frac{dx_2}{A}\cdots\frac{dx_{M-1}}{A}, \tag{2.5.8}$$

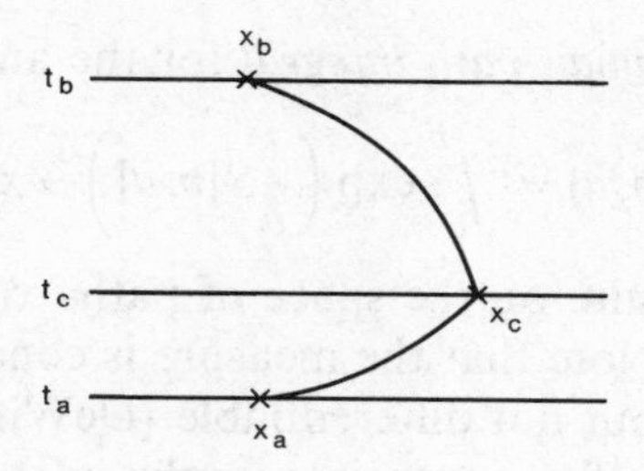

Fig. 2.10. The action is additive over adjacent regions.

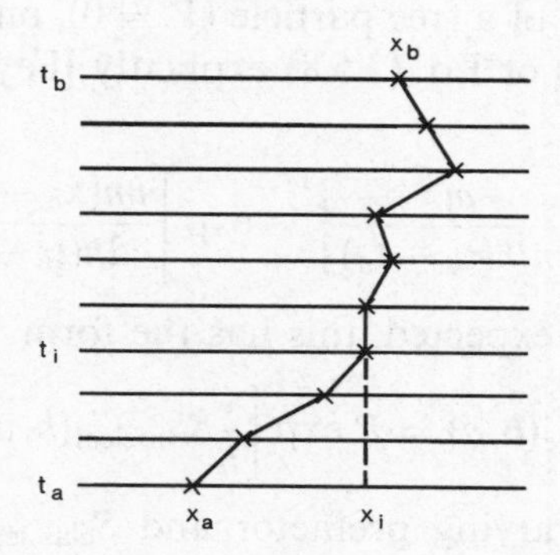

Fig. 2.11. A typical path to be summed over in the path integral.

where $S[b, a]$ is the action for the trajectory passing through the points x_i with straight sections in between (Fig. 2.11).

Here

$$A = \left(\frac{2\pi i\hbar\epsilon}{m}\right)^{1/2} \tag{2.5.9}$$

Equivalently, one can derive the path integral by using the iteration of amplitudes following Eq. (2.5.4), which gives

$$\begin{aligned} K(b, a) = \lim_{M\to\infty} \int_{x_1}\int_{x_2} \cdots \int_{x_{M-1}} & K(b, M-1) \\ \times K(M-1, M-2) & \cdots K(1, a) dx_1 dx_2 \cdots dx_{M-1}. \end{aligned} \tag{2.5.10}$$

It can be shown [Feynman & Hibbs 1965] that in the short-time-interval limit, one may approximate $K(i+1, i)$ as

$$K(i+1, i) = \frac{1}{A} \exp\left[\frac{i\epsilon}{\hbar} L\left(\frac{x_{i+1} - x_i}{\epsilon}, \frac{x_{i+1} + x_i}{2}, \frac{t_{i+1} + t_i}{2}\right)\right], \tag{2.5.11}$$

which is correct to first order in ϵ. With this approximation, one obtains an expression of the type (2.5.8), with a careful definition of the action for each segment of the path.

This defines the *Feynman path integral* for the amplitude,

$$K(b,a) = \int_a^b \exp\left(\frac{i}{\hbar} S[b,a]\right) \mathscr{D}x(t). \tag{2.5.12}$$

Here $\mathscr{D}x(t)$ is the measure on the space of paths $x(t)$, defined via the limiting processes above. Note that the measure is concentrated on paths $x(t)$ which are continuous but not differentiable [DeWitt–Morette *et al.* 1979]. Thus one sums over all appropriate paths $x(t)$, with measure $\mathscr{D}x(t)$, weighted by $\exp(i\hbar^{-1}S[b,a])$.

For the simplest case of a free particle ($V = 0$), one can do the Gaussian integrals in Eq. (2.5.12) or Eq. (2.5.8) explicitly [Feynman & Hibbs 1965], to obtain

$$K_{\text{free}}(b,a) = \left[\frac{m}{2\pi i\hbar(t_b - t_a)}\right]^{1/2} \exp\left[\frac{im(x_b - x_a)^2}{2\hbar(t_b - t_a)}\right], \tag{2.5.13}$$

the correct answer. As expected, this has the form

$$K_{\text{free}}(b,a) = P \exp\left[\frac{i}{\hbar} S_{\text{classical}}(b,a)\right], \tag{2.5.14}$$

where P is a slowly varying prefactor and $S_{\text{classical}}$ is the action of the classical straight-line path from (x_a, t_a) to (x_b, t_b).

Given the Feynman path integral, one deduces the Schrödinger equation for a wave function $\psi(x,t)$ propagated by $K(b,a)$, as in Eq. (2.5.3) [Feynman & Hibbs 1965]. One uses the approximation in Eq. (2.5.11), to obtain, for propagation through a time interval ϵ:

$$\begin{aligned}\psi(x, t+\epsilon) &= \int_{-\infty}^{\infty} \frac{1}{A} \exp\left[\frac{i\epsilon}{\hbar} L\left(\frac{x-y}{\epsilon}, \frac{x+y}{2}, t + \frac{\epsilon}{2}\right)\right] \psi(y,t)dy \\ &= \int_{-\infty}^{\infty} \frac{1}{A} \exp\left[\frac{im(x-y)^2}{2\hbar\epsilon}\right] \\ &\quad \times \exp\left[\frac{-i\epsilon}{\hbar} V\left(\frac{x+y}{2}, t + \frac{\epsilon}{2}\right)\right] \psi(y,t)dy. \end{aligned} \tag{2.5.15}$$

Let us write $y = x + \eta$; one expects destructive interference from the $\exp[im(x-y)^2/2\hbar\epsilon]$ term in Eq. (2.5.15), unless η is small, of order $(\epsilon\hbar/m)^{1/2}$. Thus

$$\begin{aligned}\psi(x, t+\epsilon) &= \int_{-\infty}^{\infty} \frac{1}{A} \exp\left(\frac{im\eta^2}{2\hbar\epsilon}\right) \\ &\quad \times \exp\left[\frac{-i\epsilon}{\hbar} V\left(x + \frac{\eta}{2}, t + \frac{\epsilon}{2}\right)\right] \psi(x+\eta, t)d\eta. \end{aligned} \tag{2.5.16}$$

One expands ψ in a power series, keeping only terms of order ϵ, and hence only terms of second order in η. Then

$$\begin{aligned}\psi(x,t)+\epsilon\frac{\partial\psi}{\partial t}+\cdots=\int_{-\infty}^{\infty}\frac{1}{A}\exp\left(\frac{im\eta^2}{2\hbar\epsilon}\right)\left[1-\frac{i\epsilon}{\hbar}V(x,t)+\cdots\right]\\ \times\left[\psi(x,t)+\eta\frac{\partial\psi}{\partial x}+\tfrac{1}{2}\eta^2\frac{\partial^2\psi}{\partial x^2}+\cdots\right]d\eta.\end{aligned} \tag{2.5.17}$$

The leading term on the right-hand side gives $\psi(x,t)$ since

$$A=(2\pi i\hbar\epsilon/m)^{1/2}$$

implies

$$\frac{1}{A}\int_{-\infty}^{\infty}\exp\left(\frac{im\eta^2}{2\hbar\epsilon}\right)d\eta=1. \tag{2.5.18}$$

One also needs

$$\frac{1}{A}\int_{-\infty}^{\infty}\exp\left(\frac{im\eta^2}{2\hbar\epsilon}\right)\left\{\begin{matrix}\eta\\ \eta^2\end{matrix}\right\}d\eta=\left\{\begin{matrix}0\\ i\hbar\epsilon/m\end{matrix}\right\}. \tag{2.5.19}$$

Hence

$$\psi+\epsilon\frac{\partial\psi}{\partial t}=\psi-\frac{i\epsilon}{\hbar}V\psi-\frac{\hbar\epsilon}{2im}\frac{\partial^2\psi}{\partial x^2}. \tag{2.5.20}$$

Hence $\psi(x,t)$ obeys the Schrödinger equation

$$i\hbar\frac{\partial\psi}{\partial t}=-\frac{\hbar^2}{2m}\frac{\partial^2\psi}{\partial x^2}+V(x,t)\psi\equiv H\psi. \tag{2.5.21}$$

One can similarly treat the parametrized one-particle model of section 2.2, which gives a closer analogy with general relativity. As in section 2.2, one uses the variables $x^\alpha=(t,x)$, their conjugate momenta p_α, and the Lagrange multiplier N. The action is

$$S[x^\alpha,p_\alpha;N]=\int d\tau(p_\alpha x^\alpha-N\mathscr{H}), \tag{2.5.22}$$

where

$$\begin{aligned}\mathscr{H}&\equiv p_0+H(x^\alpha,p_\alpha)\\ &=0\text{ classically,}\end{aligned} \tag{2.5.23}$$

with

$$H(x^\alpha,p_\alpha)=\frac{1}{2m}(p_1)^2+V(x^\alpha). \tag{2.5.24}$$

The quantum amplitude can again be evaluated in this version of the theory via a path integral. As described in Eqs. (2.2.9–10), the action (2.5.22) is invariant under arbitrary re-parametrizations of τ. It is clearly sufficient to specify $x^\alpha(\tau)$ up to re-parametrization, in specifying a path.

Further, one only needs to specify x^α, but not N, as boundary data; N can be given any value by re-parametrization, since [Eq. (2.2.10)] $N \to N' = N d\tau/d\tau'$. Thus one expects to have wave functions of the form $\psi(x^\alpha)$, as in the non-parametrized theory. One deals with the over-counting of paths by choosing a gauge condition, for example $x^0 = T(\tau)$, with T a monotonically increasing function which agrees with x^0 on the inital and final surfaces [Hartle 1986]. The amplitude is then

$$K(x_2{}^\alpha, x_1{}^\alpha) = \int \mathscr{D}x^\alpha(\tau)\mathscr{D}N(\tau)\det(\dot{T})\delta(x^0 - T(\tau)) \times \exp(i\hbar^{-1}S[x^\alpha, p_\alpha; N]). \tag{2.5.25}$$

Here $\delta(x^0 - T(\tau))$ is the gauge-fixing delta-function, and $\det(\dot{T})$ is the associated Faddeev–Popov determinant needed to obtain the correct measure [Itzykson & Zuber 1980].

It can be shown [Hartle 1986] that the expression for $K(x_2{}^\alpha, x_1{}^\alpha)$ in Eq. (2.5.25) agrees with the non-parametrized path integral

$$K(t_2, x_2; t_1, x_1) = \int_1^2 \exp(i\hbar^{-1}S_{21})\mathscr{D}x(t). \tag{2.5.26}$$

Hence the amplitude $K(x_2{}^\alpha, x_1{}^\alpha)$ constructed for the parametrized theory obeys the Schrödinger equation

$$\mathscr{H}_2 K = \left[-i\hbar\frac{\partial}{\partial t_2} - \frac{\hbar^2}{2m}\frac{\partial^2}{\partial x_2{}^2} + V(t_2, x_2)\right] K = 0, \tag{2.5.27}$$

and correspondingly with $i \to -i$ for $\mathscr{H}_1 K$ which involves derivatives at the initial surface. Similarly, any physical wave function $\psi(x^\alpha)$ (obtained from initial data by propagating with K) obeys the Schrödinger equation. This approach has provided another route to the Schrödinger equation, which previously in section 2.2 was written down following the Dirac procedure. Again, it provides a close analogy with the quantum-gravity constraints $\mathscr{H}_\perp\Psi = 0$, $\mathscr{H}_i\Psi = 0$, which will be studied in the following sections. In particular, it shows that if one regards the gravitational path integral as primary, then the quantum constraints (with some factor ordering) can be derived from the path integral.

Finally, in this discussion of one-particle non-relativistic quantum mechanics, let us consider quantum mechanics in a potential (such as in Fig. 2.12) admitting a ground state $\psi_0(x)$:

$$H\psi_0(x) = E_0\psi_0(x). \tag{2.5.28}$$

As above, the propagator or quantum amplitude is

$$\langle x'', t''|x', t'\rangle = \int \mathscr{D}x(t)\exp\left\{\frac{i}{\hbar}S[x(t)]\right\}. \tag{2.5.29}$$

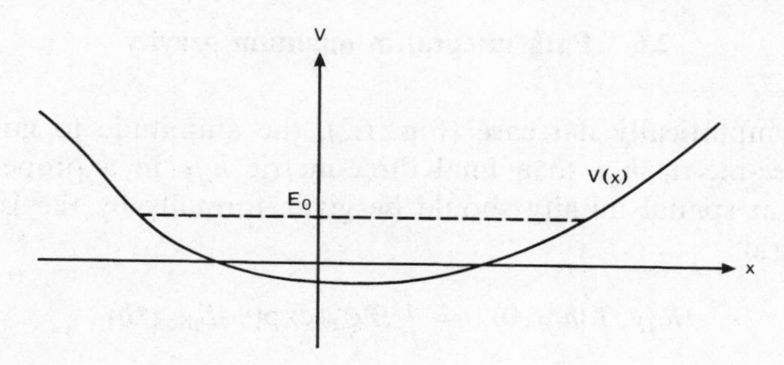

Fig. 2.12. A potential $V(x)$ admitting a ground state with energy E_0.

One fixes the initial position, arbitrarily, at $x' = 0$, and regards t'' as zero and t' as negative. Then (replacing t' by t) one expands $\langle x_0, 0|0, t\rangle$ in a complete set of energy eigenstates:

$$\begin{aligned}\langle x_0, 0|0, t\rangle &= \Sigma_n \langle x_0, 0|n\rangle\langle n|0, t\rangle \\ &= \Sigma_n e^{iE_n t/\hbar}\psi_n(x_0)\psi_n{}^*(0).\end{aligned} \tag{2.5.30}$$

Making the analytic continuation [Hartle 1986] $t \to -i\tau$, one has

$$\begin{aligned}\Sigma_n e^{E_n T/\hbar}\psi_n(x_0)\psi_n{}^*(0) &= \langle x_0, 0|0, -i\tau\rangle \\ &= \int \mathscr{D}x(\tau)\exp\left\{-\frac{1}{\hbar}I[x(\tau)]\right\},\end{aligned} \tag{2.5.31}$$

where $I[x(\tau)]$ is the *Euclidean action functional*

$$I[x(\tau)] = \int d\tau \left[\frac{m}{2}\left(\frac{dx}{d\tau}\right)^2 + V(x)\right]. \tag{2.5.32}$$

One can normalize the energy such that $E_0 = 0$. Now let $T \to -\infty$. Then only the ground state survives in the left-hand side of Eq. (2.5.31), giving

$$\text{const.}\ \psi_0(x_0) = \int \mathscr{D}x(\tau)\exp(-\hbar^{-1}I[x(\tau)]). \tag{2.5.33}$$

Thus the ground state is given by the *Euclidean path integral* on the right-hand side of Eq. (2.5.33), where the integral is over all paths which start at x_0 at Euclidean time $\tau = 0$ and proceed to $\tau = -\infty$ with $x \to 0$ (say).

Here we already see the use of the Euclidean, as opposed to the Lorentzian, path integral. The analogue in quantum cosmology of the Euclidean construction leading to Eq. (2.5.33) is the Hartle–Hawking path integral [Hartle & Hawking 1983], which will be discussed in the following sections. More generally, it will be argued in section 2.6 that all amplitudes in quantum gravity or quantum cosmology should be defined via a Euclidean path integral.

2.6 Path integral in quantum gravity

In the asymptotically flat case (Fig. 1.2), the amplitude to go from an initial three-metric h_{ijI} to a final three-metric h_{ijF} in a proper time T measured at spatial infinity should be given formally by the Lorentzian path integral

$$\langle h_{ijF}, T | h_{ijI}, 0 \rangle_L = \int \mathscr{D} g_{\mu\nu} \exp(-iS_{\text{mod}}/\hbar). \tag{2.6.1}$$

The integral is over Lorentzian four-metrics which agree with the boundary data h_{ijI}, h_{ijF} on the initial and final surfaces Σ_0, Σ_1. The action S_{mod} of Eq. (2.3.68) includes boundary terms on Σ_0 and Σ_1, and also at spatial infinity; the latter give $-MT$ [Teitelboim 1977b] where M is the four-dimensional mass of the spacetime (see section 4.4). This has the property that the action is additive over adjacent spacetime regions (section 2.3), as required in quantum theory. Strictly, one should also include gauge-fixing and Faddeev–Popov (ghost) terms in Eq. (2.6.1) [Itzykson & Zuber 1980, Vilkovisky 1984].

As in section 2.5, in the case of parametrized quantum mechanics, one can formally derive the quantum constraints from the path integral (this follows whether one uses the Lorentzian or Euclidean version). In the asymptotically flat case, the amplitude will obey the quantum constraints

$$\mathscr{H}_\perp \langle h_{klF}, T | h_{klI}, 0 \rangle = 0, \tag{2.6.2}$$
$$\mathscr{H}_i \langle h_{klF}, T | h_{klI}, 0 \rangle = 0, \tag{2.6.3}$$

where $\mathscr{H}_\perp$ and $\mathscr{H}_i$ are now operators which act (say) on the final variables h_{klF}. Here the operator version of $h_{kl}(x)$ is multiplication by $h_{kl}(x)$, and the operator version of $\pi^{kl}(x)$ is $-2i\hbar\kappa^2\delta/\delta h_{kl}(x)$. These operators obey the correct canonical commutation rules. In the asymptotically flat case, the amplitude also obeys a Schrödinger equation

$$i\hbar \frac{\partial}{\partial T} \langle h_{klF}, T | h_{klI}, 0 \rangle = \hat{H} \langle h_{klF}, T | h_{klI}, 0 \rangle, \tag{2.6.4}$$

where

$$H = \frac{1}{2\kappa^2} \int d^3x (N\mathscr{H}_\perp + N^i \mathscr{H}_i) + \text{boundary terms at spatial infinity.} \tag{2.6.5}$$

Here it is precisely the boundary terms in H at spatial infinity which contribute to the right-hand side of Eq. (2.6.4) (section 4.4).

The classical $\mathscr{H}_\perp$ is [Eq. (2.3.57)]

$$\begin{aligned} \mathscr{H}_\perp &= h^{-1/2}[\pi^{kl}\pi_{kl} - \tfrac{1}{2}(\pi^k{}_k)^2] - h^{1/2}\,{}^3R \\ &\equiv G_{ijkl}\pi^{ij}\pi^{kl} - h^{1/2}\,{}^3R, \end{aligned} \tag{2.6.6}$$

where

$$G_{ijkl} = \tfrac{1}{2}h^{-1/2}(h_{ik}h_{jl} + h_{il}h_{jk} - h_{ij}h_{kl}) \\ = G_{(ij)(kl)} \tag{2.6.7}$$

is the *DeWitt metric* [DeWitt 1967]. One can regard G_{ijkl} as a 6×6 matrix in the space of symmetric index pairs (ij).

The quantum $\mathscr{H}_\perp$ is given by

$$\mathscr{H}_\perp = -\ell^4 G_{ijkl}(x)\frac{\delta}{\delta h_{ij(x)}}\frac{\delta}{\delta h_{kl(x)}} - h^{1/2}(x)\,{}^3R(x), \tag{2.6.8}$$

up to the choice of factor ordering in the first term. Here

$$\ell^2 = 2\kappa^2\left(\frac{\hbar G}{c^3}\right) = 2\kappa^2(\text{Planck length})^2, \tag{2.6.9}$$

where the

$$\text{Planck length} \simeq 1{\cdot}6 \times 10^{-33}\ \text{cm} \tag{2.6.10}$$

gives the length scale characteristic of quantum gravity. Curvatures corresponding to this length will be attained in the very early universe or, for example, in very-high-energy graviton collisions [D'Eath 1978, D'Eath & Payne 1992a,b,c, D'Eath 1996]. The time-reverse of such collisions describes the last stages of the quantum-mechanical evaporation of a black hole, when the black hole has reached a mass of the order of the Planck mass $(\hbar c/G)^{1/2}$ [Hawking 1975] and then evaporates by tunnelling into (say) a pair of Planckian-energy gravitons. The quantum $\mathscr{H}^i$ has the form

$$\mathscr{H}^i = -4\kappa^2\,{}^3\nabla_k\left(\frac{\delta}{\delta h_{ik}}\right). \tag{2.6.11}$$

Any physical wave functional $\Psi[h_{ij}(x), T]$ should have the form

$$\Psi[h_{ij}(x), T] = \int \mathscr{D}h'_{ij}\langle h_{ij}, T|h'_{ij}, 0\rangle\Psi[h'_{ij}(x), 0] \tag{2.6.12}$$

(if $T > 0$). Hence $\Psi[h_{ij}(x), T]$ obeys the *Wheeler–DeWitt equation*

$$\mathscr{H}_\perp\Psi = 0 \tag{2.6.13}$$

and the *momentum constraints*

$$\mathscr{H}^i\Psi = 0, \tag{2.6.14}$$

together with the Schrödinger equation

$$i\hbar\frac{\partial\Psi}{\partial T} = \hat{H}\Psi. \tag{2.6.15}$$

When the spatial sections Σ are compact, rather than asymptotically flat, one does not have any time-at-infinity T. A wave functional describing a physical quantum state has the form $\Psi[h_{ij}(x)]$ obeying $\mathscr{H}_\perp\Psi = 0$ and $\mathscr{H}^i\Psi = 0$.

The momentum constraints $\mathscr{H}^i\Psi = 0$, or

$$^3\nabla_i\left(\frac{\delta\Psi}{\delta h_{ik}(x)}\right) = 0, \tag{2.6.16}$$

have a simple interpretation [Higgs 1958,1959]. Consider an infinitesimal coordinate transformation on the surface $t =$ const. , mapping

$$x^i \to x^i + \xi^i(x^k). \tag{2.6.17}$$

The spatial metric transforms as [Hawking & Ellis 1973]

$$h_{ik} \to h_{ik} + {}^3\nabla_i\xi_k + {}^3\nabla_k\xi_i. \tag{2.6.18}$$

The resulting change in Ψ is

$$\begin{aligned}\delta\Psi &= 2\int d^3x\, {}^3\nabla_{(i}\xi_{k)}\frac{\delta\Psi}{\delta h_{ik(x)}} \\ &= 0\end{aligned} \tag{2.6.19}$$

on integrating by parts and using Eq. (2.6.16). Thus $\Psi[h_{ij}(x)]$ is invariant under spatial coordinate transformations applied to its argument $h_{ij}(x)$. Hence any physical state Ψ is a functional on the space of three-geometries – the space of gauge-inequivalent metrics on Σ. This space is called 'superspace' [Wheeler 1968], where the terminology is not to be confused with that used in supersymmetric theories [Wess & Bagger 1992].

In the Wheeler–DeWitt equation (2.6.8),(2.6.13), the DeWitt metric G_{ijkl} has canonical form $(-+++++)$ at each space point [DeWitt 1967]. The Wheeler–DeWitt equation is then a 'hyperbolic' functional differential equation in superspace. The minus sign corresponds to the 'conformal direction', with $\delta h_{ij}(x) = \delta\phi(x)h_{ij}(x)$, discussed later in this section. In quantum cosmology, e.g. , for spacetime models which are perturbations of a $k = +1$ Robertson–Walker model (section 2.8), one finds that when the model is treated suitably by solving the linearized momentum constraints [Halliwell & Hawking 1985], it can be regarded as having precisely one negative direction – corresponding to the radius a of the universe, while all the physical perturbation modes give positive directions. In that case one does have a hyperbolic differential equation with an infinite number of degrees of freedom, and the Wheeler–DeWitt equation gives a time-evolution equation for Ψ.

As in section 2.3, one can include matter coupled to the gravitational field. Classically [Eqs. (2.3.63),(2.3.64)] one has the constraints

$$\mathscr{H}_\perp = \mathscr{H}_{\perp\text{grav}} + 2\kappa^2 h^{1/2}T_{\perp\perp} = 0, \tag{2.6.20}$$

$$\mathscr{H}_i = \mathscr{H}_{i\text{grav}} - 2\kappa^2 h^{1/2}T_{\lambda i}n^\lambda = 0. \tag{2.6.21}$$

One needs to reformulate the total Lagrangian $L_{\text{grav}} + L_{\text{matter}}$ in Hamiltonian form. For example, for a scalar field one can take ϕ as a coordinate, with momentum $\pi_\phi \propto \partial\phi/\partial n$ proportional to the normal derivative of ϕ. On quantization, one obtains the quantum constraints $\mathscr{H}_\perp\Psi = 0$ and $\mathscr{H}_i\Psi = 0$ for the wave function $\Psi = \Psi[h_{ij}(x), \phi(x)]$ in (e.g.) the coordinate representation. The momentum constraint $\mathscr{H}_i\Psi = 0$ has the same interpretation as above: Ψ is invariant under spatial coordinate transformations applied to $h_{ij}(x)$ and to the matter fields. As in the previous paragraph for perturbations of a Friedmann model, matter fields give further plus signs in the kinetic part of the Wheeler–DeWitt equation, so that the Wheeler–DeWitt equation remains 'hyperbolic'. Further, as in the case of pure gravity, the path integral $\int \mathscr{D}g\mathscr{D}\phi \exp(iS/\hbar)$ again gives solutions of the quantum constraints automatically.

All of the above discussion could equally well have been given using the Euclidean formulation of the theory. In section 2.4 it was pointed out that one would only have a well-posed classical boundary-value problem for the vacuum Einstein equations if one worked with Euclidean-signature four-metrics $g_{\mu\nu}$, giving an elliptic problem. Now if one uses the path integral to evaluate quantum amplitudes, one can carry out a semi-classical expansion about a classical solution joining initial and final data [DeWitt 1984a]. Indeed, in most quantum theories, this is the principal method of calculation. Since (section 2.4) the Lorentzian classical solution of the Einstein vacuum field equations does not exist in the asymptotically flat case, it seems more sensible to rotate $T \to -iT$ and study the *Euclidean amplitude*

$$\langle h_{ijF}, -iT|h_{ijI}, 0\rangle_E = \int \mathscr{D}g_{\mu\nu} \exp(-I/\hbar). \qquad (2.6.22)$$

The integral is over Riemannian (positive-definite) four-metrics $g_{\mu\nu}$ which agree with the boundary data. For completeness, gauge-fixing and Faddeev–Popov terms should again be included in the right-hand side of Eq. (2.6.22). Here I is the *Euclidean action*

$$I = -\frac{1}{2\kappa^2}\int_{\text{vol}} d^4x g^{1/2}\,{}^4R - \frac{1}{\kappa^2}\int_{\text{bdry}} d^3x h^{1/2}\text{tr}K, \qquad (2.6.23)$$

with

$$K_{ij}^{\text{Euclidean}} = {}^4\nabla_i n_j, \qquad (2.6.24)$$

n^μ being the outward normal from the boundary, with $n_\mu n^\mu = 1$.

Note that it is not clear that the Euclidean amplitude

$$\langle h_{ijF}, -iT|h_{ijI}, 0\rangle_E$$

is the analytic continuation (with respect to time) of

$$\langle h_{ijF}, T|h_{ijI}, 0\rangle_L.$$

The analytic continuation of a Riemannian metric $g_{\mu\nu}$ will not in general have a real Lorentzian section. Thus the respective path integrals are sums over quite different geometries, and one has to make a definite choice between the two possibilities. To summarize the reasons for favouring the Euclidean amplitude:

1. it works for Yang–Mills theories [Itzykson & Zuber 1980],
2. the path integral becomes real; this may improve its convergence (see later in this section),
3. the classical boundary-value problem becomes elliptic,
4. one may be able to approximate the amplitude as
$$\text{prefactor} \times \exp(-\hbar^{-1} I_{\text{classical}}).$$

Once one has the Euclidean amplitude for imaginary time $-iT$, one may rotate the time-coordinate back towards real time and obtain a value for $\langle h_{ijF}, T | h_{ijI}, 0\rangle$ as a limit from the Euclidean region. There is no apparent reason why this should agree with the Lorentzian value $\langle h_{ijF}, T | h_{ijI}, 0\rangle_L$ for typical strong-field boundary metrics, although there is agreement in the case when a comparison can be made for linearized fields and weak fields in perturbation theory.

As at the end of section 2.5, one can obtain the ground state of quantum gravity (only defined in the asymptotically flat case) by means of the limit $\lim_{T\to-\infty}\langle h_{ij}, 0 | \delta_{ij}, -iT\rangle_E$. For closed cosmologies, with compact spacelike hypersurfaces Σ, a possible analogue is the *Hartle–Hawking state* [Hawking 1982, Hartle & Hawking 1983] or *no-boundary state.* Schematically, including a generic matter field ϕ, this is defined by

$$\Psi_{HH}(h_{ij}, \phi_0) = \int \mathscr{D}g_{\mu\nu}\mathscr{D}\phi \exp\left(-\hbar^{-1} I\left[g_{\mu\nu}, \phi\right]\right). \qquad (2.6.25)$$

The integral is over Riemannian four-geometries $g_{\mu\nu}$ on compact manifolds with boundary Σ, on which the induced three-metric is h_{ij}, and over matter fields ϕ which agree with ϕ_0 on the boundary (Fig. 2.1). As usual, gauge-fixing and Faddeev–Popov fields should be understood. The state Ψ_{HH} automatically obeys the quantum constraint equations defining a physical state: $\mathscr{H}_\perp \Psi_{HH} = 0$, $\mathscr{H}_i \Psi_{HH} = 0$. The Hartle–Hawking state is thus a preferred state (being defined by a particularly simple path integral) which incorporates particular boundary conditions for the Wheeler–DeWitt equation. This will be investigated in the following section 2.7. The Hartle–Hawking state for spatially homogeneous supersymmetric mini-superspace models will be discussed in chapter 5.

Assuming that the Riemannian classical boundary-value problem has a solution, one will have a semi-classical approximation

$$\Psi_{HH} \sim \text{prefactor} \times \exp(-\hbar^{-1} I_{\text{classical}}) \tag{2.6.26}$$

where $I_{\text{classical}}$ is the Euclidean action of the infilling classical solution. This approximation should be valid provided $|I_{\text{classical}}| \gg \hbar$. For the Einstein Lagrangian with a cosmological term Λ, or coupled to a scalar field, $I_{\text{classical}}$ may be complex (section 2.7), with solutions appearing in complex conjugate pairs. It is not known whether this phenomenon also arises for the vacuum Einstein equations.

Finally, in this section, consider the question of convergence of the Euclidean path integral for quantum gravity. The Euclidean action [Eq. (2.6.23)] is

$$I[g_{\mu\nu}] = -\frac{1}{2\kappa^2}\int_{\text{vol}} d^4x g^{1/2}\, {}^4R - \frac{1}{\kappa^2}\int_{\text{bdry}} d^3x h^{1/2}(\text{tr}K). \tag{2.6.27}$$

Under a conformal transformation

$$\tilde{g}_{\mu\nu} = \Omega^2 g_{\mu\nu},\ \Omega = \Omega(x), \tag{2.6.28}$$

it can be shown [Hawking & Ellis 1973, Wald 1984] that

$$\begin{aligned} I[\tilde{g}_{\mu\nu}] = &-\frac{1}{2\kappa^2}\int_{\text{vol}} d^4x g^{1/2}(\Omega^2\, {}^4R + 6\Omega_{,\mu}\Omega^{,\mu}) \\ &-\frac{1}{\kappa^2}\int_{\text{bdry}} d^3x h^{1/2}\Omega^2(\text{tr}K). \end{aligned} \tag{2.6.29}$$

By making Ω rapidly varying, one can make I arbitrarily large and negative. Then the Euclidean amplitude $\int \mathscr{D}g_{\mu\nu} \exp(-\hbar^{-1}I)$ will not converge.

One can attempt to cure this [Gibbons *et al.* 1978]. First, for simplicity, consider the Euclidean amplitude $\langle\delta_{ij}, 0|\delta_{ij}, -i\infty\rangle_E$ corresponding to a path integral over metrics $g_{\mu\nu}$ which are asymptotically flat in all directions (i.e. in both space and time directions). Split the integral into an integration $\int \mathscr{D}\Omega()$ over the conformal factor, and an integration over metrics defining different conformal equivalence classes. Thus one writes

$$g_{\mu\nu} = \Omega^{-2}\tilde{g}_{\mu\nu} \tag{2.6.30}$$

with $\Omega \to 1$ at infinity, and as a reference condition specifying a particular member of a conformal equivalence class, one chooses Ω such that

$${}^4\tilde{R} = 0. \tag{2.6.31}$$

One must solve for Ω the equation

$$\square(\Omega^{-1}) - \frac{1}{6}\, {}^4R\Omega^{-1} = 0, \tag{2.6.32}$$

with $\Omega \to 1$ at infinity, where

$$\Box(\Omega^{-1}) = g^{\mu\nu}(\Omega^{-1})_{;\mu\nu} \tag{2.6.33}$$

with $_{;\mu\nu}$ denoting covariant differentiation in the metric $g_{\mu\nu}$. Eq. (2.6.32) follows from Eq. (2.6.31) because [Hawking & Ellis 1973, Wald 1984]

$$^4\tilde{R} = \Omega^2\, {}^4R - 6\Omega^3 \Box(\Omega^{-1}). \tag{2.6.34}$$

To make the path integral over the conformal factor Ω 'convergent', write $\Omega = 1 + Y$ and rotate $Y \to iY$, so that the positive Gaussian form of $\exp(-\hbar^{-1} I[\tilde{g}_{\mu\nu}])$ of Eq. (2.6.28) in Ω becomes a negative Gaussian expression in Y. Thus

$$I[g_{\mu\nu}] = I[\tilde{g}_{\mu\nu}] + \frac{3}{\kappa^2}\int_{\text{vol}} d^4x \tilde{g}^{1/2}(\tilde{\nabla}Y)^2 \tag{2.6.35}$$

for such complex metrics $g_{\mu\nu}$. Note that the $I[\tilde{g}_{\mu\nu}]$ term in (2.6.35) arises from the surface term in the action with $\Omega \to 1$ at infinity. Now the path integral over the conformal factor 'converges'. Actually it gives a one-loop factor, depending on the reference metric $\tilde{g}_{\mu\nu}$, of the form $[\det(-3\kappa^{-2}\tilde{\Box})]^{-1/2}$ [Hawking 1979]; this may include one-loop divergences. One can then consider the integration over reference metrics $\tilde{g}_{\mu\nu}$. The positive action theorem [Schoen & Yau 1979] shows that $I[\tilde{g}_{\mu\nu}] \geq 0$ for metrics $\tilde{g}_{\mu\nu}$ which are asymptotically flat in all directions, obeying $^4\tilde{R} = 0$. This suggests that the path integral over $\tilde{g}_{\mu\nu}$ might be convergent.

One can try to repeat the procedure in the compact case with boundary (Hartle–Hawking state) – see Fig. 1.1. One again requires $\Omega = 1$ on the boundary, to give $g_{ij} = h_{ij}$ (the specified three-metric) there. Again one solves Eq. (2.6.32) for $\Omega(x)$, so as to make $^4\tilde{R} = 0$. However, there is now no known theorem about the sign of $I[\tilde{g}_{\mu\nu}]$ under these conditions, so that one cannot form any conclusion about the convergence of the path integral for the Hartle–Hawking state.

In general, finding a contour for which the path integral converges is a major unsolved problem. In mini-superspace models (section 2.7), derived by assuming a high degree of spacetime symmetry, one can usually find a contour by *ad hoc* analysis. The Hartle–Hawking path integral $\Psi_{HH}(h_{ij})$ should still be real, as its definition suggests, but it may oscillate in certain regions, corresponding to the existence of complex classical paths (section 2.7). To emphasize the role of the contour C_0, the path integral will sometimes be denoted by $\int_{C_0} \mathscr{D}g_{\mu\nu} \exp(-I/\hbar)$. In supergravity, these problems are somewhat alleviated because of the restrictive nature of the supersymmetry constraints (see chapters 5 and 8); sometimes one can then find the Hartle–Hawking state in closed form.

2.7 Mini-superspace models

Since the full theory of quantum gravity is so hard, one instead makes progress by first studying highly symmetrical models, so that the degrees of freedom of the gravitational and any matter fields can be reduced to a finite number. For example, one can consider a Lorentzian $k = +1$ Friedmann model. (The references [Hartle & Hawking 1983, Hawking 1984, Hartle 1986] are followed in this section.) For later convenience, the metric can be written as

$$ds^2 = \sigma^2[-N^2(t)dt^2 + a^2(t)d\Omega_3{}^2], \tag{2.7.1}$$

where $d\Omega_3{}^2$ is the metric on the unit three-sphere. Here

$$\sigma^2 = \ell^2/24\pi^2 = 2/3\pi, \tag{2.7.2}$$

where $\ell^2 = 16\pi = 2\kappa^2$ [Eq. (2.6.9)], and one can, if desired, assume that the spacetime contains a homogeneous scalar field $\Phi(t)$. The Euclidean version of the metric is

$$ds^2 = \sigma^2[N^2(\tau)d\tau^2 + a^2(\tau)d\Omega_3{}^2], \tag{2.7.3}$$

containing a homogeneous scalar field $\Phi(\tau)$.

As a first example, consider a conformally invariant scalar field coupled to Einstein gravity with a cosmological constant Λ [Hartle & Hawking 1983]. (This example is not intended to be physically realistic!) Let us work in the Euclidean régime. The action I is

$$I = I_E + I_\Phi, \tag{2.7.4}$$

a sum of gravitational and matter parts. Here I_E is given by

$$\ell^2 I_E = \int_{\text{vol}} d^4x g^{1/2}({}^4R - 2\Lambda) - 2\int d^3x h^{1/2}(\text{tr}K). \tag{2.7.5}$$

Compare this with the Lorentzian Eq. (2.3.48), in simplifying the Lagrangian. The Euclidean definition of K_{ij} differs by a factor of i from the Lorentzian definition (since the definition of the normal n^μ does). Hence

$$\ell^2 I_E = -\int d^4x g^{1/2}[{}^3R + (\text{tr}K)^2 - K_{ij}K^{ij} - 2\Lambda], \tag{2.7.6}$$

where the Euclidean definition is used for K_{ij}. Now

$${}^3R = 6/(\sigma a)^2, \tag{2.7.7}$$

$$(\text{tr}K)^2 - K_{ij}K^{ij} = \frac{6}{\sigma^2}\left(\frac{da/d\tau}{aN}\right)^2, \tag{2.7.8}$$

following Eq. (2.3.24). The volume of space is [Misner *et al.* 1973]

$$\int d^3x\, h^{1/2} = 2\pi^2(\sigma a)^3. \tag{2.7.9}$$

Hence

$$I_E = \frac{1}{2}\int d\tau \left(\frac{N}{a}\right)\left[-\left(\frac{a\dot{a}}{N}\right)^2 - a^2 + H^2a^4\right], \tag{2.7.10}$$

where

$$H^2 = \frac{\sigma^2\Lambda}{3}, \quad \dot{a} = \frac{da}{d\tau}. \tag{2.7.11}$$

The Euclidean conformal scalar field has action [Hartle 1986]

$$\begin{aligned} I_\Phi = &\frac{1}{2}\int_{\text{vol}} d^4x\, g^{1/2}[(\nabla\Phi)^2 + \frac{1}{6}\,{}^4R\Phi^2] \\ &+ \frac{1}{12}\int_{\text{bdry}} d^3x\, h^{1/2}(\text{tr}K)\Phi^2. \end{aligned} \tag{2.7.12}$$

Here the boundary term has been introduced to cancel off the second normal derivative terms in $\frac{1}{12}(g^{1/2})^4R\Phi^2$. The classical field equation for Φ is

$$g^{\mu\nu}\Phi_{;\mu\nu} - \frac{1}{6}\,{}^4R\Phi = 0, \tag{2.7.13}$$

the *conformally invariant wave equation* in four dimensions. Eq. (2.7.13) is invariant [Wald 1984] under

$$\begin{aligned} g_{\mu\nu}(x) &\rightarrow \Omega^2(x)g_{\mu\nu}(x), \\ \Phi(x) &\rightarrow \Omega^{-1}(x)\Phi(x). \end{aligned} \tag{2.7.14}$$

It is convenient to make the rescaling

$$\phi = (2\pi^2\sigma^2)^{1/2}\Phi. \tag{2.7.15}$$

One obtains

$$I_\Phi = \frac{1}{2}\int d\tau \left(\frac{N}{a}\right)\left[\frac{a^4}{N^2}\left(\dot{\phi} + \frac{\dot{a}}{a}\phi\right)^2 + a^2\phi^2\right]. \tag{2.7.16}$$

Now define

$$\phi = \frac{\chi}{a}, \tag{2.7.17}$$

and note that χ is conformally invariant. Then

$$\begin{aligned} I &= I_E + I_\Phi \\ &= \frac{1}{2}\int d\tau \left(\frac{N}{a}\right)\left[-\left(\frac{a}{N}\dot{a}\right)^2 + \left(\frac{a}{N}\dot{\chi}\right)^2 + U(a,\chi)\right] \end{aligned} \tag{2.7.18}$$

where

$$U(a,\chi) = -a^2 + H^2a^4 + \chi^2. \tag{2.7.19}$$

The three terms on the right-hand side of (2.7.19) arise from the spatial curvature, the Λ-term and the matter energy. Equivalently, one could have computed the Lorentzian action first, and then obtained the Euclidean action by letting $t \to -i\tau$.

Now one can derive the Hamiltonian form. Note that

$$\pi_a = i\frac{\partial L_{\text{Euclidean}}}{\partial \dot{a}}, \pi_\chi = i\frac{\partial L_{\text{Euclidean}}}{\partial \dot{\chi}}. \tag{2.7.20}$$

Hence

$$\pi_a = -i\frac{a}{N}\dot{a}, \ \pi_\chi = i\frac{a}{N}\dot{\chi}. \tag{2.7.21}$$

One obtains the Euclidean action in the form

$$I = \int d\tau(-i\pi_a\dot{a} - i\pi_\chi\dot{\chi} + N\mathscr{H}_\perp), \tag{2.7.22}$$

where

$$\mathscr{H}_\perp = \frac{1}{2a}[-\pi_a{}^2 + \pi_\chi{}^2 + U(a,\chi)]. \tag{2.7.23}$$

Classically,

$$\mathscr{H}_\perp = 0. \tag{2.7.24}$$

The classical constraint (2.7.23),(2.7.24) can be written as

$$\frac{1}{2}G^{AB}\pi_A\pi_B + \frac{1}{2a}U(a,\chi) = 0, \tag{2.7.25}$$

with G_{AB} the hyperbolic metric on superspace:

$$G_{AB} = \begin{pmatrix} -a & 0 \\ 0 & a \end{pmatrix}, \tag{2.7.26}$$

where the momenta and coordinates are written as

$$\pi_A = (\pi_a, \pi_\chi), \ x^A = (a, \chi). \tag{2.7.27}$$

The Wheeler–DeWitt equation for the Friedmann $k = +1$ model with Λ-term and conformal scalar field is

$$\frac{1}{2}\left[-\hbar\nabla^2 + \frac{1}{a}U(a,\chi)\right]\Psi(a,\chi) = 0, \tag{2.7.28}$$

with the factor-ordering choice

$$\nabla^2 = \frac{1}{\sqrt{-G}}\frac{\partial}{\partial x^A}\left(\sqrt{-G}G^{AB}\frac{\partial}{\partial x^B}\right) \tag{2.7.29}$$

giving the covariant Laplacian $\nabla_A(G^{AB}\nabla_B(\ ))$ in the metric G_{AB}. This makes sense because the operator ∇^2 is conformally invariant in two dimensions [Wald 1984]. One needs conformal invariance of the quantum Wheeler–DeWitt operator [Misner 1972] since there is an ambiguity of

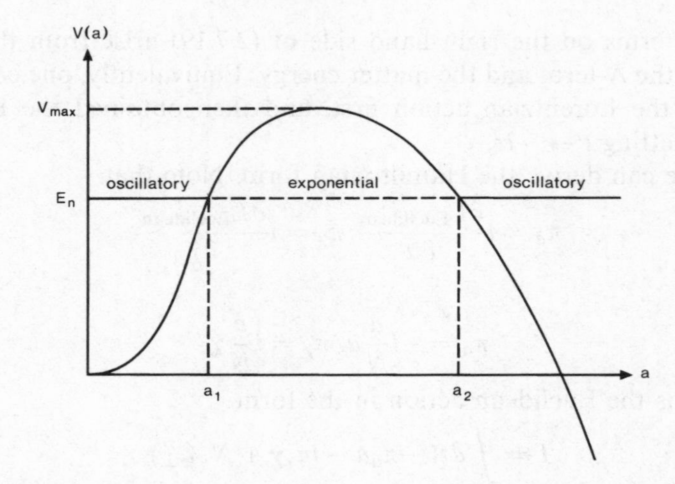

Fig. 2.13. The potential $V(a)$ for the radial dependence of the Wheeler– DeWitt equation.

multiplying the classical constraint $\mathscr{H}_\perp$ by an arbitrary function of the coordinates before quantizing; this freedom corresponds to a conformal rescaling of the superspace metric, and the quantization procedure must be invariant under this conformal transformation. Hence one obtains the Wheeler–DeWitt equation

$$\frac{1}{a}\left(\hbar^2\frac{\partial^2\Psi}{\partial a^2}-\hbar^2\frac{\partial^2\Psi}{\partial\chi^2}+U\Psi\right)=0. \tag{2.7.30}$$

Here the $\partial/\partial a$ direction is timelike; this property recurs in more general models, where all other degrees of freedom are spacelike. Because of the form (2.7.19) of $U(a,\chi)$, one can separate variables:

$$\Psi(a,\chi)=C(a)f(\chi), \tag{2.7.31}$$

where

$$\left(-\hbar^2\frac{d^2}{d\chi^2}+\chi^2\right)f=Ef. \tag{2.7.32}$$

The solutions are harmonic-oscillator states $f_n(\chi)$, with $E_n=(2n+1)\hbar\,(n=0,1,2,\ldots)$. Then $C(a)$ obeys

$$-\hbar^2\frac{d^2C}{da^2}+(a^2-H^2a^4-E_n)C=0. \tag{2.7.33}$$

This is the time-independent Schrödinger equation with potential $V(a)=a^2-H^2a^4$. This potential has a maximum value $V_{\max}$ for $a>0$, and is depicted in Fig. 2.13. The solutions $C(a)$ corresponding to the Hartle–Hawking state for different values of H are shown below.

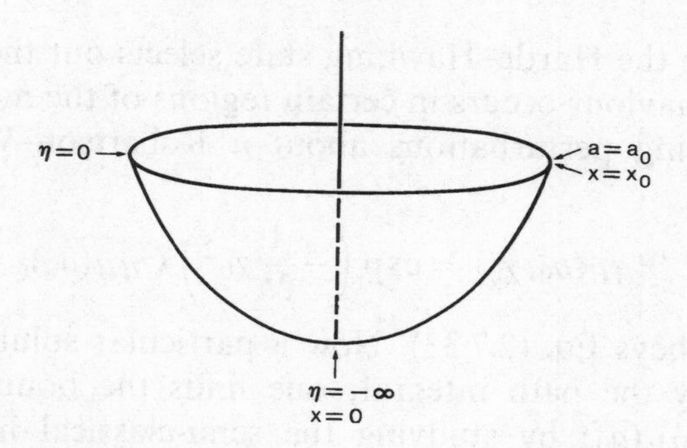

Fig. 2.14. A typical Friedmann geometry, with scalar field, contributing to the Hartle–Hawking state.

To find the Hartle–Hawking state for this model, start by fixing the gauge by taking $N = a$. Denote τ by η in this gauge. Then

$$I[a,\chi] = \tfrac{1}{2}\int_{-\infty}^{0} d\eta[-(a')^2 + (\chi')^2 + U(a,\chi)], \tag{2.7.34}$$

with

$$a' = da/d\eta,\ U = -a^2 + H^2a^4 + \chi^2. \tag{2.7.35}$$

To understand the limits on η, consider Fig. 2.14, which depicts a typical compact manifold-with-boundary contributing to the Euclidean path integral (2.6.25), with a three-sphere of radius σa_0 at the boundary, and $\chi = \chi_0$ there. Note that, at the 'South pole' $a \to 0$ of Fig. 2.14, $a^{-1}da/d\eta = N^{-1}da/d\eta$ is finite and non-zero, being d(radius)/d(proper distance). Hence $\eta \to -\infty$ as $a \to 0$. Using translation invariance for the parameter η, one can further arrange that $\eta = 0$ at the boundary $a = a_0$. Further the scalar field Φ or ϕ should be regular near $a = 0$; hence $\chi = a\phi = 0$ at $\eta = -\infty$. The Hartle–Hawking state is, for a suitable contour C_0,

$$\Psi_{HH} = \int_{C_0} \mathscr{D}a\mathscr{D}\chi \exp(-\hbar^{-1}I[a,\chi]), \tag{2.7.36}$$

which, as above, automatically obeys the Wheeler–DeWitt equation.

The path integral separates. The χ-integral is

$$f(\chi) = \int \mathscr{D}\chi \exp\left[-\frac{1}{2\hbar}\int_{-\infty}^{0} d\eta(\chi'^2 + \chi^2)\right], \tag{2.7.37}$$

with $\chi = 0$ at $\eta = -\infty$, and $\chi = \chi_0$ at $\eta = 0$. As in Eqs. (2.5.32),(2.5.33), this is just the imaginary-time path integral for the harmonic-oscillator ground state, giving

$$f(\chi) = \text{const.}\ f_0(\chi_0). \tag{2.7.38}$$

Thus in this case the Hartle–Hawking state selects out the matter ground state; similar behaviour occurs in certain regions of the models of sections 2.8, 2.9, describing perturbations about a Robertson–Walker universe. Hence

$$\Psi_{HH}(a_0, \chi_0) = \exp\left(-\frac{1}{2\hbar}\chi_0{}^2\right) C_{HH}(a_0), \tag{2.7.39}$$

where $C_{HH}(a)$ obeys Eq. (2.7.33). Now a particular solution for $C_{HH}(a_0)$ is picked out by the path integral; one finds the boundary conditions which define $C_{HH}(a_0)$ by studying the semi-classical limit (see below) [Hartle 1986]. It is natural and indeed essential (see below) to regard the path integral over a as a contour integral. One finds three possible contours, but only two of them give the correct semi-classical limit [Halliwell & Louko 1989a,b, Hawking 1995]. An advantage of supergravity may be that the quantum constraints are much more restrictive than the Wheeler–DeWitt equation, and lead to possibly unique quantum states, given suitable boundary conditions (chapter 5). Two typical solutions for $C_{HH}(a_0)$ are given in Figs. 2.15a,b, one in the case $H \simeq 1$ and the other with $H \ll 1$. The procedure for finding such Hartle–Hawking states will now be discussed.

In order to interpret a wave function physically, one needs to know the probability measure. Since the Wheeler–DeWitt operator $\mathscr{H}_\perp$ should be Hermitian, one is naturally led to define the inner product between two quantum states as [Hartle 1986, Hawking & Page 1986]

$$\langle \Psi, | \Psi \rangle = \int\int da\, d\phi (-G)^{1/2} \Psi_1^* \Psi_2, \tag{2.7.40}$$

where $G = \det(G_{AB})$ with G_{AB} given in Eq. (2.7.26). The probability of being in a small interval $(a, a + \delta a)$ and $(\phi, \phi + \delta\phi)$ is thus

$$P = (-G)^{\frac{1}{2}} |\Psi(a, \phi)|^2 \delta a \delta\phi. \tag{2.7.41}$$

Strictly, this is a relative probability, since for typical solutions $\Psi_1(a, \phi)$ of the Wheeler–DeWitt equation, $\langle \Psi_1 | \Psi_1 \rangle$ will diverge at large a. In Fig. 2.15b, with $H \ll 1$, it can be seen that the probability of finding a sphere with $a_0 < H^{-1}$ is very small. This can also be verified from the semi-classical or WKB analysis below.

Consider now the semi-classical approximation to the Hartle–Hawking path integral

$$C_{HH}(a_0) = \int_{C_0} \mathscr{D}a \exp(-\hbar^{-1} I_E[a]), \tag{2.7.42}$$

with

$$I_E(a_0) = \tfrac{1}{2} \int_{-\infty}^{0} d\eta (-a'^2 - a^2 + H^2 a^4), \tag{2.7.43}$$

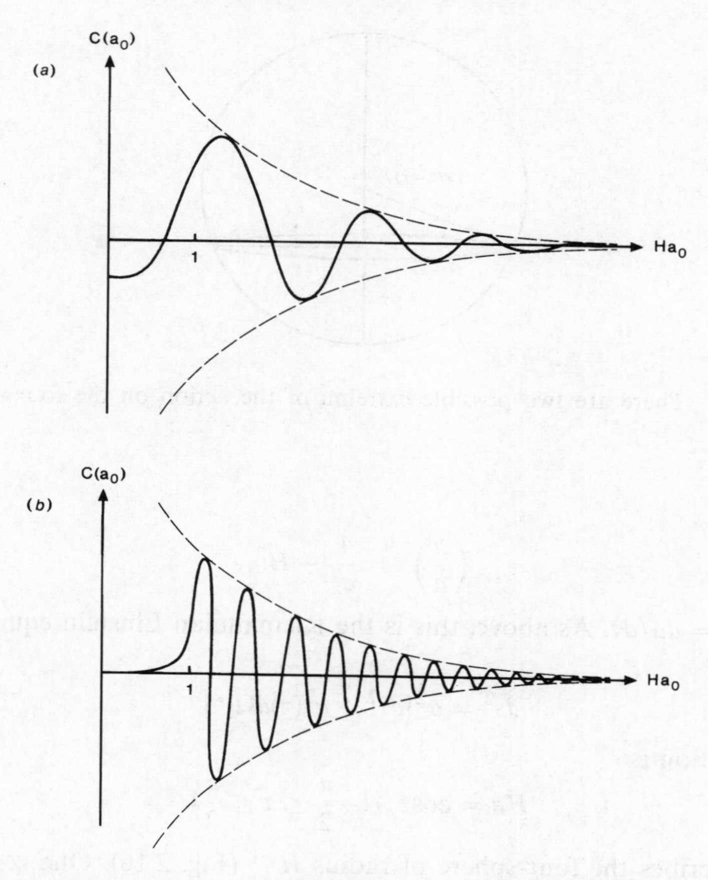

Fig. 2.15. The solution $C_{HH}(a_0)$ corresponding to the Hartle–Hawking state, in two cases: (a)$H \simeq 1$, (b)$H \ll 1$.

and (Fig. 2.14) $a(0) = a_0$, $a(-\infty) = 0$. One makes an approximation by steepest descents, finding the extrema of I_E through which the contour of integration can be distorted. First take the case $a_0 < H^{-1}$. The extrema of I_E correspond to solutions of the classical Euclidean equation of motion

$$a'' - a + 2H^2 a^3 = 0. \tag{2.7.44}$$

This has an energy integral

$$\begin{aligned}(a')^2 - a^2 + H^2 a^4 &= \text{const.} \\ &= 0\end{aligned} \tag{2.7.45}$$

since $a \to 0$ and $da/d\eta \to 0$ as $\eta \to -\infty$. Define $\sigma\tau$ to be proper distance in the corresponding Friedmann model: $d\tau = a d\eta$ (recall $\sigma^2 = 2/3\pi$).

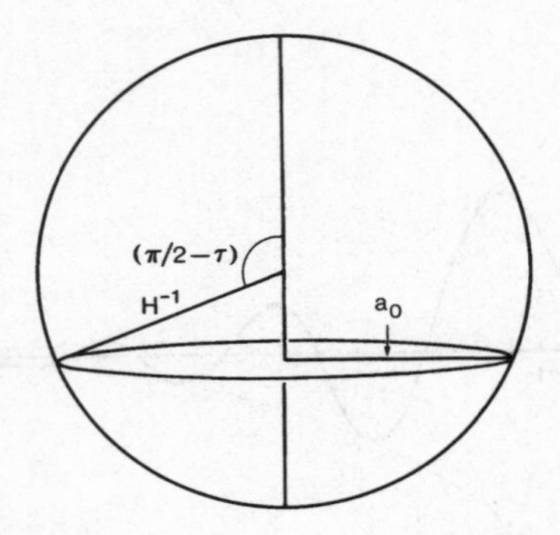

Fig. 2.16. There are two possible extrema of the action on the four-sphere of radius H^{-1}.

Then

$$\left(\frac{\dot{a}}{a}\right)^2 = \frac{1}{a^2} - H^2, \tag{2.7.46}$$

where $\dot{a} = da/d\tau$. As above, this is the Riemannian Einstein equation for the metric

$$ds^2 = \sigma^2[d\tau^2 + a^2(\tau)d\Omega_3{}^2]. \tag{2.7.47}$$

The solution is

$$Ha = \cos\tau \quad (-\frac{\pi}{2} \leq \tau \leq \frac{\pi}{2}). \tag{2.7.48}$$

This describes the four-sphere of radius H^{-1} (Fig. 2.16). One can regard $\tau = -\frac{\pi}{2}$ as the South pole and $\tau = \frac{\pi}{2}$ as the North pole.

For $\sigma_0 < 1/H$ there are two possible extrema of the action I_E, which are compact four-geometries with a three-sphere boundary of radius a_0, namely the upper and lower regions in Fig. 2.16. The corresponding classical actions are

$$I_{\text{class}} = -\frac{1}{3H^2}\left[1 \pm (1 - H^2 a_0{}^2)^{3/2}\right], \tag{2.7.49}$$

where + refers to the upper region and − to the lower region. Then I_{class} is more negative for the upper region, so that one would naively expect to get a contribution proportional to $\exp(-\hbar^{-1} I_{\text{class}})$ to the path integral from the upper region. But [Hartle & Hawking 1983, Hartle 1986] the contour C_0 in the Euclidean path integral $\int_{C_0} \mathscr{D}a \exp(-\hbar_{-1} I_E[a])$ is in the imaginary direction near the extremum, because of the rotation of the conformal factor (section 2.6). Now extrema of analytic functions are saddle points: a maximum of the action in real directions is a minimum in imaginary directions. This suggests that the steepest-descent contour comes from the lower region. In fact, the steepest-descent contour cannot be distorted to pass through the upper-region extremum – see [Hartle & Hawking 1983] for further discussion.

For $a_0 < 1/H$, taking the lower classical solution in Fig. 2.16, a standard WKB analysis [Olver 1974] gives

$$C_{HH}(a_0) \sim \mathcal{N}(-1 + a_0{}^2 - H^2 a_0{}^4)^{-1/4} \exp\left[-\frac{1}{3\hbar H^2}(1 - H^2 a_0{}^2)^{3/2}\right], \tag{2.7.50}$$

where $\mathcal{N}$ is an arbitrary normalization factor. Note that this shows an exponential decrease of $C_{HH}(a_0)$ as Ha_0 is reduced to zero – see Fig. 2.15b above. As can be seen from Eq. (2.7.50), only even powers of a_0 appear in this approximation to $C_{HH}(a_0)$. Since the path integral (2.7.42), (2.7.43) involves only even powers of a, the exact $C_{HH}(a_0)$ involves only even powers of a_0. Hence one boundary condition for $C_{HH}(a_0)$ is that $dC_{HH}/da_0 = 0$ at $a_0 = 0$.

Now take the case $a_0 > H^{-1}$. There are no real extrema since the three-sphere of radius a_0 does not fit into the four-sphere of radius H^{-1}. Instead, there are *complex* extrema. One changes $\tau \to \pm it$, and solves

$$\left(\frac{\dot{a}}{a}\right)^2 = H^2 - \frac{1}{a^2}. \tag{2.7.51}$$

This is the Lorentzian Einstein equation for the metric

$$ds^2 = \sigma^2[-dt^2 + a^2(t)d\Omega_3{}^2]. \tag{2.7.52}$$

The solution is de Sitter spacetime [Hawking & Ellis 1973] (Fig. 2.17):

$$Ha = \cosh t. \tag{2.7.53}$$

A complex solution of the Einstein field equations with Λ-term is given by taking a complex path or section of the complexified metric

$$ds^2 = \sigma^2(d\tau^2 + H^{-2}\cos^2\tau d\Omega_3{}^2) \tag{2.7.54}$$

between (say) the endpoints $\tau = -\frac{\pi}{2}$ and $\tau = it_0$, where $Ha_0 = \cosh(t_0)$ (Fig. 2.18). For example, the right-angled path in Fig. 2.18 gives the shaded region in Fig. 2.17, which is the union of a Euclidean and Lorentzian region. But one can equally well use a complex path in the geometry (2.7.54) (Fig. 2.18); since the Lagrangian density is analytic, the action between the initial and final endpoints $\tau = -\frac{\pi}{2}$ and $\tau = it_0$ is invariant. Note that complex conjugate paths (with endpoints $\pm it_0$) must contribute equally to $C_{HH}(a_0)$; the wave function $C_{HH}(a_0)$ will then be real. A WKB analysis gives, for $a_0 > H^{-1}$,

$$C_{HH}(a_0) \sim 2\mathcal{N}(H^2 a_0{}^4 - a_0{}^2 + 1)^{-1/4} \cos\left[\frac{1}{3\hbar H^2}(1 - H^2 a_0{}^2)^{3/2} - \frac{\pi}{4}\right]. \tag{2.7.55}$$

This may again be compared with Figs. 2.15a,b giving $C_{HH}(a_0)$.

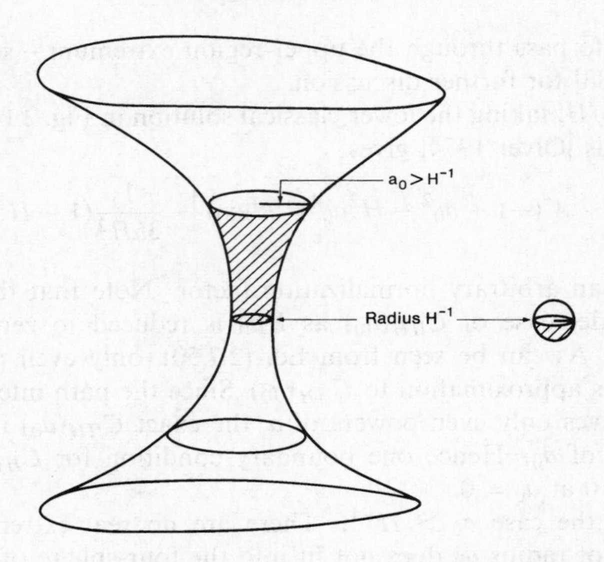

Fig. 2.17. A classical solution appearing in the Hartle–Hawking state, in which a portion of Lorentzian de Sitter spacetime is joined onto half of Riemannian de Sitter space (S^4).

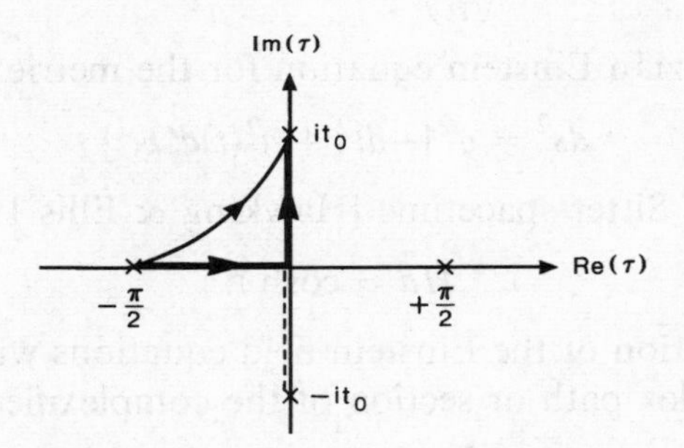

Fig. 2.18. Complex paths joining $\tau = it_0$ (corresponding to a radius $a > H^{-1}$) to the South pole $\tau = -\pi/2$ of S^4.

In the full Hartle–Hawking state

$$\Psi(a_0, \phi_0) = \exp\left[-\frac{(a_0\phi_0)^2}{2\hbar}\right] C_{HH}(a_0), \qquad (2.7.56)$$

using the probability measure (2.7.41), note that large values of Φ_0 are correlated with small values of a_0, and vice versa. This would give an experimental test of the model, were it physically realistic.

In a classically allowed region ($a_0 > H^{-1}$) corresponding to a Lorentzian classical solution, the Hartle–Hawking wave function is oscillatory, being proportional to $\mathrm{Re}[\exp(iS_{\mathrm{class}}/\hbar)]$, where S_{class} is the complex action of a complex classical solution filling in smoothly from radius a_0 down to radius 0. As above, this corresponds to de Sitter spacetime. In fact the prefactor $(H^2 a_0{}^4 - a_0{}^2 + 1)^{-1/4}$ in Eq. (2.7.55) gives the envelopes in Figs. 2.15a,b, which when squared [Eq. (2.7.41)] approximate the distribution of spheres in de Sitter spacetime. The 'classical' probability of finding a three-sphere with radius between a and $a + \delta a$ is proportional to the amount of proper time elapsed, which can be evaluated for the geometry (2.7.52), (2.7.53). In the simpler case of the harmonic oscillator, an example of an oscillating wave function and its 'classical' envelope is depicted in Fig. 11 of [Schiff 1968].

In the classically forbidden region ($a_0 < H^{-1}$), the Hartle–Hawking wave function is exponential, being proportional to $\exp(-I_{\text{class}}/\hbar)$, corresponding to a Riemannian solution – the four-sphere.

Consider now a second, more realistic example of a $k = +1$ Robertson–Walker universe, containing a scalar field Φ of mass M, with $\Lambda = 0$. It will turn out that, in the early universe, the mass term mimics the effect of the previous Λ-term [Linde 1987]. The Euclidean action is

$$\begin{aligned} I &= I_E + I_\Phi, \\ I_\Phi &= \tfrac{1}{2} \int_{\text{vol}} d^4x\, g^{1/2} [(\nabla\Phi)^2 + M^2\Phi^2], \end{aligned} \tag{2.7.57}$$

where I_E is given by Eq. (2.6.23). It is convenient to make the rescalings

$$\begin{aligned} \phi &= (2\pi^2\sigma^2)^{1/2}\Phi, \\ m &= \sigma M, \end{aligned} \tag{2.7.58}$$

where $\sigma^2 = 2/3\pi$. Then, for the Robertson–Walker universe,

$$I = \tfrac{1}{2} \int d\tau \left(\frac{N}{a}\right) \left[-\left(\frac{a\dot{a}}{N}\right)^2 + \frac{a^4}{N^2}\dot{\phi}^2 + U(a,\phi) \right], \tag{2.7.59}$$

where

$$U(a,\phi) = -a^2 + m^2 a^4 \phi^2. \tag{2.7.60}$$

The Hamiltonian constraint equation is then

$$\mathscr{H}_\perp = \frac{1}{2a} [-\pi_a{}^2 + a^{-2}\pi_\phi{}^2 + U(a,\phi)] = 0. \tag{2.7.61}$$

The metric on superspace is

$$G_{AB} = \begin{pmatrix} -a & 0 \\ 0 & a^3 \end{pmatrix}. \tag{2.7.62}$$

The configuration space ($a > 0, -\infty < \phi < \infty$) is conformal to the interior of the forward light cone in two-dimensional Minkowski spacetime.

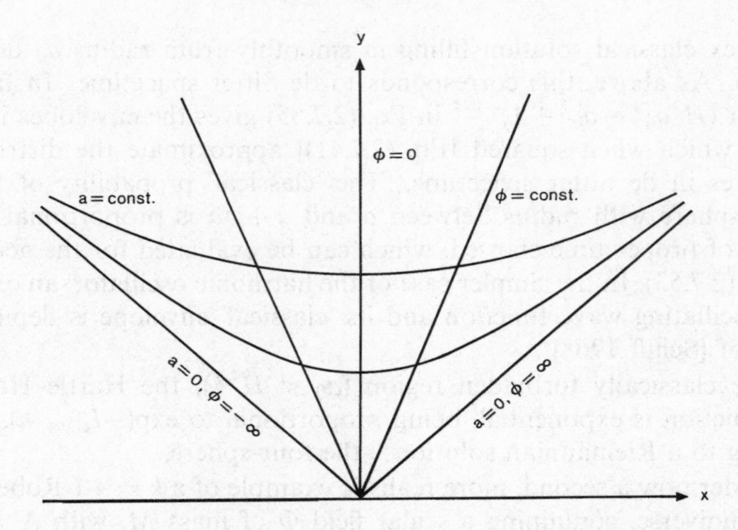

Fig. 2.19. The coordinates (a, ϕ) depicted on the xy-plane.

This can be seen by introducing new coordinates (Fig. 2.19)

$$\begin{aligned} x &= a \sinh\phi, \\ y &= a \cosh\phi, \end{aligned} \tag{2.7.63}$$

chosen such that $y \pm x$ are null coordinates. In xy coordinates,

$$G_{A'B'} = (y^2 - x^2)^{1/2} \begin{pmatrix} 1 & 0 \\ 0 & -1 \end{pmatrix}. \tag{2.7.64}$$

The future null cone of the origin $(x = 0, y = 0)$ consists of the regions (Fig. 2.19) $a = 0$, $\phi = -\infty$ and $a = 0$, $\phi = +\infty$. Lines of constant $a > 0$ are rectangular hyperbolas. Lines of constant ϕ are straight lines through the origin. Note that the metric $G_{A'B'}$ of Eq. (2.7.64) is conformal to a flat metric, and the covariant two-dimensional wave operator is conformally invariant (a general property of two-dimensional metrics, as mentioned after Eq. (2.7.29)). Hence one finds the Wheeler–DeWitt equation

$$\left[-\hbar^2 \frac{\partial^2}{\partial y^2} + \hbar^2 \frac{\partial^2}{\partial x^2} - \hat{U}(x, y) \right] \Psi = 0, \tag{2.7.65}$$

where

$$\hat{U}(x, y) = x^2 - y^2 + m^2 (y^2 - x^2)^2 \left[\tanh^{-1} \left(\frac{x}{y} \right) \right]^2. \tag{2.7.66}$$

The regions where $\hat{U} < 0$ and $\hat{U} > 0$ are shown in the xy plane in Fig. 2.20. From Eq. (2.7.60), these correspond to $a < (m|\phi|)^{-1}$ and $a > (m|\phi|)^{-1}$. By comparison with the previous Λ-term example, the regions where $\hat{U} < 0$ are expected to correspond to Euclidean classical solutions, and

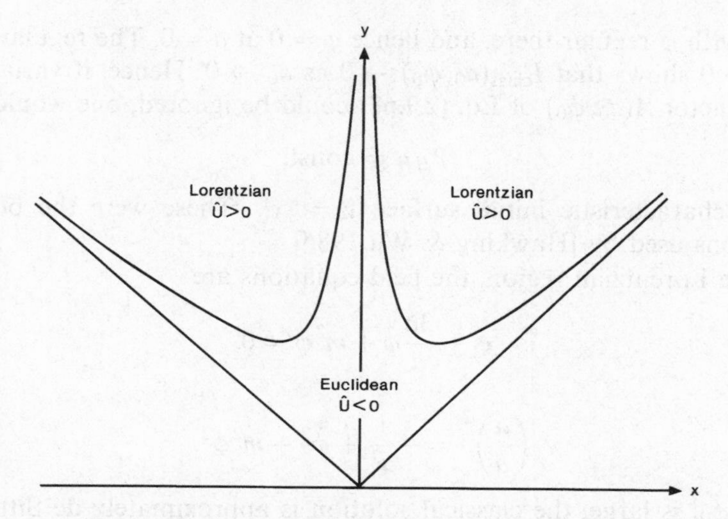

Fig. 2.20. The Lorentzian regions $\hat{U} > 0$ and the Euclidean region $\hat{U} < 0$.

the regions with $\hat{U} > 0$ to Lorentzian classical solutions. The region with $\hat{U} < 0$ is mostly at small a, except for a narrow region along the y-axis, which corresponds to large nearly-flat geometries.

The Wheeler–DeWitt equation (2.7.65) is a wave equation, and one might try to solve it by setting initial data on the light cone $y = |x|$, the data being given in principle by the path integral [Hawking & Wu 1985]. Alternatively, one can study the semi-classical approximation to

$$\Psi_{HH}(a_0, \phi_0) = \int_{C_0} \mathscr{D}a\mathscr{D}\phi \exp(-\hbar^{-1} I\,[a, \phi]). \tag{2.7.67}$$

In the 'Euclidean region' above, approximately where $\hat{U} < 0$, one expects semi-classically

$$\Phi_{HH} \sim A(a_0, \phi_0) \exp[-\hbar^{-1} I_{\text{class}}(a_0, \phi_0)], \tag{2.7.68}$$

where $A(a_0, \phi_0)$ is a quantum or WKB prefactor.

Consider the corresponding Riemannian classical solution (Fig. 2.14). In the gauge $N = 1$, the Euclidean field equations are

$$\ddot{\phi} + \frac{3\dot{a}}{a}\dot{\phi} - m^2\phi^2 = 0, \tag{2.7.69}$$

$$\left(\frac{\dot{a}}{a}\right)^2 = \frac{1}{a^2} + \dot{\phi}^2 - m^2\phi^2. \tag{2.7.70}$$

Here Eq. (2.7.70) is the $\mathscr{H}_\perp = 0$ constraint, and on differentiation leads to an equation for $\ddot{a}$. As in Fig. 2.14, the boundary conditions are that $(a, \phi) = (a_0, \phi_0)$ at the boundary, and that the geometry is smooth near

$a = 0$, with ϕ regular there, and hence $\dot{\phi} = 0$ at $a = 0$. The regularity of ϕ as $a_0 \to 0$ shows that $I_{\rm class}(a_0, \phi_0) \to 0$ as $a_0 \to 0$. Hence, if variations in the prefactor $A(a_0, \phi_0)$ of Eq. (2.7.68) could be ignored, one would have

$$\Psi_{HH} = \text{const.} \tag{2.7.71}$$

on the characteristic initial surface $y = |x|$. These were the boundary conditions used by [Hawking & Wu 1985].

In the Lorentzian region, the field equations are

$$\ddot{\phi} + \frac{3\dot{a}}{a}\dot{\phi} + m^2\phi = 0, \tag{2.7.72}$$

$$\left(\frac{\dot{a}}{a}\right)^2 = -\frac{1}{a^2} + \dot{\phi}^2 + m^2\phi^2. \tag{2.7.73}$$

When $|\phi_0|$ is large, the classical solution is approximately de Sitter, with $\phi \simeq$ const. and $\dot{\phi}$ being small because of the damped oscillation (2.7.72), and

$$\left(\frac{\dot{a}}{a}\right)^2 \simeq m^2\phi^2. \tag{2.7.74}$$

Thus the term $m^2\phi^2$ behaves as an effective cosmological constant $\sigma^2\Lambda/3$ (compare U in Eq. (2.7.60) with Eq. (2.7.19)). The same holds in the Euclidean region: for large $|\phi|$, the classical solution is approximately a portion of a four-sphere with radius $1/(m|\phi|)$. As in the earlier part of this section, the Euclidean and Lorentzian solutions can be joined together at the equator of the Euclidean four-sphere. In the Lorentzian region, where $U > 0$, the classical solutions which dominate the path integral are complex, and the wave function oscillates in this region. For large $|\phi|$, the classical solutions are approximately inflationary for some time (see Fig. 2.21). Eventually, at later times, the matter field decays and oscillates. The universe becomes matter-dominated, large and nearly flat. Then it re-collapses. Thus one can already obtain many features of a realistic quantum model of cosmology from a mini-superspace model. For a more detailed model, one must include bosonic and fermionic fluctuations about a Friedmann model, as discussed in the following two sections.

2.8 Bosonic structure in the universe

One can extend the massive scalar-field model of the previous section to allow for perturbations in the scalar and gravitational fields [Halliwell & Hawking 1985]. One can then compute the Hartle–Hawking state and examine its physical properties.

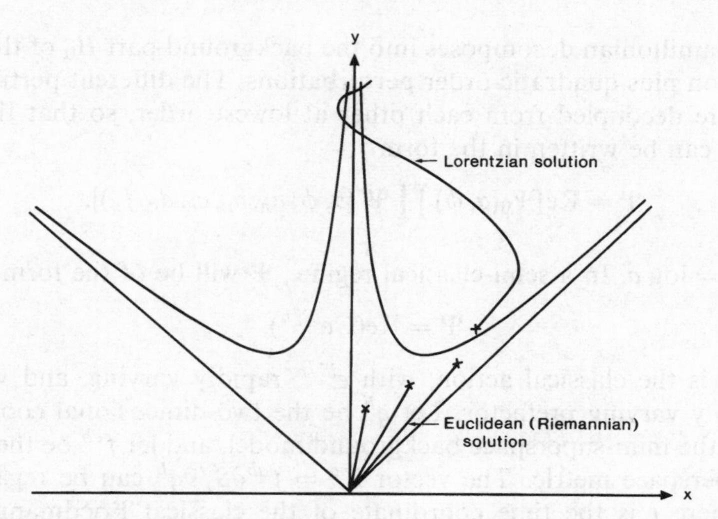

Fig. 2.21. Classical Euclidean and Lorentzian solutions of the massive scalar field model. For small radius a and large $|\phi|$, they are approximately inflationary.

The background Lorentzian geometry is

$$ds^2 = \sigma^2[-dt^2 + a^2(t)d\Omega_3{}^2]. \tag{2.8.1}$$

One perturbs this by taking the three-metric of the form

$$h_{ij} = \sigma^2 a^2(\Omega_{ij} + \epsilon_{ij}), \tag{2.8.2}$$

where Ω_{ij} is the metric of the unit three-sphere, and the perturbation ϵ_{ij} is expanded in hyperspherical harmonics [Lifschitz & Khalatnikov 1963, Gerlach & Sengupta 1978]. They describe the tensor (gravitational-wave), vector (gauge) and scalar modes. These will not be written out in detail. Their coefficients are denoted

$$a_{nlm}, b_{nlm}, c^o_{nlm}, c^e_{nlm}, d^o_{nlm}, d^e_{nlm},$$

and are functions of time t, but not of the space coordinates x^i. The wave function is then of the form

$$\Psi = \Psi(a; a_{nlm}, b_{nlm}, c^o_{nlm}, c^e_{nlm}, d^o_{nlm}, d^e_{nlm}, f_{nlm}), \tag{2.8.3}$$

where the scalar field has the expression

$$\Psi = \sigma^{-1}\left(\frac{1}{2^{1/2}\pi}\phi(t) + \sum_{nlm} f_{nlm} Q^n_{lm}\right), \tag{2.8.4}$$

where the scalar harmonics Q^n_{lm} obey

$$Q^n_{lm|k}{}^{|k} = -(n^2-1)Q^n_{lm},\ n = 1, 2, 3, \ldots. \tag{2.8.5}$$

One also expands out the lapse N and shift N^i in harmonics.

The Hamiltonian decomposes into the background part H_0 of the previous section plus quadratic-order perturbations. The different perturbation modes are decoupled from each other at lowest order, so that the wave function can be written in the form

$$\Psi = \mathrm{Re}[\Psi_0(\alpha, \phi) \prod_n \Psi^n(\alpha, \phi; a_n, b_n, c_n, d_n, f_n)], \tag{2.8.6}$$

where $\alpha = \log a$. In a semi-classical régime, Ψ will be of the form

$$\Psi = \mathrm{Re}(Ce^{iS/\hbar}), \tag{2.8.7}$$

where S is the classical action, with $e^{iS/\hbar}$ rapidly varying, and where C is a slowly varying prefactor. Let q^b be the two-dimensional coordinates (α, ϕ) of the mini-superspace background model, and let f^{ab} be the inverse mini-superspace metric. The vector $X^a = f^{ab}\partial S/\partial q^b$ can be regarded as $\partial/\partial t$, where t is the time coordinate of the classical Friedmann metric corresponding to $\Psi = \mathrm{Re}(Ce^{iS/\hbar})$. Each perturbation wave function Ψ^n obeys a time-dependent Schrödinger equation along an integral curve of the vector field X^a:

$$i\hbar \frac{\partial \Psi^n}{\partial t} = H_2^n \Psi^n. \tag{2.8.8}$$

Thus a concept of time is recovered in this semi-classical limit. Further, energy is fed back from the perturbation modes into the homogeneous background model, through the term $J.J$ in the background Wheeler–DeWitt equation

$$(-\tfrac{1}{2}\nabla^2 + e^{-3\alpha}V + \tfrac{1}{2}J.J)\Psi_0 = 0, \tag{2.8.9}$$

where $J = \sum_n \frac{\nabla \Psi^n}{\Psi^n}$, ∇ is the two-dimensional covariant derivative in the background, and $V = \frac{1}{2}(e^{6\alpha}m^2\phi^2 - e^{4\alpha})$. For the scalar-gravity model considered here, $J.J$ will be infinite, although one might expect that this difficulty would be removed by working with a supergravity theory, as described later in this book. In the present model, one makes J finite by subtractions [Halliwell & Hawking 1985]. First, one subtracts out the ground state energies of the H_2^n; this corresponds to a renormalization of the cosmological constant Λ. Then one renormalizes the gravitational constant G and also makes a further subtraction corresponding to a curvature-squared counterterm in the action. One can then assume that $J.J$ is finite.

One can then consider the Hartle–Hawking state. The boundary conditions for the Euclidean path integral are that, as $a \to 0$ with $t \to 0$, regularity of the four-metric and scalar field implies that a_n, b_n, c_n, d_n, f_n vanish there. One finds that the tensor (gravitational-wave) and scalar (coupled gravity and scalar-field) modes start out in their ground states in the Euclidean region. The remaining vector modes are pure gauge and so can be ignored.

Returning to the background geometry, one finds that the classical solutions with large nearly-constant values of $|\phi|$ are not damped by the Euclidean action $e^{-I/\hbar}$. These classical trajectories are all roughly equally probable. The classical solutions have a long inflationary period, before they start to oscillate (Fig. 2.21) and make a transition to a matter-dominated expansion. In general, if one took a more realistic model which also contained other fields of low rest mass, the matter energy which is available in the oscillations of the massive scalar field would decay into the lighter particles, giving a radiation-dominated universe.

One can follow the quantum state of each mode from the Euclidean into the Lorentzian region by considering the Schrödinger equation (2.8.8). The tensor (gravitational-wave) modes obey an oscillator equation with a time-dependent frequency $v = (n^2 - 1)^{1/2} e^{-\alpha}$. The spin-1/2 analogue of this will be seen in more detail in the spin-1/2-gravity case (section 2.9). At early times the wave function Ψ^n will be in the ground state, with the frequency v obeying $v \gg \dot{\alpha}$, the characteristic inverse time-scale of the background geometry. The wave function initially remains in its ground state. When $v \simeq \dot{\alpha}$, the adiabatic approximation will break down, since the wavelength of the gravitational mode becomes equal to the inflationary horizon scale. The wave function then 'freezes' – i.e. it becomes independent of time. Eventually the mode re-enters the horizon in the matter- or radiation-dominated epoch. At this time it will be in a superposition of highly excited states. Thus the ground-state fluctuations in the gravitational waves have been strongly amplified by the cosmological expansion [Grishchuk 1974,1977, Starobinsky 1980, Rubakov *et al.* 1982]. A similar description applies to the scalar modes.

Halliwell and Hawking further calculate the spectrum of perturbations in the microwave background radiation resulting from the Hartle–Hawking state. This can be compared with the observed results [Smoot *et al.* 1992, White *et al.* 1994]. The fluctuations in the microwave background temperature are found to obey

$$\left\langle \left(\frac{\nabla T}{T} \right)^2 \right\rangle \simeq \left(\frac{m}{m_{\rm Planck}} \right)^2. \tag{2.8.10}$$

This gives an approximately scale-free spectrum of perturbations, which is compatible with observations if the mass m of the scalar field is approximately $5 \times 10^{-5} \times$ (the Planck mass), or about 10^{14} GeV. The perturbation spectrum has Gaussian statistics, as confirmed by the observations.

To summarize, the Hartle–Hawking proposal yields a quantum state which is strongly peaked around isotropy and homogeneity. As expected, it yields a ground state at early times for the perturbation wave functions.

Further, it makes testable predictions about the microwave background radiation, which are in agreement with observation for $m \simeq 10^{14}$ GeV.

A more detailed treatment will be given in the next section of a model with fermionic perturbations.

2.9 Spin-1/2 fermions in quantum cosmology

In addition to studying the bosonic structure in the universe, described in the previous section in the context of the Hartle–Hawking state, one would also like to understand the structure of matter in a simplified model, given by spin-1/2 fermions coupled to the gravitational and scalar fields [D'Eath & Halliwell 1987]. This will also serve as an introduction to some of the ideas and properties which recur in supergravity, where spin-3/2 matter is coupled to gravity in a locally supersymmetric fashion.

First, in subsection 2.9.1, two-component spinors are described; these are needed to describe the gravitational and spin-1/2 fields, and of course later the gravitino field in supergravity. Subsection 2.9.2 deals with the Hamiltonian treatment of the coupled Einstein–Dirac system [Nelson & Teitelboim 1978]. For simplicity of exposition, the scalar field is not included in this section. One only needs its homogeneous mode $\phi(t)$ for the background, described by Hawking's massive scalar field model [Hawking 1984] (end of section 2.7). There are inhomogeneous perturbations of Φ, coupled to matter, but these do not couple at lowest order to the fermionic perturbations which will be studied here. The Hamiltonian treatment of gravity-spin-1/2 parallels that of gravity-spin-3/2 (supergravity) in section 3.2.

For quantum cosmology with compact spatial surfaces, as here with an S^3 surface, it is appropriate (subsection 2.9.3) to split the spin-1/2 field into its positive- and negative-frequency parts. In the case when there are two bounding three-surfaces, one takes spectral boundary conditions by regarding the negative-frequency parts as being fixed on the initial surface and the positive-frequency parts as fixed on the final surface. In the context of the Hartle–Hawking state, one would like to find the classical action for the case with data fixed on one boundary, with a compact Riemannian space inside. In the case $m = 0$ of a massless spin-1/2 field the classical solution will be regular only if one specifies the positive-frequency part of the spinor field on the boundary. Spectral boundary conditions are thus essential in quantum cosmology. In the case of scattering theory in supergravity, spectral conditions are also needed in order to be able to describe incoming and outgoing gravitinos (section 8.2).

The quantization of the spin-1/2 modes is described in subsection 2.9.4. As in the bosonic case (section 2.8), when the background wave function has the WKB form $C\exp(iS/\hbar)$, the wave function for each perturbation mode obeys a Schrödinger evolution equation. The boundary conditions for the Hartle–Hawking state are described, and found to describe the ground state at the beginning of the inflationary epoch. In the subsequent evolution, following from the Schrödinger equation, spin-1/2 particles are produced, and their back–reaction can be estimated. Further, one finds that, in the inflationary régime, where the background geometry is de Sitter, the Green's functions for the fermionic field are invariant under the de Sitter isometry group (e.g. the rotations of S^4 in the Euclidean case). Correspondingly, a fermionic particle detector moving along a geodesic measures a thermal spectrum [Gibbons & Hawking 1977]. Thus in the inflationary period the universe is in thermal equilibrium at the de Sitter temperature

$$T = \frac{H}{2\pi} = \frac{1}{2\pi}\left(\frac{\Lambda}{3}\right)^{1/2}, \tag{2.9.1}$$

where $H = m|\phi|$.

2.9.1 Two-component spinors

Here we describe the conventions for two-component spinors used here, and some of their properties. In order to define spinors in curved space–time one must start by considering the pseudo-orthonormal tetrad $e^a{}_\mu$, which gives the spacetime metric $g_{\mu\nu}$ as

$$g_{\mu\nu} = \eta_{ab} e^a{}_\mu e^b{}_\nu \tag{2.9.1.1}$$

and obeys the relation

$$g^{\mu\nu} e^a{}_\mu e^b{}_\nu = \eta^{ab}, \tag{2.9.1.2}$$

where $\eta_{ab} = \mathrm{diag}(-1,1,1)$ is the Minkowski metric. Latin tetrad indices $a, b, ...,$ which are raised and lowered using η^{ab}, η_{ab}, run from 0 to 3, as do Greek spacetime indices $\mu, \nu, ...,$ which are raised and lowered using $g^{\mu\nu}, g_{\mu\nu}$. Thus, Lorentzian spinor conventions are being used. One can nevertheless move to a Riemannian geometry by rotating to suitable complex $e^a{}_\mu$, e.g. by rotating the lapse function by $N \to -iN$.

The basis one-forms $e^a{}_\mu$ correspond to spinor-valued one-forms

$$e^{AA'}{}_\mu = e^a{}_\mu \sigma_a{}^{AA'}, \tag{2.9.1.3}$$

where unprimed spinor indices $A, B, ...$ take the values 0,1, and primed spinor indices $A', B', ...$ take the values $0', 1'$. The Infeld–van der Waerden translation symbols $\sigma_a{}^{AA'}$ are taken to be

$$\sigma_0 = -\frac{1}{\sqrt{2}} I, \ \sigma_i = \frac{1}{\sqrt{2}} \Sigma_i \, (i = 1,2,3), \tag{2.9.1.4}$$

where Σ_i are the Pauli spin matrices. Spinor indices are raised and lowered with the alternating spinors $\epsilon^{AB}, \epsilon_{AB}, \epsilon^{A'B'}, \epsilon_{A'B'}$, each of which is given by the matrix

$$\begin{bmatrix} 0 & 1 \\ -1 & 0 \end{bmatrix},$$

according to

$$\begin{aligned} \rho^{\dots A \dots} &= \epsilon^{AB} \rho^{\dots}{}_{B}{}^{\dots}, \quad \rho_{\dots A \dots} = \rho_{\dots}{}^{B}{}_{\dots} \epsilon_{BA}, \\ \rho^{\dots A' \dots} &= \epsilon^{A'B'} \rho^{\dots}{}_{B'}{}^{\dots}, \quad \rho_{\dots A' \dots} = \rho_{\dots}{}^{B'}{}_{\dots} \epsilon_{B'A'}. \end{aligned} \tag{2.9.1.5}$$

The inverse of Eq. (2.9.1.3) is then

$$e^{a}{}_{\mu} = -\sigma^{a}{}_{AA'} e^{AA'}{}_{\mu}. \tag{2.9.1.6}$$

The minus signs in this and several subsequent equations appear because of our spacelike metric convention, with signature +2, as opposed to the timelike convention more usually adopted when spinors are considered [Itzykson & Zuber 1980, Penrose & Rindler 1984]. Corresponding to Eqs. (2.9.1.1) and (2.9.1.2) one has

$$g^{\mu\nu} e^{AA'}{}_{\mu} e^{BB'}{}_{\nu} = -\epsilon^{AB} \epsilon^{A'B'}, \tag{2.9.1.7}$$

$$g_{\mu\nu} = -\epsilon_{AB} \epsilon_{A'B'} e^{AA'}{}_{\mu} e^{BB'}{}_{\nu}. \tag{2.9.1.8}$$

The $e^{AA'}{}_{\mu}$ also obey the Dirac relations

$$\begin{aligned} 2e_{AA'(\mu} e^{BA'}{}_{\nu)} &= -g_{\mu\nu} \epsilon_{A}{}^{B}, \\ 2e_{AA'(\mu} e^{AB'}{}_{\nu)} &= -g_{\mu\nu} \epsilon_{A'}{}^{B'}. \end{aligned} \tag{2.9.1.9}$$

Corresponding to an infinitesimal local Lorentz transformation acting on the tetrad index of $e^{a}{}_{\mu}$, the $e^{AA'}{}_{\mu}$ change by

$$\delta e^{AA'}{}_{\mu} = N^{A}{}_{B} e^{BA'}{}_{\mu} + \overline{N}^{A'}{}_{B'} e^{AB'}{}_{\mu}, \tag{2.9.1.10}$$

where $N_{AB} = N_{(AB)}$ is symmetric and $\overline{N}_{A'B'}$ is its complex conjugate. Under this transformation any spinor $\rho^{A \dots A' \dots}$ changes by

$$\delta \rho^{A \dots A' \dots} = N^{A}{}_{B} \rho^{B \dots A' \dots} + \cdots + \overline{N}^{A'}{}_{B'} \rho^{A \dots B' \dots} + \cdots. \tag{2.9.1.11}$$

In particular, any tensor $T^{\mu_1 \dots \mu_n}$ defines a spinor

$$T^{AA' \dots ZZ'} = e^{AA'}{}_{\mu_1} \cdots e^{ZZ'}{}_{\mu_n} T^{\mu_1 \dots \mu_n}, \tag{2.9.1.12}$$

where inversely

$$T^{\mu_1 \dots \mu_n} = (-1)^n e_{AA'}{}^{\mu_1} \cdots e_{ZZ'}{}^{\mu_n} T^{AA' \dots ZZ'}. \tag{2.9.1.13}$$

With our conventions, a tensor index contraction corresponds to spinor index contraction with a minus sign, e.g. ,

$$T^{\mu}{}_{\mu} = -T^{AA'}{}_{AA'}. \tag{2.9.1.14}$$

In a Hamiltonian treatment one considers the spinor-valued spatial one-forms $e^{AA'}{}_i$ defined on a spacelike hypersurface $t = \text{const.}$, where Latin spatial indices $i, j, ...$ run from 1 to 3. The intrinsic spatial metric is

$$h_{ij} = -e_{AA'i}e^{AA'}{}_j = g_{ij}, \tag{2.9.1.15}$$

and h^{ij}, h_{ij} are used to raise and lower spatial indices, where h^{ij} is the inverse of h_{ij}. If n^μ is the future-pointing unit timelike normal to the surface, then the spinor $n^{AA'}$ obeys

$$n_{AA'}e^{AA'}{}_i = 0, \tag{2.9.1.16}$$

$$n_{AA'}n^{AA'} = 1. \tag{2.9.1.17}$$

These conditions determine $n^{AA'}$ in terms of the $e^{AA'}{}_i$. Together $n^{AA'}$ and the $e^{AA'}{}_i$ form a basis for the space of spinors with one unprimed and one primed index. The relations

$$n_{AA'}n^{AB'} = \frac{1}{2}\epsilon_{A'}{}^{B'}, \tag{2.9.1.18}$$

$$n_{AA'}n^{BA'} = \frac{1}{2}\epsilon_A{}^B, \tag{2.9.1.19}$$

$$e_{AA'i}e^{AB'}{}_j = -\frac{1}{2}h_{ij}\epsilon_{A'}{}^{B'} - i\epsilon_{ijk}h^{1/2}n_{AA'}e^{AB'k}, \tag{2.9.1.20}$$

$$e_{AA'i}e^{BA'}{}_j = -\frac{1}{2}h_{ij}\epsilon_A{}^B + i\epsilon_{ijk}h^{1/2}n_{AA'}e^{BA'k}, \tag{2.9.1.21}$$

$$e_{AA'i}e_{BB'}{}^i = -\epsilon_{AB}\epsilon_{A'B'} + n_{AA'}n_{BB'} \tag{2.9.1.22}$$

are useful in expanding out spinorial expressions.

2.9.2 Hamiltonian treatment of the Einstein–Dirac theory

In the Lorentzian régime, the Einstein–Dirac action has the form

$$S = S_V + S_B, \tag{2.9.2.1}$$

where S_V is a volume term and S_B is a surface term. For S_V one takes

$$S_V = \frac{1}{2\kappa^2}\int d^4x\, eR - \frac{i}{2}\int d^4x\, e(\bar{\phi}^{A'}e_{AA'}{}^\mu D_\mu\phi^A + \bar{\chi}^{A'}e_{AA'}{}^\mu D_\mu\chi^A) + \text{H.c.}$$
$$- \frac{m}{\sqrt{2}}\int d^4x\, e(\chi_A\phi^A + \bar{\phi}^{A'}\bar{\chi}_{A'}). \tag{2.9.2.2}$$

Here $e = \det(e^a{}_\mu)$ and R is the Ricci scalar of the four-metric $g_{\mu\nu} = \eta_{ab}e^a{}_\mu e^b{}_\nu$. The Dirac field consists of the spinor fields (ϕ^A, χ^A) with Hermitian conjugates $(\bar{\phi}^{A'}, \bar{\chi}^{A'})$. These are odd Grassmann quantities, anticommuting under multiplication among themselves, but commuting with even quantities such as the tetrad $e^{AA'}{}_\mu$. The derivative $D_\mu\phi^A$ is defined by

$$D_\mu\phi^A = \partial_\mu\phi^A + \omega^A{}_{B\mu}\phi^B, \tag{2.9.2.3}$$

where the torsion-free connection forms $\omega^{ab}{}_\mu$ are given spinorially as

$$\omega^{AA'BB'}{}_\mu = \omega^{AB}{}_\mu \epsilon^{A'B'} + \bar{\omega}^{A'B'}{}_\mu \epsilon^{AB}, \tag{2.9.2.4}$$

where

$$\omega^{AB}{}_\mu = \omega^{(AB)}{}_\mu$$

and the conjugate $\bar{\omega}^{A'B'}{}_\mu$ obeys

$$\bar{\omega}^{A'B'}{}_\mu = \bar{\omega}^{(A'B')}{}_\mu.$$

Here $\omega^{ab}{}_\mu$ is given by [D'Eath 1984]

$$\omega^{ab}{}_\mu = e^{av}\partial_{[\mu}e^b{}_{v]} - e^{bv}\partial_{[\mu}e^a{}_{v]} - e^{av}e^{b\rho}e_{c\mu}\partial_{[v}e^c{}_{\rho]}. \tag{2.9.2.5}$$

Classically, the boundary terms S_B in the action are needed in order that, when data are posed on the boundaries, the variational equation $\delta S = 0$ should correctly give the classical solution. The detailed form of S_B depends on the boundary data chosen. Consider, for example, the scattering case in which there is an initial boundary Σ_I and a final boundary Σ_F. It will become apparent in the Hamiltonian treatment that one natural 'local' choice of boundary data involves specifying the $e^{AA'}{}_i$ and half of the fermionic variables $\phi^A, \chi^A, \tilde{\phi}^{A'}, \tilde{\chi}^{A'}$ on each surface. Here, as is natural in the Hamiltonian approach, the variables $\tilde{\phi}^{A'}$ and $\tilde{\chi}^{A'}$ are freed from being the Hermitian conjugates of ϕ^A and χ^A, and will be written with a tilde. There are different types of 'local' choice of boundary conditions, but suppose for example that one specifies $e^{AA'}{}_i, \tilde{\phi}^{A'}, \tilde{\chi}^{A'}$ on Σ_I and $e^{AA'}{}_i, \phi^A, \chi^A$ on Σ_F. Then

$$\begin{aligned} S_B = & \frac{1}{\kappa^2}\left[\int_{\Sigma_F} - \int_{\Sigma_I}\right] d^3x\, h^{1/2}\mathrm{tr}K \\ & + \frac{i}{2}\left[\int_{\Sigma_F} + \int_{\Sigma_I}\right] d^3x\, h^{1/2}(\tilde{\phi}^{A'} n_{AA'}\phi^A + \tilde{\chi}^{A'} n_{AA'}\chi^A). \end{aligned} \tag{2.9.2.6}$$

With the above boundary data, the classical variational equation $\delta S = 0$ gives the Dirac equations

$$e_{AA'}{}^\mu D_\mu \phi^A = i\frac{m}{\sqrt{2}}\tilde{\chi}_{A'}, \tag{2.9.2.7}$$

$$e_{AA'}{}^\mu D_\mu \chi^A = i\frac{m}{\sqrt{2}}\tilde{\phi}_{A'}, \tag{2.9.2.8}$$

$$e_{AA'}{}^\mu D_\mu \tilde{\phi}^{A'} = -i\frac{m}{\sqrt{2}}\chi_A, \tag{2.9.2.9}$$

$$e_{AA'}{}^\mu D_\mu \tilde{\chi}^{A'} = -i\frac{m}{\sqrt{2}}\phi_A, \tag{2.9.2.10}$$

as well as the Einstein equations with energy–momentum tensor formed quadratically from $(\phi^A, \chi^A, \tilde{\phi}^{A'}, \tilde{\chi}^{A'})$ and their derivatives. Note that the classical solution $e^{AA'}{}_{\mu}$ will in general not be Hermitian, because (ϕ^A, χ^A) and $(\tilde{\phi}^A, \tilde{\chi}^{A'})$ are no longer Hermitian conjugates. In fact, when one studies the full model with gravity coupled also to a scalar field, it has been seen in section 2.7 that the classical $e^{AA'}{}_{\mu}$ may be non-Hermitian (being Euclidean or complex) even in the absence of fermions.

One could instead have specified (say) the fields $(e^{AA'}{}_i, \phi^A, \chi^A)$ on Σ_I and $(e^{AA'}{}_i, \tilde{\phi}^{A'}, \tilde{\chi}^{A'})$ on Σ_F. Then the fermionic boundary term in Eq. (2.9.2.6) would appear with its sign reversed. When the chirality of a fermionic boundary variable is altered, there is a sign change in the corresponding fermionic boundary term. So far, only local fermionic boundary conditions have been mentioned. However, as described above, and discussed in detail in the following subsection 2.9.3, the most natural boundary conditions for the Hartle–Hawking state, which must be set on a single surface Σ, are non-local or spectral, being based on the eigenfunction decomposition of fields on the three-sphere S^3. Again, there are corresponding boundary terms in the action. Scattering amplitudes are also usually expressed by making a positive- and negative-frequency decomposition.

Quantum-mechanically, following section 2.3, one needs the boundary terms S_B so that the amplitude to go from given data on Σ_I to data on Σ_F can be equivalently obtained by taking an intermediate surface Σ_J and composing the amplitudes to go from Σ_I to Σ_J and Σ_J to Σ_F, by summing over a complete set of states on Σ_J.

Having described the Lagrangian theory, one obtains the Hamiltonian version [Nelson & Teitelboim 1978] by decomposing the theory with respect to a family of spacelike hypersurfaces $t = x^0 = \text{const.}$ The gravitational field is decomposed into $e^{AA'}{}_0$ and $e^{AA'}{}_i$. One may adopt the convention that spatial indices are lowered and raised using h_{ij} and h^{ij}. The quantities $e^{AA'}{}_0$ can be expanded as

$$e^{AA'}{}_0 = Nn^{AA'} + N^i e^{AA'}{}_i, \tag{2.9.2.11}$$

where N is the lapse and N^i is the shift (section 2.3).

One can calculate the momentum $p_{AA'}{}^i = \delta S/\delta \dot{e}_i^{AA'}$, and one also finds that $p_{AA'}{}^0 = 0$. The fermionic momenta conjugate to $\phi^A, \chi^A, \tilde{\phi}^{A'}, \tilde{\chi}^{A'}$ are defined as $\pi_{\phi A} = \delta S/\delta \dot{\phi}^A$, etc. Here one adopts the convention that an odd variable such as $\dot{\phi}^A$ must first be brought to the left using anti-commutation; then the differentiation is carried out [Wess & Bagger 1992]. The fermionic momenta are given by

$$\pi_{\phi A} = -\frac{i}{2} h^{1/2} n_{AA'} \tilde{\phi}^{A'}, \tag{2.9.2.12}$$

$$\pi_{\chi A} = -\frac{i}{2} h^{1/2} n_{AA'} \tilde{\chi}^{A'}, \tag{2.9.2.13}$$

with $\pi_{\tilde{\phi} A'}$ $\pi_{\tilde{\chi} A'}$ being given by minus the 'Hermitian conjugates' of $\pi_{\phi A}$

and $\pi_{\chi A}$. The momenta are related to the original variables because the Lagrangian is of first order in derivatives of fermionic fields. The coordinate and momentum variables are also related by the primary constraints [Dirac 1965]

$$J_{AB} \approx 0, \tilde{J}_{A'B'} \approx 0, \tag{2.9.2.14}$$

which follow from the invariance of the action under local Lorentz transformations. Explicitly,

$$J_{AB} = e_{(A}{}^{A'i} p_{B)A'i} + \phi_{(A}\pi_{\phi B)} + \chi_{(A}\pi_{\chi B)}, \tag{2.9.2.15}$$

and correspondingly for $\tilde{J}_{A'B'}$. The variables also obey the Hamiltonian and momentum constraints $\mathscr{H}_{\perp} = 0, \mathscr{H}_i = 0$.

Poisson brackets for systems with both bosonic and fermionic variables have been defined by [Casalbuoni 1976]. The constraints (2.9.2.14) and $\mathscr{H}_{\perp}, \mathscr{H}_i = 0$ are first-class [Dirac 1965, Hanson *et al.* 1976, Nelson & Teitelboim 1978] in the sense that their brackets are zero when the constraints are satisfied. The set of fermionic constraints (2.9.2.12), (2.9.2.13) and their 'conjugates' is second-class [Dirac 1965, Hanson *et al.* 1976, Nelson & Teitelboim 1978] in that their Poisson brackets do not vanish when the constraints are imposed. These brackets are proportional to the normal spinor $n_{AA'}$. In general, given a set λ_I of second-class constraints, one defines [Hanson *et al.* 1976]

$$C_{IJ} = [\lambda_I, \lambda_J], \tag{2.9.2.16}$$

and C^{IJ} to be the inverse of C_{IJ}. Then one defines a Dirac bracket corresponding to these second-class constraints,

$$[F, G]^* = [F, G] - [F, \lambda_I]C^{IJ}[\lambda_J, G]. \tag{2.9.2.17}$$

As described by [Nelson & Teitelboim 1978], it is valid to set $\lambda_I = 0$ provided that one uses the Dirac bracket $[F, G,]^*$, since

$$[\lambda_I, F]^* = 0 \tag{2.9.2.18}$$

for any function F of the dynamical variables. Further,

$$[F, G]^* \approx [F, G] \tag{2.9.2.19}$$

if one of the functions F, G is first-class. Here $\approx$ means that the quantities are equal subject to the constraint equations. Hence, given the form of the Hamiltonian H (see below), which is linear in first-class constraints, one has

$$[F, H]^* \approx [F, H], \tag{2.9.2.20}$$

so that the equations of motion using the Dirac bracket are the same as those using the original Poisson bracket.

Thus the basic dynamical variables of Einstein–Dirac theory may be taken to be $e^{AA'}{}_i, p_{AA'}{}^i, \phi^A, \chi^A, \tilde{\phi}^{A'}$ and $\tilde{\chi}^{A'}$. One may assume that the fermionic momentum variables have been eliminated using Eqs. (2.9.2.12),(2.9.2.13) and 'conjugates'. The Hamiltonian is of the form

$$H = \int d^3x(N\mathscr{H}_\perp + N^i\mathscr{H}_i + M_{AB}J^{AB} + \tilde{M}_{A'B'}\tilde{J}^{A'B'}). \tag{2.9.2.21}$$

Here $M_{AB} = M_{(AB)}$ and $\tilde{M}_{A'B'} = \tilde{M}_{(A'B')}$ have been introduced as Lagrange multipliers for the constraints $J^{AB} = 0, \tilde{J}^{A'B'} = 0$. The set $N, N^i, M_{AB}, \tilde{M}_{A'B'}$ of Lagrange multipliers for the first-class constraints may be freely specified during the classical dynamical evolution. As usual, N and N^i give the amount of normal and tangential displacement applied to dynamical data per unit time, and $M_{AB}, \tilde{M}_{A'B'}$ give the amount of local Lorentz transformation.

The generator $\mathscr{H}_\perp$ is

$$\begin{aligned}\mathscr{H}_\perp = {} & 2\kappa^2 h^{-1/2}\left[\pi_{ij}\pi^{ij} - 1/2(\mathrm{tr}\pi)^2\right] - \frac{1}{2\kappa^2}h^{1/2}\,{}^3R \\ & + \frac{i}{2}h^{1/2}e_{AA'}{}^i(\tilde{\phi}^{A'}\,{}^3D_i\phi^A + \tilde{\chi}^{A'}\,{}^3D_i\chi^A) + \mathrm{H.c.} \\ & + \frac{m}{\sqrt{2}}h^{1/2}(\chi_A\phi^A + \tilde{\phi}^{A'}\tilde{\chi}_{A'}),\end{aligned} \tag{2.9.2.22}$$

and $\mathscr{H}_i$ is

$$\begin{aligned}\mathscr{H}_i = {} & -2h_{ij}\pi^{jk}{}_{|k} \\ & + \frac{i}{2}h^{1/2}n_{AA'}(\tilde{\phi}^{A'}\,{}^3D_i\phi^A + \tilde{\chi}^{A'}\,{}^3D_i\chi^A) + \mathrm{H.c.} \\ & + \frac{1}{4}h_{ik}\partial_j[(\tilde{\phi}^{A'}\phi^A + \tilde{\chi}^{A'}\chi^A)\epsilon^{kjl}e_{AA'l}],\end{aligned} \tag{2.9.2.23}$$

where

$$\pi^{ij} = \tfrac{1}{2}e^{AA'(i}p_{AA'}{}^{j)} = -\frac{h^{1/2}}{2\kappa^2}(K^{ij} - h^{ij}\mathrm{tr}K) \tag{2.9.2.24}$$

and $\mathrm{tr}\pi = h_{ij}\pi^{ij}$. The spatial covariant deriviative on spinors is (e.g.)

$${}^3D_i\phi^A = \partial_i\phi^A + {}^3\omega^A{}_{B_i}\phi^B. \tag{2.2.9.25}$$

Here ${}^3\omega^{AB}{}_i$ corresponds [D'Eath 1984] to

$$\begin{aligned}{}^3\omega^{ab}{}_i = \Big(& e^{bj}\partial_{[j}e^a{}_{i]} - \tfrac{1}{2}e^{aj}e^{bk}e^c{}_i\partial_j e_{ck} \\ & - \tfrac{1}{2}e^{aj}n^b n^c\partial_j e_{ci} - \tfrac{1}{2}n^a\partial_i n^b\Big) - (a \leftrightarrow b).\end{aligned} \tag{2.9.2.26}$$

For a classical solution, the generators $\mathscr{H}_\perp$ and $\mathscr{H}_i$ vanish:

$$\mathscr{H}_\perp \approx 0, \mathscr{H}_i \approx 0. \tag{2.9.2.27}$$

The simplest Hamiltonian form of the Einstein–Dirac theory is found in the time gauge [Nelson & Teitelboim 1978], where one imposes

$$e^0{}_i = 0, \tag{2.9.2.28}$$

or equivalently

$$n^a = \delta^a_0. \tag{2.9.2.29}$$

The $e^a{}_i$ are now described by $e^\alpha{}_i$ where $\alpha, \beta, \ldots = 1, 2, 3$. One also eliminates the conjugate momenta $p_0{}^i$ by setting them to zero:

$$p_0{}^i = 0. \tag{2.9.2.30}$$

Hence the generators $J_{*i} = n^a e^b{}_i J_{ab}$ of Lorentz boosts are set to zero, where $J_{ab} \leftrightarrow J_{AB}\epsilon_{A'B'} + \bar{J}_{A'B'}\epsilon_{AB}$ are the generators of local Lorentz transformations. The local triad rotation generators $J_{\alpha\beta}$ or J_{ij} remain, and vanish classically:

$$J_{\alpha\beta} \approx 0. \tag{2.9.2.31}$$

The dynamical variables are most conveniently taken to be the triad $e^\alpha{}_i$, its conjugate momentum $p_\alpha{}^i$, and the rescaled fermionic variables

$$\begin{aligned} \psi^A &= h^{1/4}\phi^A, \ \lambda^A = h^{1/4}\chi^A, \\ \tilde{\psi}^{A'} &= h^{1/4}\tilde{\phi}^{A'}, \ \tilde{\lambda}^{A'} = h^{1/4}\tilde{\chi}^{A'}. \end{aligned} \tag{2.9.2.32}$$

The only non-zero Dirac brackets are

$$[e^\alpha{}_i(x), p_\beta{}^j(x')]^* = \delta^\alpha_\beta \delta^j_i \delta(x, x'), \tag{2.9.2.33}$$

$$[\psi^A(x), \tilde{\psi}^{A'}(x')]^* = -2in^{AA'}\delta(x, x'), \tag{2.9.2.34}$$

$$[\lambda^A(x), \tilde{\lambda}^{A'}(x')]^* = -2in^{AA'}\delta(x, x'). \tag{2.9.2.35}$$

The Hamiltonian is given by Eqs. (2.9.2.21–23),(2.9.2.15), with $e^0{}_i$ and $p_0{}^i$ set to zero.

In quantizing, one picks a maximal (anti)commuting set of variables, such as $e^\alpha{}_i(x)$, $\psi^A(x)$, and $\lambda^A(x)$, which one regards as 'coordinate' variables. The remaining variables, here $p_\alpha{}^i(x)$, $\tilde{\psi}^{A'}(x)$, and $\tilde{\lambda}^{A'}(x)$, are 'momentum-like'. Quantum states are represented by Grassmann algebra-valued wave functionals $\Psi(e^\alpha{}_i(x), \psi^A(x), \lambda^A(x))$. As will be described for supergravity in section 3.3, a formal inner product between wave functions can be given (compare [Faddeev & Slavnov 1980]). In the case that the wave functions do not depend on any extraneous fermionic variables (such as might be introduced through boundary data), the inner product is complex-valued. Dirac brackets $[\,,\,]^*$ must be replaced by commutators $[\,,\,]_-$ or anticommutators $[\,,\,]_+$ according to [Casalbuoni 1976]:

$$\begin{aligned} [E_1, E_2]_- &= i\hbar[E_1, E_2]^*, \\ [O, E]_- &= i\hbar[O, E]^*, \\ [O_1, O_2]_+ &= i\hbar[O_1, O_2]^*. \end{aligned} \tag{2.9.2.36}$$

Then the remaining variables $p_\alpha{}^i, \bar{\psi}^{A'}, \bar{\lambda}^{A'}$ are represented by

$$p_\alpha{}^i(x) \to -i\hbar \frac{\delta}{\delta e^\alpha{}_i(x)}, \tag{2.9.2.37}$$

$$\bar{\psi}^{A'}(x) \to 2\hbar n^{AA'} \frac{\delta}{\delta \psi^A(x)}, \tag{2.9.2.38}$$

$$\bar{\lambda}^{A'}(x) \to 2\hbar n^{AA'} \frac{\delta}{\delta \lambda^A(x)}. \tag{2.9.2.39}$$

The quantum fermionic operators $\bar{\psi}^{A'}(x)$ and $\bar{\lambda}^{A'}(x)$ are written with a bar, since they are the Hermitian adjoints of $\psi^A(x), \lambda^A(x)$ with the inner product. Also, $p_\alpha{}^i(x)$ is Hermitian.

The classical constraints (2.9.2.27),(2.9.2.14) become quantum constraints on physically allowed wave functionals:

$$\mathscr{H}_\perp \Psi = 0, \tag{2.9.2.40}$$

$$\mathscr{H}_i \Psi = 0, \tag{2.9.2.41}$$

$$J_{\alpha\beta} \Psi = 0. \tag{2.9.2.42}$$

The constraints $\mathscr{H}_i\Psi = 0$ and $J_{\alpha\beta}\Psi = 0$ are of first order in momenta, and are naturally ordered with momentum operators on the right. One can show that Eqs. (2.9.2.41),(2.9.2.42) imply that the wave function Ψ is invariant when spatial coordinate transformations and local rotations are applied to its arguments $e^\alpha{}_i, \psi^A$ and λ^A. These properties still hold in different fermionic representations. There is no obvious factor ordering for the second-order operator $\mathscr{H}_\perp$; for example, one apparently cannot achieve a closed algebra under (anti)commutation for the set of constraint generators $\mathscr{H}_\perp$, $\mathscr{H}_i$ and $J_{\alpha\beta}$. This difficulty suggests that one should turn to supergravity models, as in chapter 3 and following, where there is a preferred ordering for $\mathscr{H}_\perp$ and $\mathscr{H}_i$, consequent on a preferred ordering for the supersymmetry generators S_A and $\bar{S}_{A'}$. For the present purpose in quantum cosmology, in the order of approximation to be used, the ordering of $\mathscr{H}_\perp$ does not make a substantial difference to the results, and will be chosen for convenience.

In this cosmological example, the action and Hamiltonian can be expanded out to second order in gravitational variables about the $k = +1$ Friedmann background, and to second order in fermionic fields. The bosonic and fermionic perturbations are non-interacting at this order, and can be treated separately. To recover the metric representation for the gravitational field, one fixes the local $SO(3)$ rotation freedom; the wave function for the bosonic perturbations will be that of [Halliwell & Hawking 1985], as outlined in section 2.8. One takes a normalized left-invariant basis $E^{\alpha}{}_i$ of one-forms on S^3 [Misner *et al.* 1973], with $E^{\alpha}{}_i E_{\alpha j}$ giving the metric of the unit three-sphere. Letting a be the radius of the background three-sphere, one writes

$$e^{\alpha}{}_i = a(E^{\alpha}{}_i + \epsilon^{\alpha}{}_i), \tag{2.9.2.43}$$

where $\epsilon^{\alpha}{}_i$ are triad perturbations. One then imposes the gauge condition

$$E^{\alpha}{}_i \epsilon_{\alpha j} - E^{\alpha}{}_j \epsilon_{\alpha i} = 0, \tag{2.9.2.44}$$

whence the constraints $J_{\alpha\beta}$ become second-class. As described above, any second-class constraints can be eliminated by the Dirac procedure. Once this is done, the gravitational variables can be taken to be h_{ij} and π^{ij}, with the same fermionic variables as above. At lowest order in perturbations, the only non-zero Dirac brackets are the standard brackets [DeWitt 1967] between h_{ij} and π^{ij}, and the fermionic brackets (2.9.2.34–35). Hence, given the results of [Halliwell & Hawking 1985], it is valid to study only the fermionic perturbations.

2.9.3 *Spectral boundary conditions*

Consider now the expansion of the fermion fields $\phi_A, \tilde{\phi}_{A'}, \chi_A, \tilde{\chi}_{A'}$ or their weighted versions $\psi_A, \lambda_A, \tilde{\psi}_{A'}, \tilde{\lambda}_{A'}$ in harmonics on the three-sphere. As described in [D'Eath & Halliwell 1987], one may construct a complete set of spinor harmonics $\rho_A^{np}(x), \bar{\sigma}_A^{np}(x)$ and $\bar{\rho}_{A'}^{np}(x), \sigma_{A'}^{np}(x)$ for the expansion of a pair such as $(\phi_A, \tilde{\phi}_{A'})$. Here $n = 0, 1, 2, \ldots$ and $p = 1, 2, \ldots, (n+1)(n+2)$. The harmonic equations are

$$e^{AA'j}\,{}^3D_j \rho_A = -(n + \tfrac{3}{2})\,{}_e n^{AA'}\,\rho_A, \tag{2.9.3.1}$$

$$e^{AA'j}\,{}^3D_j \sigma_{A'} = -(n + \tfrac{3}{2})\,{}_e n^{AA'}\,\sigma_{A'}, \tag{2.9.3.2}$$

$$e^{AA'j}\,{}^3D_j \bar{\rho}_{A'} = (n + \tfrac{3}{2})\,{}_e n^{AA'}\,\bar{\rho}_{A'}, \tag{2.9.3.3}$$

$$e^{AA'j}\,{}^3D_j \bar{\sigma}_A = (n + \tfrac{3}{2})\,{}_e n^{AA'}\,\bar{\sigma}_A. \tag{2.9.3.4}$$

Here

$${}_e n^{AA'} = -i n^{AA'} \tag{2.9.3.5}$$

is the Euclidean normal spinor. Thus ρ_A and $\sigma_{A'}$ are the positive-frequency harmonics, and $\bar{\rho}_{A'}, \bar{\sigma}_A$ are the negative-frequency harmonics. They obey the relations

$$\int d\mu\, \rho_A^{np} n^{AA'} \bar{\rho}_{A'}^{mq} = \delta^{nm} H_n^{pq}, \tag{2.9.3.6}$$

$$\int d\mu\, \rho_A^{np} \epsilon^{AB} \rho_B^{mq} = \delta^{nm} A_n^{pq}, \qquad (2.9.3.7)$$

$$\int d\mu\, \bar{\sigma}_A^{np} n^{AA'} \sigma_{A'}^{mq} = \delta^{nm} H_n^{pq}, \qquad (2.9.3.8)$$

$$\int d\mu\, \sigma_{A'}^{np} \epsilon^{A'B'} \sigma_{B'}^{mq} = \delta^{nm} A_n^{pq}, \qquad (2.9.3.9)$$

where $d\mu$ is the measure on the unit three-sphere. Here, by suitable choice of the harmonics, one may take $H_n^{pq} = \delta^{pq}$, and $A_n^{pq} = \sqrt{2} C_n^{pq}$, where C_n^{pq} is block-diagonal with blocks $\begin{pmatrix} 0 & 1 \\ -1 & 0 \end{pmatrix}$. Integrals analogous to (2.9.3.6–9), involving one ρ and one σ, possibly conjugated, are zero by virtue of the eigenvalue equations (2.9.3.1–4).

One may then expand out

$$\phi_A = \frac{e^{-3\alpha/2}}{2\pi} \sum_{np} \sum_{q} \alpha_n^{pq} [m_{np}(t) \rho_A^{nq}(x) + \tilde{r}_{np}(t) \bar{\sigma}_A^{nq}(x)], \qquad (2.9.3.10)$$

$$\tilde{\phi}_{A'} = \frac{e^{-3\alpha/2}}{2\pi} \sum_{np} \sum_{q} \alpha_n^{pq} [\tilde{m}_{np}(t) \bar{\rho}_{A'}^{nq}(x) + r_{np}(t) \sigma_{A'}^{nq}(x)], \qquad (2.9.3.11)$$

$$\chi_A = \frac{e^{-3\alpha/2}}{2\pi} \sum_{np} \sum_{q} \beta_n^{pq} [s_{np}(t) \rho_A^{nq}(x) + \tilde{t}_{np}(t) \bar{\sigma}_A^{nq}(x)], \qquad (2.9.3.12)$$

$$\tilde{\chi}_{A'} = \frac{e^{-3\alpha/2}}{2\pi} \sum_{np} \sum_{q} \beta_n^{pq} [\tilde{s}_{np}(t) \bar{\rho}_{A'}^{nq}(x) + t_{np}(t) \sigma_{A'}^{nq}(x)]. \qquad (2.9.3.13)$$

Here the weight factor $e^{-3\alpha/2}$ corresponds (up to a constant) to $h^{1/4}$ in Eq. (2.9.2.32); it is simplest to set boundary conditions for the weighted fields $\psi_A, \lambda_A, \tilde{\psi}_{A'}, \tilde{\lambda}_{A'}$, which obey the simplest bracket relations. The index q runs from 1 to $(n+1)(n+2)$, just as p does. The coefficients α_n^{pq} and β_n^{pq} are introduced to avoid coupling between different values of p in the harmonic expansion of the action. For a given n, α_n^{pq} and β_n^{pq} may be regarded as block-diagonal matrices α_n, β_n of dimension $(n+1)(n+2)$, with blocks $\begin{pmatrix} 1 & 1 \\ 1 & -1 \end{pmatrix}$ for α_n and $\begin{pmatrix} 1 & -1 \\ -1 & -1 \end{pmatrix}$ for β_n. The set of time-dependent coefficients $m_{np}, r_{np}, t_{np}, s_{np}$ and $\tilde{m}_{np}, \tilde{r}_{np}, \tilde{t}_{np}, \tilde{s}_{np}$ may be taken to be the fermionic variables, each an odd element of a Grassmann algebra.

As mentioned at the beginning of this section 2.9, one can use spectral boundary conditions for the fermions, instead of taking ψ_A and λ_A as fixed on a final surface Σ_F and $\tilde{\psi}_{A'}, \tilde{\lambda}_{A'}$ as fixed on an initial surface Σ_I. In particular, one can specify the positive-frequency coefficients $m_{np}, r_{np}, t_{np}, s_{np}$ on Σ_F and negative-frequency coefficients $\tilde{m}_{np}, \tilde{r}_{np}, \tilde{t}_{np}, \tilde{s}_{np}$ on Σ_I. Were we studying scattering theory, rather than quantum cosmology, this would be appropriate. But it is also essential for the Hartle–Hawking state in the case that the fermion mass m is zero, where one shrinks the initial surface Σ_I to nothing. As will be seen below [Eqs. (2.9.3.20–23)], in the massless case only the positive-frequency fermionic data on Σ_F admit a regular classical solution in the classical Riemannian geometry interior to Σ_F. Thus only positive-frequency fermionic data allow even a semi-classical approximation $\exp(-I_E/\hbar)$ to the Hartle–Hawking state, where I_E is the classical Euclidean action.

We now decompose the Einstein–Dirac action into harmonics. From the above information, one finds

$$S = S_0 + S_f + S_{fB}. \tag{2.9.3.14}$$

Here S_0 is the action of the background model. The volume fermionic contribution S_f is

$$S_f = S_f^{(1)}[m, s, \tilde{m}, \tilde{s}] + S_f^{(2)}[t, r, \tilde{t}, \tilde{r}]. \tag{2.9.3.15}$$

Here

$$S_f^{(1)} = \sum_{np} \int dtN \Big[\frac{i}{2N}(\tilde{m}_{np}\dot{m}_{np} + m_{np}\dot{\tilde{m}}_{np} + \tilde{s}_{np}\dot{s}_{np} + s_{np}\dot{\tilde{s}}_{np}$$
$$+ e^{-\alpha}(n + \tfrac{3}{2})(\tilde{m}_{np}m_{np} + \tilde{s}_{np}s_{np}) - m(s_{np}m_{np} + \tilde{m}_{np}\tilde{s}_{np})\Big] \tag{2.9.3.16}$$

and $S_f^{(2)}$ is identical, with $m_{np}, s_{np}, \tilde{m}_{np}, \tilde{s}_{np}$ replaced by $t_{np}, r_{np}, \tilde{t}_{np}, \tilde{r}_{np}$. The appropriate boundary term for the above spectral boundary conditions, with $m_{np}, s_{np}, t_{np}, r_{np}$ fixed on Σ_F and $\tilde{m}_{np}, \tilde{s}_{np}, \tilde{t}_{np}, \tilde{r}_{np}$ fixed on Σ_I, is

$$S_{fB} = \frac{i}{2}\sum_{np}(\tilde{m}_{np}m_{np} + \tilde{s}_{np}s_{np} + \tilde{t}_{np}t_{np} + \tilde{r}_{np}r_{np})_{\Sigma_F}$$
$$+ \frac{i}{2}\sum_{np}(\tilde{m}_{np}m_{np} + \tilde{s}_{np}s_{np} + \tilde{t}_{np}t_{np} + \tilde{r}_{np}r_{np})_{\Sigma_I}. \tag{2.9.3.17}$$

If one uses the Grassmann variables x and y to denote m_{np} and s_{np}, or t_{np} and r_{np}, then the fermionic action, including boundary terms, is the sum over n, p of actions

$$S_n(x, \tilde{x}, y, \tilde{y}) = \int dtN \Big[\frac{i}{2N}(\tilde{x}\dot{x} + x\dot{\tilde{x}} + \tilde{y}\dot{y} + y\dot{\tilde{y}})$$
$$+ e^{-\alpha}(n + \tfrac{3}{2})(\tilde{x}x + \tilde{y}y) - m(yx + \tilde{x}\tilde{y})\Big]$$
$$+ \frac{i}{2}(\tilde{x}x + \tilde{y}y)_{\Sigma_F} + \frac{i}{2}(\tilde{x}x + \tilde{y}y)_{\Sigma_I}. \tag{2.9.3.18}$$

The total fermionic action has the form

$$S_f + S_{fB} = \sum_{np} [S_n(m_{np}, \tilde{m}_{np}, s_{np}, \tilde{s}_{np}) + S_n(t_{np}, \tilde{t}_{np}, r_{np}, \tilde{r}_{np})]. \qquad (2.9.3.19)$$

Here (2.9.3.18–19) represents $x, \tilde{x}$ and $y, \tilde{y}$ time-dependent Fermi oscillators, coupled through the mass term. If $e^{-\alpha}$ did not depend on time, one would have a time-independent Fermi oscillator, treated in [DeWitt 1984b].

The Lorentzian classical fermionic field equations, derived from the action of Eq. (2.9.3.18), are

$$i\frac{\dot{x}}{N} + vx - m\tilde{y} = 0, \qquad (2.9.3.20)$$

$$i\frac{\dot{\tilde{x}}}{N} - v\tilde{x} + my = 0, \qquad (2.9.3.21)$$

$$i\frac{\dot{y}}{N} + vy + m\tilde{x} = 0, \qquad (2.9.3.22)$$

$$i\frac{\dot{\tilde{y}}}{N} - v\tilde{y} - mx = 0 \qquad (2.9.3.23)$$

where $v = e^{-\alpha}(n + \frac{3}{2})$. These are the components of the Dirac equation (2.9.2.7–10). There is also a decoupled second-order form: x and y each obey the equation

$$\frac{1}{N}\frac{d}{dt}\left(\frac{\dot{x}}{N}\right) + \left(\frac{\dot{v}}{iN} + v^2 + m^2\right) x = 0, \qquad (2.9.3.24)$$

while $\tilde{x}$ and $\tilde{y}$ each obey

$$\frac{1}{N}\frac{d}{dt}\left(\frac{\dot{\tilde{x}}}{N}\right) + \left(\frac{-\dot{v}}{iN} + v^2 + m^2\right) \tilde{x} = 0. \qquad (2.9.3.25)$$

Proceeding along the lines of subsection 2.9.2, one finds the Hamiltonian

$$H_n(x, \tilde{x}, y, \tilde{y}) = N[v(x\tilde{x} + y\tilde{y}) + m(yx + \tilde{x}\tilde{y})], \qquad (2.9.3.26)$$

where the only non-zero Dirac-bracket relations are

$$[x, \tilde{x}]^* = -i, \ [y, \tilde{y}]^* = -i. \qquad (2.9.3.27)$$

The total fermionic Hamiltonian is

$$H_f = \sum_{np} H_{np} = \sum_{np} [H_n(m_{np}, \tilde{m}_{np}, s_{np}, \tilde{s}_{np}) + H_n(t_{np}, \tilde{t}_{np}, r_{np}, \tilde{r}_{np})]; \quad (2.9.3.28)$$

this is the expansion of the fermionic part of Eq. (2.9.2.22) in harmonics. Since the bosonic perturbations decouple from the fermionic perturbations, the total Hamiltonian constructed from the background Hamiltonian (2.7.60), (2.7.61) and the fermionic Hamiltonian H_f of Eq. (2.9.3.28) must vanish by the Hamiltonian constraint

$$H_0 + H_f = 0. \qquad (2.9.3.29)$$

The use of the coefficients m_{np}, s_{np}, etc., instead of $\phi_A, \chi_A, \tilde{\phi}_{A'}, \tilde{\chi}_{A'}$ has a further advantage. The $m_{np}, s_{np}, \ldots$ are invariant under local Lorentz transformations and spatial diffeomorphisms. Hence the J_{ab} and $\mathscr{H}_i$ constraints are automatically satisfied, and one only needs to satisfy the Hamiltonian constraint (2.9.3.29).

One can now quantize the model consisting of the background coupled to the fermions. A wave function is taken to have the form

$$\Psi(\alpha, \phi, m_{np}, s_{np}, t_{np}, r_{np}).$$

This obeys the Wheeler–DeWitt equation

$$(H_0 + H_f)\Psi = 0. \tag{2.9.3.30}$$

Here one chooses quantum operators such that [Casalbuoni 1976]

$$[E_1, E_2]_- = i\hbar[E_1, E_2]^*, \tag{2.9.3.31}$$

$$[O, E]_- = i\hbar[O, E]^*, \tag{2.9.3.32}$$

$$[O_1, O_2]_+ = i\hbar[O_1, O_2]^*, \tag{2.9.3.33}$$

where E denotes an even element and O denotes an odd element. In particular, for the fermions one has

$$[x, \bar{x}]_+ = 1, \; [y, \bar{y}]_+ = 1. \tag{2.9.3.34}$$

One chooses the holomorphic representation [Faddeev & Slavnov 1980]

$$\bar{x} \to \hbar\frac{\partial}{\partial x}, \; \bar{y} \to \hbar\frac{\partial}{\partial y}, \tag{2.9.3.35}$$

using left differentiation [Wess & Bagger 1992]; e.g. before performing the operation $\hbar\partial/\partial x$, each appearance of x must be brought to the left using anticommutation and then the differentiation must be performed.

The factor ordering of the background operator H_0 can be chosen as usual (section 2.7). Among the fermionic terms H_f, there is an ambiguity in the terms $x\bar{x}$. The Weyl ordering [Berezin & Marinov 1977, Henneaux & Teitelboim 1982]

$$x\bar{x} \to \frac{\hbar}{2}\left[x\frac{\partial}{\partial x} - \frac{\partial}{\partial x}x\right] \tag{2.9.3.36}$$

will be used. Then the fermionic Hamiltonian H_n becomes

$$H_n = N\left[-v + v\left(x\frac{\partial}{\partial x} + y\frac{\partial}{\partial y}\right) + m\left(yx + \frac{\partial^2}{\partial x \partial y}\right)\right], \tag{2.9.3.37}$$

where here and for the remainder of this section $\hbar$ is set to 1. One needs the eigenstates of the operator $N^{-1}H_n$, obeying $N^{-1}H_n\psi = E\psi$. The eigenstates and eigenvalues are

$$\psi^{(0)} = N_0\left[1 + \frac{m}{(v+\omega)}xy\right], \qquad E_0 = -\omega, \tag{2.9.3.38}$$

$$\psi^{(1)} = N_1 x, \qquad E_1 = 0, \tag{2.9.3.39}$$

$$\psi^{(2)} = N_2 y, \qquad E_2 = 0, \tag{2.9.3.40}$$

$$\psi^{(3)} = N_3\left[1 + \frac{m}{(v-\omega)}xy\right], \qquad E_3 = \omega, \tag{2.9.3.41}$$

where $\omega = (v^2 + m^2)^{1/2}$ and $N_0, \ldots, N_3$ are normalization terms. Note that these four eigenstates are orthogonal: the inner product is that appropriate to the holomorphic representation [Berezin 1966, Faddeev & Slavnov 1980]

$$(f,g) = \int \overline{f(x,y)}g(x,y)e^{-x\bar{x}-y\bar{y}}dx d\bar{x} dy d\bar{y}. \tag{2.9.3.42}$$

The fermionic integration follows the Berezin rules [Berezin 1966]

$$\int dx = 0, \int x dx = 1, \int d\bar{x} = 0, \int \bar{x} d\bar{x} = 1, \tag{2.9.3.43}$$

and analogously for y and $\bar{y}$. If the eigenstates (2.9.3.38–41) are normalized with respect to the inner product, one finds

$$N_0 = \left(\frac{\omega + v}{2\omega}\right)^{1/2}, N_1 = N_2 = 1, N_3 = \left(\frac{\omega - v}{2\omega}\right)^{1/2}. \tag{2.9.3.44}$$

Note that the operators x and $\partial/\partial x$ are adjoints with respect to (2.9.3.42), as are y and $\partial/\partial y$. One can further check that H_n in Eq. (2.9.3.37) is self-adjoint.

One can define creation and annihilation operators among the states (2.9.3.38–41):

$$a = \frac{1}{(2\omega)^{1/2}}\left[(v+\omega)^{1/2}\frac{\partial}{\partial y} + \frac{m}{(v+\omega)^{1/2}}x\right], \tag{2.9.3.45}$$

$$b = \frac{1}{(2\omega)^{1/2}}\left[(v+\omega)^{1/2}\frac{\partial}{\partial x} - \frac{m}{(v+\omega)^{1/2}}y\right], \tag{2.9.3.46}$$

and their adjoints

$$a^\dagger = \frac{1}{(2\omega)^{1/2}}\left[(v+\omega)^{1/2}y + \frac{m}{(v+\omega)^{1/2}}\frac{\partial}{\partial x}\right] \tag{2.9.3.47}$$

$$b^\dagger = \frac{1}{(2\omega)^{1/2}}\left[(v+\omega)^{1/2}x - \frac{m}{(v+\omega)^{1/2}}\frac{\partial}{\partial y}\right]. \tag{2.9.3.48}$$

The only non-zero anticommutators among these are

$$[a, a^\dagger]_+ = 1, \ [b, b^\dagger]_+ = 1. \tag{2.9.3.49}$$

One has, for example,

$$\begin{aligned} a\psi^{(0)} &= 0, \qquad a^\dagger\psi^{(0)} = \psi^{(2)}, \\ a\psi^{(2)} &= \psi^{(0)}, \qquad a^\dagger\psi^{(2)} = 0. \end{aligned} \tag{2.9.3.50}$$

The operators a and $a^\dagger$ step between $\psi^{(0)}$ and $\psi^{(2)}$ and between $\psi^{(1)}$ and $\psi^{(3)}$; b and $b^\dagger$ step between $\psi^{(0)}$ and $\psi^{(1)}$ and between $\psi^{(2)}$ and $\psi^{(3)}$.

Number operators N_a and N_b can be defined by $N_a = a^\dagger a$ and $N_b = b^\dagger b$. The Hamiltonian can be written as

$$N^{-1}H_n = \omega(N_a + N_b) - \omega. \tag{2.9.3.51}$$

The states (2.9.3.38–41) are eigenstates of N_a and N_b:

$$\begin{aligned} N_a\psi^{(0)} &= 0, \qquad N_a\psi^{(1)} = 0, \\ N_a\psi^{(2)} &= \psi^{(2)}, \qquad N_a\psi^{(3)} = \psi^{(3)}. \end{aligned} \tag{2.9.3.52}$$

Applying N_b gives analogous results, with $\psi^{(1)}$ and $\psi^{(2)}$ interchanged.

This approach to particle states uses instantaneous Hamiltonian diagonalization [Birrell & Davies 1982], depending on the time-dependent quantities $\nu = e^{-\alpha}(n + \frac{3}{2})$ and $\omega = (\nu^2 + m^2)^{1/2}$. In this context one may interpret the states $\psi^{(0)}, \psi^{(1)}, \psi^{(2)}$ and $\psi^{(3)}$. One may regard $a^\dagger$ and $b^\dagger$ as creation operators of a particle and an antiparticle, respectively, in the mode np. Then $\psi^{(0)}$ gives the vacuum state, $\psi^{(1)}$ gives a one-antiparticle state, $\psi^{(2)}$ gives a one-particle state, and $\psi^{(3)}$ gives a one-particle–one-antiparticle state. The freedom of choosing x, y in Eqs. (2.9.3.38–41) to be either m_{np}, s_{np} or t_{np}, r_{np} allows for the two helicity states of the particle. The use of Hamiltonian diagonalization gives different definitions of what one means by a particle state for different values of the scale factor. The Hamiltonian definition of a particle state will not in general agree with the definition given by measurements taken by a detector moving on a geodesic [Birrell & Davies 1982], as will be outlined in the following subsection 2.9.4.

One would like to understand the evolution of a fermionic quantum state. This is most easily done when the background quantum state is semi-classical, corresponding to the Lorentzian case $\Psi_0 = C\exp(iS/\hbar)$, say. Using the mini-superspace metric G_{AB} of section 2.7 to define the inner product in the following equation, one writes

$$\frac{\nabla\Psi_0}{\Psi_0}\cdot\nabla = i\nabla S\cdot\nabla = i\frac{\partial}{\partial t}. \tag{2.9.3.53}$$

The integral curves of $\partial/\partial t$ are the classical trajectories which correspond to the classical limit of the mini-superspace background model. One finds [D'Eath & Halliwell 1987] that, in this limit, the fermionic wave function in each mode obeys the time-dependent Schrödinger equation along a trajectory:

$$(H_{np} - \tfrac{1}{2}\Omega_{np})\Psi_{np} = i\hbar \frac{\partial \Psi_{np}}{\partial t}. \qquad (2.9.3.54)$$

The quantity Ω_{np} will be chosen later, in order to subtract off the vacuum energies of the perturbation modes. In particular, one can solve the Schrödinger equation [D'Eath & Halliwell 1987] in the case of an exact de Sitter background. One uses (e.g.) the Hartle–Hawking proposal to set initial data at the boundary of the Euclidean and Lorentzian region in (α, ϕ) space, and then evolves the quantum state to the future in the Lorentzian region, using Eq. (2.9.3.54). One can also arrive at similar conclusions by working with a large-n (adiabatic) approximation, as in the following subsection 2.9.4.

2.9.4 Quantum state

Consider now the Hartle–Hawking state

$$\Psi(\alpha, \phi, m_{np}, s_{np}, t_{np}, r_{np})$$
$$= \int_C \mathscr{D}e^a{}_\mu \mathscr{D}\phi_A \mathscr{D}\tilde{\phi}_{A'} \mathscr{D}\chi_A \mathscr{D}\tilde{\chi}_{A'} \exp(-I/\hbar). \qquad (2.9.4.1)$$

Here I is the Euclidean action, obtained by choosing the lapse function N to be negative imaginary, and taking $I = -iS$. Strictly, one should include gauge-fixing and Faddeev–Popov ghost terms in the path integral (2.9.4.1) [Itzykson & Zuber 1980]. The class C of paths in the space with coordinates $(\alpha, \phi, m_{np}, \tilde{m}_{np}, s_{np}, \tilde{s}_{np}, t_{np}, \tilde{t}_{np}, r_{np}, \tilde{r}_{np})$ describes compact four-geometries with boundary at radius e^α, with scalar field ϕ at the boundary (Fig. 2.14), and spin-1/2 fields $\phi_A, \tilde{\phi}_{A'}, \chi_A, \tilde{\chi}_{A'}$ which are bounded. Writing $\tau = \int iN dt$ for the proper Euclidean time coordinate, the conditions of regularity for α and ϕ at the 'South pole' $\tau = 0$ are

$$e^\alpha = 0, \ \frac{de^\alpha}{d\tau} = 1, \ \phi = \phi_0, \ \frac{d\phi}{d\tau} = 0, \qquad (2.9.4.2)$$

at $\tau = 0$. Boundedness of the spin-1/2 fields $\phi_A, \tilde{\phi}_{A'}, \chi_A, \tilde{\chi}_{A'}$ at the South pole $\tau = 0$, and the weight factors $e^{-3\alpha/2}$ in Eqs. (2.9.3.10–13), imply that

$$m_{np} = s_{np} = t_{np} = r_{np} = 0,$$
$$\tilde{m}_{np} = \tilde{s}_{np} = \tilde{t}_{np} = \tilde{r}_{np} = 0, \qquad (2.9.4.3)$$

at $\tau = 0$. At the boundary, $\tau = \tau'$ say, $m_{np}, s_{np}, t_{np}, r_{np}$ match the prescribed final values in the argument of Ψ in Eq. (2.9.4.1), while the variables $\tilde{m}_{np}, \tilde{s}_{np}, \tilde{t}_{np}$ and $\tilde{r}_{np}$ are unrestricted.

It is helpful to consider the boundary conditions for the fermions in the classical approximation. Note that the fermionic action is quadratic, so that the path integral for the fermionic modes is indeed semi-classical. In the decomposition (2.9.3.10–13), one can write each of the fields $\phi_A, \chi_A, \tilde{\phi}_{A'}$ and $\tilde{\chi}_{A'}$ in the form $\phi_A = \phi_A^{(+)} + \phi_A^{(-)}$, etc., where the (+) part corresponds to the modes with unbarred coefficients and the (−) part corresponds to the modes with barred coefficients. Thus the classical boundary-value problem involves finding the classical (Euclidean–Dirac) solution with $\phi_A^{(+)}, \chi_A^{(+)}, \tilde{\phi}_{A'}^{(+)}$ and $\tilde{\chi}_{A'}^{(+)}$ matching prescribed values on the final surface Σ_F, and $\phi_A, \chi_A, \tilde{\phi}_{A'}, \tilde{\chi}_{A'}$ being regular in the interior of Σ_F. The Euclidean version of the Dirac equation (2.9.2.7–10) or of Eqs. (2.9.3.20–23) leads to second-order equations for the variables $\phi_A^{(+)}, \chi_A^{(+)}, \tilde{\phi}_{A'}^{(+)}, \tilde{\chi}_{A'}^{(+)}$. One solves these in the interior of Σ_F subject to prescribed boundary data on Σ_F, to find $\phi_A^{(+)}, \chi_A^{(+)}, \tilde{\phi}_{A'}^{(+)}, \tilde{\chi}_{A'}^{(+)}$ uniquely in the interior of Σ_F. In the massive case $m \neq 0$, one can then use Eqs. (2.9.3.20–23) to obtain $\phi_A^{(-)}, \chi_A^{(-)}, \tilde{\phi}_{A'}^{(-)}, \tilde{\chi}_{A'}^{(-)}$ from the (+) variables. Now the (+) fields will be analytic, being solutions of elliptic boundary-value problems. Hence the (−) fields will also be analytic, from Eqs. (2.9.3.20–23). One can then check that the (+) and (−) fields together provide a solution of the classical Euclidean Dirac equations, derived from Eqs. (2.9.2.7–10) or Eqs. (2.9.3.20–23). Thus one has a unique solution of the classical fermionic field equations, subject to the boundary conditions with prescribed $\phi_A^{(+)}, \chi_A^{(+)}, \tilde{\phi}_{A'}^{(+)}, \tilde{\chi}_{A'}^{(+)}$, showing that the boundary-value problem is well posed.

The one case excluded from the argument of the previous paragraph is the massless case, with $m = 0$. In this case it is essential (not just a matter of choice) to specify the (+) variables on the surface Σ_F, and not to specify any of the (−) variables there. One can see this from the Euclidean version of the differential equations (2.9.3.20–23), which decouple when $m = 0$. The solutions for the unbarred variables x and y are regular at $\tau = 0$, corresponding to regular (+) solutions. But the solutions for $\tilde{x}$ and $\tilde{y}$ are divergent at $\tau = 0$, if the boundary data for $\tilde{x}$ and $\tilde{y}$ are non-zero. Hence, when $m = 0$, one obtains a regular classical solution in the interior for the (+) variables, when these are specified on the boundary, and one takes the (−) variables to be zero on the boundary and hence in the interior.

The above treatment is a special case of the more general boundary-value problem, where a general three-geometry (Σ_F, h_{ij}) bounds a compact four-geometry $(\mathscr{M}, g_{\mu\nu})$. One encounters here the general theory of [Atiyah *et al.* 1975] on the spectral theory of elliptic operators, where one splits the fermionic data into (+) and (−) parts, using the spatial projection of the Dirac operator.

Now consider the wave function which is found in the semi-classical approximation. Since the perturbation modes do not couple to each other, a natural Ansatz is

$$\Psi = \Psi_0(q^a) \prod_{np} \Psi_{np}(q^a, m_{np}, s_{np}, t_{np}, r_{np}). \tag{2.9.4.4}$$

Here $q^a = (\alpha, \phi)$. The background wave function Ψ_0 obeys the background Wheeler–DeWitt equation (2.7.65) modified by a term representing the sum of the energies of the fermion modes [D'Eath & Halliwell 1987]. Since it can be shown that this energy term makes only a small correction, one may assume that $\Psi_0(q^a)$ is of the type described in section 2.7.

The path integral for the fermions takes the form

$$\int \mathscr{D}x \mathscr{D}\tilde{x} \mathscr{D}y \mathscr{D}\tilde{y} \exp(-I_n/\hbar). \tag{2.9.4.5}$$

The integral involves the paths $(x(\tau), \tilde{x}(\tau), y(\tau), \tilde{y}(\tau))$ with $\tilde{x} = \tilde{y} = 0$ at $\tau = 0$ and $x = x', y = y'$ at $\tau = \tau'$, where (x', y') are the arguments of the wave function. Since the action I_n is quadratic in the fermionic fields, the path integral (2.9.4.5) may be evaluated to give

$$A \exp(-I_n^{\text{cl}}.) \tag{2.9.4.6}$$

Here A is a one-loop prefactor, which depends only on the background (α, ϕ), and may be absorbed into $\Psi_0(\alpha, \phi)$. The classical Euclidean action I_n^{cl} contains the (x', y') dependence. One finds that the volume term in I_n^{cl} vanishes at a classical solution, giving just a boundary contribution

$$I_n^{\text{cl}} = \tfrac{1}{2}[\tilde{x}(\tau')x' + \tilde{y}(\tau')y']. \tag{2.9.4.7}$$

This action depends on the classical background solution for (α, ϕ). To obtain the wave function $\psi_n(\alpha', \phi', x', y')$, one must perform a functional integration of Eq. (2.9.4.6) over paths $(\alpha(\tau), \phi(\tau))$ which obey the initial conditions (2.9.4.2) and agree with the boundary data (α', ϕ') at $\tau = \tau'$. The main contribution is expected to come from paths close to solutions of the classical Euclidean field equations. For these paths, one may use the adiabatic approximation, where α is taken to be a slowly varying function of τ. Precisely, one assumes

$$\left|\frac{d\alpha}{d\tau}\right| \ll (n + \tfrac{3}{2})e^{-\alpha}. \tag{2.9.4.8}$$

This equation holds for all paths near $\tau = 0$, because of the initial conditions (2.9.4.2). One can further verify that, where $\alpha(\tau)$ is the solution of the field equations subject to the initial data (2.9.4.2), the expression (2.9.4.8) holds in the entire Euclidean region. Here, (2.9.4.8) is the condition that, in the Lorentzian region, the mode labelled by n is inside the horizon of the de Sitter spacetime. The approximation (2.9.4.8) will be used to solve the Dirac equation, which will lead to the exponent in Eqs. (2.9.4.6–7).

Following Eqs. (2.9.4.6–7), one needs $\tilde{x}(\tau')$ and $\tilde{y}(\tau')$ in terms of x' and y'. The Euclidean second-order equation, corresponding to Eq. (2.9.3.24), is

$$\frac{d^2x}{d\tau^2} - \left(\frac{\partial v}{\partial \tau} + v^2 + m^2\right) x = 0, \tag{2.9.4.9}$$

where $v = e^{-\alpha}(n + \frac{3}{2})$. The adiabatic approximation (2.9.4.8) reads $|dv/d\tau| \ll v^2$. Hence the solution to Eq. (2.9.4.9) obeying

$$x(0) = 0, \quad x(\tau') = x'$$

is approximately

$$x(\tau) = \frac{\sinh(\omega\tau)}{\sinh(\omega\tau')} x'. \tag{2.9.4.10}$$

Now $x(\tau)$ and $\tilde{y}(\tau)$ are related by the Euclidean version of Eq. (2.9.3.20):

$$-\frac{dx}{d\tau} + vx - m\tilde{y} = 0. \tag{2.9.4.11}$$

Hence

$$\tilde{y}(\tau') = \frac{1}{m}[v - \omega \coth(\omega\tau')]x'. \tag{2.9.4.12}$$

Similarly for y and $\tilde{x}$, one finds

$$\tilde{x}(\tau') = \frac{1}{m}[-v + \omega \coth(\omega\tau')]y'. \tag{2.9.4.13}$$

Hence the Euclidean action I_n^{cl} of Eq. (2.9.4.7) is

$$I_n^{\text{cl}} = \frac{1}{m}[v - \omega \coth(\omega\tau')]x'y'. \tag{2.9.4.14}$$

For large n, one has $\coth(\omega\tau') \simeq 1$, giving

$$I_n^{\text{cl}} \simeq \frac{(v-m)}{m} x'y' = \frac{-m}{(\omega + v)} x'y'. \tag{2.9.4.15}$$

Hence the perturbation wave functions are given approximately by

$$\begin{aligned} \psi_n(\alpha', \phi', x', y') &\simeq \exp\left[\frac{mx'y'}{(\omega+v)}\right] \\ &\simeq 1 + \frac{mx'y'}{(\omega+v)}, \end{aligned} \tag{2.9.4.16}$$

up to a prefactor, which does not depend on x' and y'. This can be compared with Eqs. (2.9.3.38–41) for the eigenstates of the fermionic Hamiltonian; one sees that the Hartle–Hawking wave function gives the lowest-energy (ground) eigenstate for the fermionic modes. These results can be confirmed by solving the fermionic equations in an exact de Sitter background [D'Eath & Halliwell 1987].

One may calculate particle creation during the de Sitter phase. Define $\langle N_a \rangle$ to be the average number of particles in the state ψ_n at fixed α and ϕ, and $\langle N_b \rangle$ to be the average number of antiparticles. One can calculate (see below) [D'Eath & Halliwell 1987]

$$\langle N_a \rangle = \frac{(\psi_n, N_a \psi_n)}{(\psi_n, \psi_n)} = \langle N_b \rangle. \tag{2.9.4.17}$$

Then the total number Γ of particles and antiparticles created is

$$\begin{aligned}\Gamma &= 2 \sum_{np} (\langle N_a \rangle + \langle N_b \rangle) \\ &= 4 \sum_{n=0}^{\infty} (n+1)(n+2) \langle N_a \rangle. \end{aligned} \tag{2.9.4.18}$$

In (say) the Euclidean de Sitter régime, one has

$$\phi(\tau) \simeq \text{const.}\,, \ \mathrm{e}^{\alpha(\tau)} \simeq \frac{1}{H} \sin(H\tau), \tag{2.9.4.19}$$

where H is the Hubble constant. One needs to distinguish three different regions for the sum over n:

(i)	$n \ll n_1 = m e^{\alpha}$	(i.e. $\nu \ll m$),
(ii)	$n_1 \ll n \ll n_2 = H e^{\alpha}$	(i.e. $m \ll \nu \ll H$),
(iii)	$n_2 \ll n$	(i.e. $\nu \gg H$).

For region (iii) ('very short wavelengths') the modes are inside the horizon; indeed in the exact model where de Sitter evolution is followed by a matter-dominated phase, there are an infinite number of very short wavelength modes which never leave the horizon, and which are approximately in the ground state (2.9.4.16). One can check from the estimate [D'Eath & Halliwell 1987]

$$\langle N_a \rangle \simeq \frac{m^2}{2\omega(\omega+\nu)} \frac{H^2}{[(\omega+\nu)^2 + H^2 + m^2]} \tag{2.9.4.20}$$

that the contribution of the large-n region (iii) to Γ in Eq. (2.9.4.18) is finite. The contribution to Γ from region (iii) is of order $m^2 e^{\alpha}/H$. In region (i), one finds $\langle N_a \rangle \simeq 1$, giving a contribution to Γ of order $m^3 e^{3\alpha}$. In region (ii), $\langle N_a \rangle \simeq \frac{1}{2}$, and the contribution of this region to Γ is of order $H^3 e^{3\alpha}$. This is the largest of the three contributions. Thus, in summary, the particle production during the de Sitter epoch is finite: the main contribution is from modes with $m \ll \nu \ll H$, giving an expression for Γ of order $H^3 e^{3\alpha}$.

In the massless case $m = 0$, the variables $x, \bar{x}$ decouple from $y, \bar{y}$. The Hamiltonian operator H_{np} for each mode is a sum of four terms, each of the form $-v/2 + vx\partial/\partial x$. The eigenstates are 1 and x, with eigenvalues $-v/2$ and $v/2$, respectively. Now the Hartle–Hawking state is, as usual, the lowest-energy eigenstate, so that the initial condition for the Schrödinger equation (2.9.3.54) is $\Psi_{np} = 1$. One can check that $\Psi_{np} = 1$ for all time is the solution of the Schrödinger equation, with $\Omega_{np} = -4v$. Thus, for the massless model with $m = 0$, the fermionic modes are in their ground state at all times, so that there is no particle production. This is to be expected, since the massless Dirac action is conformally invariant [Birrell & Davies 1982], and the Friedmann–Robertson–Walker universes studied here are conformal to the Einstein static universe [Hawking & Ellis 1973], in which there is no particle production.

One can estimate the back-reaction of the spin-1/2 fermions on the Friedmann background geometry [D'Eath & Halliwell 1987]. This is infinite, and needs to be regularized. One finds that the back-reaction is a small correction to the potential $V(\alpha, \phi)$, and hence that the calculation of particle production treated above is consistent, since the particles produced do not substantially change the gravitational field which caused their production.

One can also examine the fermionic quantum state, corresponding to the Hartle–Hawking state, in terms of the response of a particle detector moving along a geodesic [D'Eath & Halliwell 1987]. It is simplest to work in the approximation in which the Wheeler–DeWitt equation reduces to a set of time-dependent Schrödinger equations along the classical trajectories of the homogeneous background modes. This is equivalent to studying quantum field theory on a fixed spacetime for the fermionic fields. The background is naturally taken to be de Sitter space, obtained formally by holding the homogeneous scalar field ϕ at a constant value. The Green's functions (expectation values in the Hartle–Hawking state of products of terms such as $\phi_A(x_1)$ and $\bar{\phi}_{A'}(x_2)$) are invariant under isometries of de Sitter space. For consider the Euclidean de Sitter space – the four-sphere S^4 of radius H^{-1}. One can study the quantum state of the field on a three-sphere S^3 of radius $e^\alpha < H^{-1}$, given by a path integral over the fermionic fields. The state depends on the geometry only through the radius of the S^3, not on the location of the S^3 in the S^4. This corresponds to $SO(5)$ invariance of the state and its Green's function, in the Euclidean context. In the Lorentzian case, this becomes $SO(4,1)$ de Sitter invariance. There is an analogous de Sitter invariance of the perturbation state in the bosonic model of section 2.8 [Halliwell & Hawking 1985]. The relevant de Sitter-invariant Green's function is known [Allen & Lütken 1987], and can be used to obtain transition rates for the fermionic particle detector. One finds that the detector experiences a thermal spectrum at the de Sitter temperature [Gibbons & Hawking 1977]

$$T = \frac{H}{2\pi} = \frac{1}{2\pi}\left(\frac{\Lambda}{3}\right)^{1/2}. \tag{2.9.4.21}$$

The thermal spectrum has a distribution of the Fermi–Dirac form, but the distribution does not have the correct density-of-states factor to be precisely Planckian. The same thermality will hold for the wave function describing perturbations in the bosonic model of section 2.8.

In summary, the Hartle–Hawking state for this model yields the ground state for the fermionic oscillators at the boundary between the Euclidean and Lorentzian de Sitter regions. One can evolve the fermionic quantum state to the future along a classical trajectory of the Lorentzian background, using the Schrödinger equation (2.9.3.54), and obtain the (finite) amount of particle creation, measured using Hamiltonian diagonalization. The back-reaction of the fermions on the Friedmann background is almost negligible. Further, since there is no coupling at quadratic order between the inhomogeneous modes of the metric and the fermionic field modes, one finds that the fermion modes have a negligible effect on the isotropy of the microwave background radiation. One might well expect that the underlying reason for both these properties is the exclusion principle, which prevents too much matter from being present in one place. In the de Sitter region, a fermionic particle detector experiences a thermal spectrum at the de Sitter temperature $T = H/2\pi$.

As remarked earlier in this subsection, the spectral boundary conditions, used here in quantum cosmology, are also the most natural boundary conditions in the context of scattering [Itzykson & Zuber 1980]. For example, if one is working in the Euclidean régime, and one sets up an 'in' hypersurface Σ_I and an 'out' hypersurface Σ_F, then a natural choice of fermionic boundary conditions is to specify the (+) positive-frequency modes on Σ_F and the (−) negative-frequency modes on Σ_I. This can also be done in supergravity, and spectral boundary conditions in supergravity will be studied in section 8.2.

3

Hamiltonian supergravity and canonical quantization

3.1 Introduction

Before moving on from quantum cosmology – based on the canonical quantization of general relativity – to supersymmetric quantum cosmology, one should, for a fuller understanding, study the canonical quantization of supergravity described in this chapter, and elaborated in the following chapter 4.

This chapter first describes the Hamiltonian formulation of supergravity, and then treats the canonical quantization of the theory [D'Eath 1984]. The original Hamiltonian treatments were by [Deser *et al.* 1977a, Fradkin & Vasiliev 1977, Pilati 1978]. This chapter follows principally Pilati, since his treatment preserves the local Lorentz invariance of the theory, allowing for the full local invariance of $N = 1$ supergravity. The subsequent quantization is best described in terms of two-component spinors, as described in subsection 2.9.1. Accordingly the Hamiltonian form is re-derived using two-component spinors in section 3.2. The basic canonical variables are the even (commuting) Hermitian spatial tetrad $e^{AA'}{}_i$ ($i = 1, 2, 3$) and its conjugate momentum $p_{AA'}{}^i$, together with the gravitino variables $\psi^A{}_i$, $\bar{\psi}^{A'}{}_i$, which are odd elements of a Grassmann algebra. Here $e^{AA'}{}_i$ gives the three-metric according to $h_{ij} = -e^{AA'}{}_i e_{AA'j}$. It is given by $e^{AA'}{}_i = e^a{}_i \sigma^{AA'}{}_a$ [Eq. (2.9.1.3)], where $e^a{}_i$ are the spatial tetrad components, and $\sigma^{AA'}{}_a$ are the Infeld–van der Waerden translation symbols (2.9.1.4).

The Hamiltonian takes a form of the standard type for theories with gauge invariances:

$$H = \int d^3x \Big(N\mathscr{H}_\perp + N^i \mathscr{H}_i + \psi^A{}_0 S_A + \bar{S}_{A'} \bar{\psi}_0^{A'} - \omega_{AB0} J^{AB} - \bar{\omega}_{A'B'0} \bar{J}^{A'B'} \Big), \tag{3.1.1}$$

plus terms at spatial infinity [Teitelboim 1977a,b]. The quantities

$$N,\ N^i,\ \psi^A{}_0,\ \bar{\psi}^{A'}{}_0,\ \omega_{AB0},\ \bar{\omega}_{A'B'0}$$

appear as Lagrange multipliers for the generators

$$\mathscr{H}_\perp,\ \mathscr{H}_i,\ S_A,\ \bar{S}_{A'},\ J^{AB},\ \bar{J}^{A'B'},$$

which are formed from the basic variables $e^{AA'}{}_i, \psi^A{}_i, \bar{\psi}^{A'}{}_i$ above. Here N and N^i are the lapse function and shift vector, $\psi^A{}_0$ and $\bar{\psi}^{A'}{}_0$ are the zero components of the four-dimensional gravitino field $(\psi^A{}_\mu, \bar{\psi}^{A'}{}_\mu)$, and $\omega_{AB0}, \bar{\omega}_{A'B'0}$ are the zero components of the connection forms [D'Eath 1984]. $\mathscr{H}_\perp$ gives the generator of (modified) normal displacements applied to the basic Hamiltonian variables, $\mathscr{H}_i$ gives the generator of (modified) spatial coordinate transformations, S_A and $\bar{S}_{A'}$ are the generators of supersymmetry transformations, and J^{AB} and $\bar{J}^{A'B'}$ are the generators of local Lorentz transformations [Teitelboim 1977a]. Here 'modified' indicates that a certain amount of field-dependent supersymmetry and local Lorentz transformations has been added to the coordinate transformation. One can alter this modification by re-defining the Lagrange multipliers in Eq. (3.1.1) in a field-dependent way. The form (3.1.1) has been given here since the Lagrange multipliers have simple interpretations. Classically the dynamical variables obey the constraints

$$\mathscr{H}_\perp = 0,\ \mathscr{H}_i = 0,\ S_A = 0,\ \bar{S}_{A'} = 0,\ J^{AB} = 0,\ \bar{J}^{A'B'} = 0. \tag{3.1.2}$$

As in section 2.9 for spin-1/2 fields, there are second-class constraints relating $\psi^A{}_i$ and $\bar{\psi}^{A'}{}_i$ to their momenta. These are eliminated by analogy with subsection 2.9.2, leading to Dirac brackets. The constraints are first class: the Dirac bracket $[\ ,\ ]^*$ of any pair can be written as a sum of terms which vanish when the constraints (3.1.2) hold [Teitelboim 1977a, Pilati 1978].

In the quantization, one replaces Dirac brackets by (anti)commutators [Casalbuoni 1976], as in subsection 2.9.2, and then looks for a quantum representation of this system (section 3.4). The Dirac bracket $[\psi^A{}_i, \bar{\psi}^{A'}{}_j]$ is non-zero, so that if one regards (say) $\psi^A{}_i(x)$ as a coordinate variable, then $\bar{\psi}^{A'}{}_j(x)$ must be regarded as a momentum variable. In this case, one is led to describe quantum states by wave functions

$$f\left(e^{AA'}{}_i(x),\ \psi^A{}_i(x)\right),$$

for example. The variables $\bar{\psi}^{A'}{}_i(x)$ and $p_{AA'}{}^i(x)$ can then be represented by functional differential operators acting on f, such that the (anti)commutation relations following from the Dirac brackets among the basic variables hold with suitable factor ordering. A formal inner product can be defined such that $p_{AA'}{}^i$ is Hermitian and that ψ^A_i, $\bar{\psi}^{A'}{}_i$ are Hermitian adjoints. This is the generalization to supergravity of the finite-dimensional inner product (2.9.3.42) for the odd variables $x, y, \tilde{x}, \tilde{y}$, which arose from the spin-1/2 field. Given a wave function $f(e, \psi)$, one can also construct the corresponding function $\tilde{f}(e, \tilde{\psi})$ in a representation based on eigenstates of $\tilde{\psi}^{A'}{}_i(x)$, using a functional Fourier transform.

Following the Dirac procedure [Dirac 1965] or the path-integral approach [Hawking 1987], in the quantum theory the constraints become conditions on physically allowed wave functions. These are discussed in section 3.4. The constraints $J^{AB} f = 0$, $\bar{J}^{A'B'} f = 0$ simply describe the invariance of physically allowed wave functions under local Lorentz transformations. With a suitable choice of factor ordering, the constraint $\bar{S}_{A'} f = 0$ describes the transformation of a physical wave functional $f(e, \psi)$ under an infinitesimal supersymmetry transformation of $e^{AA'}{}_i$, $\psi^A{}_i$ parametrized by $\tilde{\epsilon}^{A'}(x)$. The wave functional is not invariant under $\tilde{\epsilon}^{A'}$ supersymmetry, since the action changes by a boundary term. Instead, the wave functional transforms in a well-defined way, depending on $\tilde{\epsilon}^{A'}, e^{AA'}{}_i$ and $\psi^A{}_i$. Here we free $e^{AA'}{}_i$ from the requirement of being Hermitian, and $(\psi^A{}_i, \tilde{\psi}^{A'}{}_i)$ and $(\epsilon^A, \tilde{\epsilon}^{A'})$ from being Hermitian conjugates. This is natural even in the Lorentzian context studied here, since (see section 3.5) in evaluating a quantum amplitude one naturally sets data $(e^{AA'}{}_i, \tilde{\psi}^{A'}{}_i)_I$ on an initial surface and independently sets $(e^{AA'}{}_i, \psi^A{}_i)_F$ on a final surface. Further, the action of supergravity is invariant under independent primed and unprimed local supergravity transformations. The constraint $S_A f = 0$, involving the adjoint operator S_A, also describes the transformation property of the corresponding functional $\tilde{f}(e, \tilde{\psi})$ in the $(e, \tilde{\psi})$ basis, under an infinitesimal supersymmetry transformation parametrized by $\epsilon^A(x)$. These transformation properties are all that is required of a physical state. Instead of

$$\mathscr{H}_{AA'} = -n_{AA'}\mathscr{H}_{\perp} + e_{AA'}{}^i \mathscr{H}_i, \tag{3.1.3}$$

where $n^{AA'}$ is the spinor version of the unit timelike normal n^μ, one can use ${}_2\mathscr{H}_{AA'}$ with the anticommutator $[S_A, \bar{S}_{A'}]_* \approx {}_2\mathscr{H}_{AA'}$. One automatically has ${}_2\mathscr{H}_{AA'} f = 0$. (Classically, ${}_2\mathscr{H}_{AA'}$ differs from $\mathscr{H}_{AA'}$ by terms linear in J^{AB} and $\bar{J}^{A'B'}$ [Teitelboim 1977a, Henneaux 1983].)

In section 3.5, we consider the quantum amplitude K to go from an initial configuration $(e^{AA'}{}_i, \tilde{\psi}^{A'}{}_i)_I$ to a final configuration $(e^{AA'}{}_i, \psi^A{}_i)_F$, possibly with some boundary data at spatial infinity. These are the natural local boundary data for fermions; one cannot specify both $\psi^A{}_{iI}$ and $\psi^A{}_{iF}$, since the classical evolution of $\psi^A{}_\mu$ is via the first-order Rarita–Schwinger equation, and $\psi^A{}_{iI}$ would evolve into a final field which would in general not equal the prescribed $\psi^A{}_{iF}$ (even up to gauge). The boundary data used here agree with that in the holomorphic representation for fermions [Faddeev 1976, Faddeev & Slavnov 1980]. All physical wave functionals can be constructed from the quantum amplitude K

by superposition. K must obey the quantum constraints, and also the Schrödinger equation in the asymptotically flat case. The amplitude can be defined as a Feynman path integral, where the action S includes boundary terms at the initial and final surfaces and at spatial infinity. This provides an alternative treatment of the invariance or transformation properties of K, allowing (for example) another derivation of the transformation property of K under a primed supersymmetry transformation at the final surface, found previously in section 3.4 from the constraint $\bar{S}_{A'}K = 0$.

3.2 Hamiltonian formulation of supergravity

The essential features of the Hamiltonian treatment of classical supergravity, needed for the subsequent discussion of quantization, are summarized in this section. Two-component spinors (subsection 2.9.1) will be necessary in the subsequent quantum treatment, and are used throughout. The basic quantities which appear in the action of supergravity are the tetrad, given by the spinor-valued one-form $e^{AA'}{}_{\mu}$, where $e^{AA'}{}_{\mu} = \bar{e}^{AA'}_{\mu}$ is Hermitian, and the spin-3/2 field, given by the spinor-valued one-form $\psi^{A}{}_{\mu}$ and its Hermitian conjugate $\bar{\psi}^{A'}{}_{\mu}$. The variables $e^{AA'}{}_{\mu}$ are even, commuting with all other variables, whereas $\psi^{A}{}_{\mu}$ are odd Grassmann quantities anticommuting among themselves.

In second-order formalism, $e^{AA'}{}_{\mu}, \psi^{A}{}_{\mu}, \tilde{\psi}^{A'}{}_{\mu}$ determine the connection forms $\omega^{ab}{}_{\mu} = \omega^{[ab]}{}_{\mu}$, or equivalently their spinorial version

$$\omega^{AA'BB'}{}_{\mu} = \omega^{AB}{}_{\mu}\epsilon^{A'B'} + \bar{\omega}^{A'B'}_{\mu}\epsilon^{AB}, \tag{3.2.1}$$

where $\omega^{AB}{}_{\mu} = \omega^{(AB)}{}_{\mu}$ is symmetric [van Nieuwenhuizen 1981]. To express this, define a derivative operator D_{μ} using $\omega^{AB}{}_{\mu}$ which acts on spinor-valued forms and only notices their spinor indices, but not their spacetime indices, so that, for example,

$$D_{\mu}e^{AA'}{}_{\nu} = \partial_{\mu}e^{AA'}{}_{\nu} + \omega^{A}{}_{B\mu}e^{BA'}{}_{\nu} + \bar{\omega}^{A'}{}_{B'\mu}e^{AB'}{}_{\nu}, \tag{3.2.2}$$

$$D_{\mu}\psi^{A}{}_{\nu} = \partial_{\mu}\psi^{A}{}_{\nu} + \omega^{A}{}_{B\mu}\psi^{B}{}_{\nu}. \tag{3.2.3}$$

Then the spinor-valued two-form $D_{[\mu}e^{AA'}{}_{\nu]}$ is given by

$$D_{[\mu}e^{AA'}{}_{\nu]} = S^{AA'}{}_{\mu\nu}, \tag{3.2.4}$$

where $S^{AA'}{}_{\mu\nu}$ is the torsion. This is a standard definition of the torsion, used when one is working with a pseudo-orthonormal frame $e^{a}{}_{\mu}$ (the first of Cartan's structure equations) – see Eq. (3.1) of [Eguchi *et al.* 1980]. It can be related to the alternative definition in terms of Christoffel symbols

$S^{\alpha}{}_{\mu\nu} = \frac{1}{2}(\Gamma^{\alpha}{}_{\mu\nu} - \Gamma^{\alpha}{}_{\nu\mu})$. With our normalization for the action below, one has

$$S^{AA'}{}_{\mu\nu} = \frac{i\kappa^2}{2}\bar{\psi}^{A'}{}_{[\mu}\psi^{A}{}_{\nu]}, \tag{3.2.5}$$

where $\kappa^2 = 8\pi$. Eq. (3.2.5) is found [van Nieuwenhuizen 1981] from variation of the action (3.2.12) below with respect to the connection forms $\omega_{ab\mu}$. This Palatini procedure [Misner *et al.* 1973, Wald 1984] gives the connection forms explicitly in terms of the tetrad $e^{AA'}{}_{\mu}$ and gravitino field $(\psi^{A}{}_{\mu}, \bar{\psi}^{A'}{}_{\mu})$, as below. In particular, it gives the torsion (3.2.5).

Equation (3.2.4) can be solved for $\omega^{AB}{}_{\mu}$. Let ${}^{s}\omega^{ab}{}_{\mu}$ be the torsion-free connection forms obtained from Eq. (3.2.4) with zero right-hand side, given by

$${}^{s}\omega^{ab}{}_{\mu} = e^{av}\partial_{[\mu}e^{a}{}_{\nu]} - e^{bv}\partial_{[\mu}e^{a}{}_{\nu]} - e^{av}e^{b\rho}e_{c\mu}\partial_{[\nu}e^{c}{}_{\rho]}. \tag{3.2.6}$$

Then

$$\omega^{AB}{}_{\mu} = {}^{s}\omega^{AB}{}_{\mu} + \kappa^{AB}{}_{\mu}, \tag{3.2.7}$$

where $\kappa^{\nu\rho}{}_{\mu} = \kappa^{[\nu\rho]}{}_{\mu}$ is the contorsion tensor, of which the spinor version $\kappa^{AA'BB'}{}_{\mu} = e^{AA'}{}_{\nu}e^{BB'}{}_{\rho}\kappa^{\nu\rho}{}_{\mu}$ is written

$$\kappa^{AA'BB'}{}_{\mu} = \kappa^{AB}{}_{\mu}\epsilon^{A'B'} + \bar{\kappa}^{A'B'}{}_{\mu}\epsilon^{AB} \tag{3.2.8}$$

with $\kappa^{AB}{}_{\mu} = \kappa^{(AB)}{}_{\mu}$. In terms of the torsion tensor

$$S^{\rho}{}_{\mu\nu} = -e_{AA'}{}^{\rho}S^{AA'}{}_{\mu\nu},$$

the contorsion is given by

$$\kappa_{\nu\rho\mu} = S_{\nu\mu\rho} + S_{\rho\nu\mu} + S_{\mu\nu\rho}. \tag{3.2.9}$$

The spinor-valued curvature two-forms are then defined as

$$R^{AB}{}_{\mu\nu} = 2\left(\partial_{[\mu}\omega^{AB}{}_{\nu]} + \omega^{A}{}_{C[\mu}\omega^{CB}{}_{\nu]}\right) \tag{3.2.10}$$

and its conjugate $\bar{R}^{A'B'}{}_{\mu\nu}$, where $R^{AB}{}_{\mu\nu} = R^{(AB)}{}_{\mu\nu}$ (Cartan's second structure equation [Eguchi *et al.* 1980]). From this one obtains the curvature scalar

$$R = e_{AA'}{}^{\mu}e_{B}{}^{A'\nu}R^{AB}{}_{\mu\nu} + \text{H.c.} \tag{3.2.11}$$

The action of supergravity is taken to be

$$S = \int d^4x\left[\frac{1}{2\kappa^2}(\det e)R + \frac{1}{2}\epsilon^{\mu\nu\rho\sigma}\left(\bar{\psi}^{A'}{}_{\mu}e_{AA'\nu}D_{\rho}\psi^{A}{}_{\sigma} + \text{H.c.}\right)\right], \tag{3.2.12}$$

supplemented by boundary terms at spatial infinity and on any bounding surfaces. Here we are using geometrical units with $c = G = 1$. Note that in taking the Hermitian conjugate of a product of fermionic variables, one should reverse their order, so that e.g. the conjugate of $\bar{\psi}^{A'}{}_{\mu}\psi^{A}{}_{\nu}$ is $\bar{\psi}^{A'}{}_{\nu}\psi^{A}{}_{\mu}$. Strictly one could also include in the action the contribution of the auxiliary fields of supergravity [van Nieuwenhuizen 1981, Wess & Bagger 1992]. However, since these vanish classically, we have dropped them here. The auxiliary fields will however play an important role in section 4.3 and in chapter 8.

In the absence of boundary surfaces, the action (3.2.12) is invariant under three kinds of local symmetry transformations: local Lorentz, coordinate, and supersymmetry transformations. Under an infinitesimal local Lorentz transformation,

$$\begin{aligned} \delta e^{AA'}{}_{\mu} &= N^{A}{}_{B}e^{BA'}{}_{\mu} + \bar{N}^{A'}{}_{B'}e^{AB'}{}_{\mu}, \\ \delta\psi^{A}{}_{\mu} &= N^{A}{}_{B}\psi^{B}{}_{\mu}, \ \delta\bar{\psi}^{A'}{}_{\mu} = \bar{N}^{A'}{}_{B'}\bar{\psi}^{B'}{}_{\mu} \end{aligned} \tag{3.2.13}$$

where $N^{AB} = N^{(AB)}$ is symmetric. Under an infinitesimal coordinate transformation any spinor-valued form such $e^{AA'}{}_{\mu}$ or $\psi^{A}{}_{\mu}$ changes by its Lie derivative [Misner *et al.* 1973], e.g. ,

$$\begin{aligned} \delta e^{AA'}{}_{\mu} &= \xi^{\nu}\partial_{\nu}e^{AA'}{}_{\mu} + e^{AA'}{}_{\nu}\partial_{\mu}\xi^{\nu}, \\ \delta\psi^{A}{}_{\mu} &= \xi^{\nu}\partial_{\nu}\psi^{A}{}_{\mu} + \psi^{A}{}_{\nu}\partial_{\mu}\xi^{\nu}. \end{aligned} \tag{3.2.14}$$

Under an infinitesimal local supersymmetry transformation,

$$\begin{aligned} \delta e^{AA'}{}_{\mu} &= -i\kappa\left(\epsilon^{A}\bar{\psi}^{A'}{}_{\mu} + \bar{\epsilon}^{A'}\psi^{A}{}_{\mu}\right), \\ \delta\psi^{A}{}_{\mu} &= 2\kappa^{-1}D_{\mu}\epsilon^{A}, \ \delta\bar{\psi}^{A'}{}_{\mu} = 2\kappa^{-1}D_{\mu}\bar{\epsilon}^{A'}, \end{aligned} \tag{3.2.15}$$

where ϵ^{A} and its conjugate $\bar{\epsilon}^{A'}$ are spacetime-dependent odd (anticommuting) fields.

The Hamiltonian formulation of supergravity can then be developed following standard methods for systems with constraints [Dirac 1965, Hanson *et al.* 1976]. This has been done in [Fradkin & Vasiliev 1977, Pilati 1978] and also, in a special gauge where the Lorentz rotation freedom is fixed, in [Deser *et al.* 1977a]. It will clarify the structure of the theory to keep all its gauge freedom at this stage, and we follow principally the treatment of Pilati.

In the Hamiltonian decomposition the tetrad variables are split into $e^{AA'}{}_{0}$ and $e^{AA'}{}_{j}$ $(i, j, ... = 1, 2, 3)$. The spatial spinor-valued forms $e^{AA'}{}_{i}$ determine the spatial metric $h_{ij} = -e_{AA'i}e^{AA'}{}_{j}$ [Eq. (2.9.1.5)], which is then used together with its inverse h^{ij} to lower and raise spatial indices $i, j, ...$. The spinor version $n^{AA'}$ of the unit timelike future-directed normal n^{μ} to a surface $t =$ const. can be used together with the $e_{i}{}^{AA'}$ to make a basis for the space of spinors with one unprimed and one primed index. It is determined by the $e_i^{AA'}$ through the conditions [Eqs. (2.9.1.16) and (2.9.1.17)]

$$n_{AA'}e^{AA'}{}_i = 0, \tag{3.2.16}$$

$$n_{AA'}n^{AA'} = 1. \tag{3.2.17}$$

Then the remaining variables $e_0{}^{AA'}$ can be expanded out as

$$e^{AA'}{}_0 = Nn^{AA'} + N^i e^{AA'}{}_i, \tag{3.2.18}$$

where N is the lapse function and N^i is the shift vector. Similarly the spin-3/2 variables are split into $\psi^A{}_0$, $\psi^A{}_i$ and their conjugates.

The momenta conjugate to the variables $e^{AA'}{}_0$, $\psi^A{}_0$ and $\bar{\psi}^{A'}{}_0$ are zero. Moreover the momenta conjugate to $\psi_i{}^A$ and $\bar{\psi}^{A'}{}_i$ can be expressed in terms of $\psi_i{}^A$, $\bar{\psi}^{A'}{}_i$ and $e^{AA'}{}_i$, since the Lagrangian involves only first-order derivatives of fermionic variables. Defining $\pi_A{}^i = \delta S/\delta\dot{\psi}^A_i$, $\tilde{\pi}_{A'}{}^i = \delta S/\delta\dot{\bar{\psi}}^{A'}{}_i$, with the convention that odd variables must be brought to the left using anticommutation rules before functional differentiation is carried out [Wess & Bagger 1992], one finds

$$\begin{aligned} \pi_A{}^i &= -\tfrac{1}{2}\epsilon^{ijk}\bar{\psi}^{A'}{}_j e_{AA'k}, \\ \tilde{\pi}_{A'}{}^i &= \tfrac{1}{2}\epsilon^{ijk}\psi^A{}_j e_{AA'k}, \end{aligned} \tag{3.2.19}$$

We have used the notation $\tilde{\pi}_{A'}{}^i$ since this quantity is minus the conjugate of $\pi_A{}^i$. The basic dynamical variables in the theory can then be reduced to $e^{AA'}{}_i$, $p_{AA'}{}^i$, $\psi^A{}_i$, and $\bar{\psi}^{A'}{}_i$, where $p_{AA'}{}^i = \delta S/\delta\dot{e}^{AA'}{}_i$ is the momentum conjugate to $e^{AA'}{}_i$. Because of the invariance of the Langrangian density in Eq. (3.2.12) under local Lorentz transformations, these variables obey the primary constraints

$$J_{AB} = 0, \quad \bar{J}_{A'B'} = 0, \tag{3.2.20}$$

where

$$J_{AB} = e_{(A}{}^{A'i}p_{B)A'i} + \psi_{(A}{}^i\pi_{B)i}. \tag{3.2.21}$$

The remaining variables $N, N^i, \psi_0{}^A$, and $\bar{\psi}^{A'}{}_0$ are not subject to dynamical equations, but can be freely specified in the dynamical evolution of Hamiltonian data.

The momentum $p_{AA'}{}^i$ can be expressed in terms of the second fundamental form K_{ij} on surfaces $t =$ const. Note that knowledge of $p_{AA'}{}^i$ is equivalent to knowledge of its projections

$$p^{ji} = -e^{AA'k}p_{AA'}{}^i, \quad p^{\perp i} = n^{AA'}p_{AA'i}, \tag{3.2.22}$$

respectively, a spatial tensor and vector density. One finds that the symmetric part of p^{ji} is given by the same relation

$$\pi^{ji} = -\frac{h^{1/2}}{2\kappa^2}\left(K^{(ij)} - \text{tr}K\, h^{ij}\right) \tag{3.2.23}$$

as in the pure gravitational case, where

$$\pi^{ij} = -\tfrac{1}{2}p^{(ij)}. \tag{3.2.24}$$

Here the spatial tensor K_{ij} is defined as minus the spatial projection of the covariant derivative of n_μ:

$$K_{ij} = -e^a{}_i \partial_j n_a + n_a \omega^{ab}{}_j e_{bi}. \tag{3.2.25}$$

The symmetric part of K_{ij} contains information about the normal derivative of the spatial metric h_{ij}:

$$K_{(ij)} = \frac{1}{2N}\left(N_{i\|j} + N_{j\|i} - h_{ij,0}\right) - 2S_{(ij)\perp}, \tag{3.2.26}$$

where $\|$ denotes a spatial covariant derivative without torsion and $S_{ij\perp} = n^\mu S_{ij\mu}$. In the presence of torsion, K_{ij} also has an antisymmetric part

$$K_{[ij]} = S_{\perp ij} = n^\mu S_{\mu ij}. \tag{3.2.27}$$

The remaining parts $p^{[ij]}$ and $p^{\perp i}$ of $p_{AA'}{}^i$ can be expressed in terms of projections of

$$J_{ab} \leftrightarrow J_{AB}\epsilon_{A'B'} + \bar{J}_{A'B'}\epsilon_{AB}$$

together with torsion terms.

As in subsection 2.9.2, Poisson brackets can be defined in a classical theory containing both bosons and fermions [Casalbuoni 1976] The bracket between two even elements or an odd and an even element is as usual antisymmetric, but the bracket between two odd elements is symmetric under interchange. After $\pi_A{}^i$ and $\tilde{\pi}_{A'}{}^i$ are eliminated as dynamical variables through the second-class constraints (3.2.19), and one finds the following Dirac brackets $[\, ,\,]^*$ among the basic variables $e^{AA'}{}_i$, $p_{AA'}{}^i$, $\psi^A{}_i$, and $\bar{\psi}^{A'}{}_i$:

$$\left[e^{AA'}{}_i(x), e^{BB'}{}_j(x')\right]^* = 0, \tag{3.2.28}$$

$$\left[e^{AA'}{}_i(x), p_{BB'}{}^j(x')\right]^* = \epsilon^A{}_B \epsilon^{A'}{}_{B'} \delta_i{}^j \delta(x, x'), \tag{3.2.29}$$

$$\left[p_{AA'}{}^i(x), p_{BB'}{}^j(x')\right]^* = \frac{1}{4}\epsilon^{jln}\psi_{Bn}D_{AB'kl} \times \epsilon^{ikm}\bar{\psi}_{A'm}\delta(x, x') + \text{H.c.}, \tag{3.2.30}$$

$$\left[\psi^A{}_i(x), \psi^B{}_j(x')\right]^* = 0, \tag{3.2.31}$$

$$\left[\psi_i^A(x), \bar{\psi}^{A'}{}_j(x')\right]^* = -D^{AA'}{}_{ij}\delta(x,x'), \tag{3.2.32}$$

$$\left[e^{AA'}{}_i(x), \psi^B{}_j(x')\right]^* = 0, \tag{3.2.33}$$

$$\left[p_{AA'}{}^i(x), \psi^B{}_j(x')\right]^* = \frac{1}{2}\epsilon^{ikl}\psi_{Al}D^B{}_{A'jk}\delta(x,x'), \tag{3.2.34}$$

and conjugate relations, where

$$D^{AA'}{}_{jk} = -2ih^{-1/2}e^{AB'}{}_k e_{BB'j}n^{BA'}. \tag{3.2.35}$$

The evolution of a dynamical variable A with no explicit time dependence is given by

$$\frac{dA}{dt} = [A, H]^*, \tag{3.2.36}$$

where the bracket can be expanded out by using the distributive laws given in [Casalbuoni 1976]. The Hamiltonian H is found to be

$$H = \int d^3x \Big(N\, {}_1\mathcal{H}_\perp + N^i\, {}_1\mathcal{H}_i + \psi^A{}_0\, {}_1S_A + {}_1\bar{S}_{A'}\bar{\psi}^{A'}{}_0 + M_{AB}J^{AB} + \bar{M}_{A'B'}\bar{J}^{A'B'} \Big), \tag{3.2.37}$$

together with surface terms at spatial infinity [Teitelboim 1977b]. Here $N, N^i, \psi^A{}_0$, and $\bar{\psi}^{A'}{}_0$ occur as Lagrange multipliers together with the additional multipliers $M_{AB} = M_{(AB)}$ and $\bar{M}_{A'B'}$, which are introduced to allow for the primary constraints $J_{AB} = 0$, $\bar{J}_{A'B'} = 0$ [Dirac 1965]. The quantities ${}_1\mathcal{H}_\perp$, ${}_1\mathcal{H}_i$, and ${}_1S_A$ are functions only of the basic variables, given by

$$\begin{aligned} {}_1\mathcal{H}_\perp = &-\frac{1}{2}\kappa^2 h^{1/2}\, {}^3R + 2\kappa^2 h^{-1/2}\left[\pi_{ij}\pi^{ij} - \frac{1}{2}(tr\pi)^2\right] \\ &+ \frac{\kappa^2}{8}h^{1/2}n_{AA'}\bar{\psi}^{A'}{}_{[i}\psi^A{}_{k]}n^{BB'}\bar{\psi}_{B'}{}^{[i}\psi_B{}^{k]} \\ &+ \left(\frac{1}{2}\epsilon^{ijk}\bar{\psi}^{A'}{}_i n_{AA'}\, {}^3D_j\psi^A{}_k + \text{H.c.}\right), \end{aligned} \tag{3.2.38}$$

$$\begin{aligned} {}_1\mathcal{H}_i = &-2h_{ij}\pi^{jk}{}_{\|k} + \frac{i}{2}h^{1/2}h_{ij}\left(n^{AA'}\bar{\psi}_{A'}^{[j}\psi_A{}^{k]}\right)_{|k} \\ &+ \frac{\kappa^2}{2}h^{1/2}n_{AA'}\bar{\psi}^{A'}{}_{[j}\psi^A{}_{i]}e^{BB'}{}_k\bar{\psi}_{B'}{}^{[j}\psi_B{}^{k]} \\ &- 2i\kappa^2\pi^{jk}e_{AA'j}\bar{\psi}^{A'}{}_{[i}\psi^A{}_{k]} \\ &+ \left(\frac{1}{2}\epsilon^{jkl}\bar{\psi}^{A'}{}_j e_{AA'i}\, {}^3D_k\psi^A{}_l + \text{H.c.}\right), \end{aligned} \tag{3.2.39}$$

$$
\begin{aligned}
{}_1S_A = {} & \epsilon^{ijk} e_{AA'i}\, {}^3D_j \bar{\psi}^{A'}{}_k + i\kappa^2 \pi^{ij} e_{AA'i} \bar{\psi}^{A'}{}_j \\
& + \frac{1}{4}\kappa^2 h^{1/2} e_{AA'i} \bar{\psi}^{A'}{}_j n^{BB'} \bar{\psi}_{B'}{}^{[i} \psi_B{}^{j]} \\
& - \frac{1}{4} i\kappa^2 \epsilon^{ijk} n_{AA'} \bar{\psi}^{A'}{}_j n_{BB'} \bar{\psi}^{B'}{}_k \psi^B{}_i .
\end{aligned} \tag{3.2.40}
$$

Here the derivative ${}^3D_i$ acts on spinor-valued spatial forms according (for example) to

$$
{}^3D_i \psi^A{}_j = \partial_i \psi^A{}_j + {}^3\omega^A{}_{Bi} \psi^B{}_j . \tag{3.2.41}
$$

The spatial connection forms ${}^3\omega^{AB}{}_i$ can be expressed as

$$
{}^3\omega^{AB}{}_i = {}^{3s}\omega^{AB}{}_i + {}^3\kappa^{AB}{}_i , \tag{3.2.42}
$$

where the torsion-free spatial connection forms ${}^{3s}\omega^{AB}{}_i$ are found from

$$
\begin{aligned}
{}^{3s}\omega^{ab}{}_i = {} & \left(e^{bj} \partial_{[j} e^a{}_{i]} - \frac{1}{2} e^{aj} e^{bk} e^c{}_i \partial_j e_{ck} - \frac{1}{2} e^{aj} n^b n^c \partial_j e_{ci} - \frac{1}{2} n^a \partial_i n^b \right) \\
& - (a \leftrightarrow b),
\end{aligned} \tag{3.2.43}
$$

while the ${}^3\kappa^{AB}{}_i$ are found from the spatial contorsion ${}^3\kappa_{jki} = \kappa_{jki}$ and the definition

$$
\begin{aligned}
{}^3\kappa_{AA'BB'i} & = {}^3\kappa_{ABi}\, \epsilon_{A'B'} + {}^3\bar{\kappa}_{A'B'i}\, \epsilon_{AB} \\
& = e_{AA'}{}^j e_{BB'}{}^k\, {}^3\kappa_{jki} .
\end{aligned} \tag{3.2.44}
$$

In Eq. (3.2.38), ${}^3R$ is the spatial curvature scalar allowing for torsion S_{ijk}, and in Eq. (3.2.39) the spatial covariant derivative with torsion is denoted by a bar $|$.

During the evolution of data with the Hamiltonian H, the momenta conjugate to $N, N^i, \psi^A{}_0$, and $\bar{\psi}^{A'}{}_0$ remain zero, leading to the secondary constraints

$$
{}_1\mathscr{H}_\perp = 0, \quad {}_1\mathscr{H}^i = 0, \quad {}_1S_A = 0, \quad {}_1\bar{S}_{A'} = 0, \tag{3.2.45}
$$

which must hold classically in addition to the primary constraints $J_{AB} = 0$, $\bar{J}_{A'B'} = 0$.

The symmetries of the theory in the Hamiltonian form are most easily understood by rewriting H so that the Lagrange multipliers of J^{AB}, $\bar{J}^{A'B'}$ are minus the components ω_{AB0}, $\bar{\omega}_{A'B'0}$ of the connection forms [Pilati 1978], just as the multipliers of

$$
{}_1\mathscr{H}_{AA'} = -n_{AA'}\, {}_1\mathscr{H}_\perp + e_{AA'}{}^i\, {}_1\mathscr{H}_i \tag{3.2.46}
$$

and ${}_1S_A$, ${}_1\bar{S}_{A'}$ are the zero components $-e^{AA'}{}_0$, $\psi^A{}_0$ and $\bar{\psi}^{A'}{}_0$. This gives

$$
\begin{aligned}
H = \int d^3x \Big(& N\mathscr{H}_\perp + N^i \mathscr{H}_i + \psi^A{}_0 S_A + \bar{S}_{A'} \bar{\psi}^{A'}{}_0 \\
& - \omega_{AB0} J^{AB} - \bar{\omega}_{A'B'0} \bar{J}^{A'B'} \Big),
\end{aligned} \tag{3.2.47}
$$

plus terms at spatial infinity, where $\mathscr{H}_\perp$, $\mathscr{H}_i$, S_A, $\bar{S}_{A'}$ differ from ${}_1\mathscr{H}_\perp$ and from ${}_1\mathscr{H}_i$, ${}_1S_A$ and ${}_1\bar{S}_{A'}$ by terms proportional to projections of J_{ab} or its derivatives. In particular, we shall need the explicit form of

$$S_A = \epsilon^{ijk} e_{AA'i}\, {}^3D_j \bar{\psi}^{A'}{}_k - \frac{1}{2} i\kappa^2 p_{AA'}{}^i \bar{\psi}^{A'}{}_i + \frac{1}{2}\kappa^2 h^{1/2} n_{AA'} \bar{\psi}^{A'}{}_i e^{BB'}{}_i \bar{\psi}_{B'}{}^{[j} \psi_B{}^{i]} - \frac{1}{4} i\kappa^2 \epsilon^{ijk} n_{AA'} \bar{\psi}^{A'}{}_j n_{BB'} \bar{\psi}^{B'}{}_k \psi^B{}_i \,, \tag{3.2.48}$$

Again, the secondary constraints

$$\mathscr{H}_\perp = 0, \quad \mathscr{H}^i = 0, \qquad S_A = 0, \quad \bar{S}_{A'} = 0, \tag{3.2.49}$$

must hold classically. As described in [Teitelboim 1977a], S_A and $\bar{S}_{A'}$ are the generators of supersymmetry transformations acting on the basic dynamical variables, J_{AB} and $\bar{J}_{A'B'}$ are the generators of local Lorentz transformations, while $\mathscr{H}_{AA'}$ is the projection along the tetrad of a modified surface deformation generator in which the usual coordinate transformation has to be supplemented by a suitable supersymmetry transformation and Lorentz transformation. Thus, in addition to the geometrical interpretation of the lapse N and shift N^i, one can interpret $\psi^A{}_0$, $\bar{\psi}^{A'}{}_0$ as giving the amount of supersymmetry transformation and ω_{AB0}, $\bar{\omega}_{A'B'0}$ the amount of Lorentz rotation applied per unit time to Hamiltonian data evolved using H in Eq. (3.2.47) with $N = 0$, $N^i = 0$.

The constraints (3.2.49) and $J_{AB} = 0$, $\bar{J}_{A'B'} = 0$ are first-class, i.e., the Dirac brackets of $\mathscr{H}_{AA'}$, S_A, $\bar{S}_{A'}$, J_{AB}, and $\bar{J}_{A'B'}$ are zero when the constraints hold. Although the Dirac brackets among a related set of generators were found in [Fradkin & Vasiliev 1977], there is as yet no complete treatment of the brackets among $\mathscr{H}_{AA'}$, S_A, and $\bar{S}_{A'}$. (Brackets involving the Lorentz generators J_{AB}, $\bar{J}_{A'B'}$ are standard.) The approach used in [Teitelboim 1977a] can only give these Dirac brackets up to terms proportional to the primary constraint generators J_{AB}, $\bar{J}_{A'B'}$, as remarked in [Henneaux 1983]. In particular, the Dirac brackets among the supersymmetry generators are

$$[S_A(x), S_B(x')]^* = 0, \quad [\bar{S}_{A'}(x), \bar{S}_{B'}(x')]^* = 0\,, \qquad [S_A(x), \bar{S}_{A'}(x')]^* = \frac{1}{2} i\kappa^2 \mathscr{H}_{AA'}(x)\delta(x, x') + \text{terms proportional to } J \text{ and } \bar{J}. \tag{3.2.50}$$

There are no extra terms in the brackets $[S, S]^*$ and $[\bar{S}, \bar{S}]^*$, but such terms are certainly present in $[S, \bar{S}]^*$, as may be checked by computing the p^2 terms arising from the $[p\bar{\psi}, p\psi]^*$ contribution, using the brackets in (3.2.32), and verifying that they only agree with the combination

$$2\kappa^2 h^{-1/2}\left[\pi_{ij}\pi^{ij} - \frac{1}{2}(tr\pi)^2\right]$$

appearing in $\mathscr{H}_\perp$ when the constraint $J_{ab} = 0$ is imposed.

Finally, we need to describe the boundary terms at spatial infinity which must be included in the Hamiltonian (3.2.47) in the asymptotically flat case. Suppose that we wish to allow N and N^i to tend to non-zero values $-\alpha^0$, $-\alpha^i$ as $r \to \infty$, describing time and space translations at infinity, and that $\psi^A{}_0, \bar{\psi}^{A'}{}_0$ tend to non-zero values $\psi^A{}_0(\infty), \bar{\psi}^{A'}{}_0(\infty)$, describing a supersymmetry transformation at infinity. Then, as shown in [Teitelboim 1977b], the volume term H_0 in Eq. (3.2.47) must be supplemented by surface terms at infinity to give the complete Hamiltonian

$$H = H_0 + \alpha^a P_a + \psi^A{}_0(\infty) Q_A + \bar{Q}_{A'}\bar{\psi}^A{}_0(\infty), \tag{3.2.51}$$

where

$$P^0 = \frac{1}{2\kappa^2}\int_{r\to\infty} dS_i\,(h_{ij,j} - h_{jj,i}) \ ,$$
$$P^i = -2\int_{r\to\infty} dS_j \pi^{ij}, \tag{3.2.52}$$
$$Q_A = -\int_{r\to\infty} dS_j \epsilon^{ijk} e_{AA'i}\bar{\psi}^{A'}{}_k,$$

and dS_i is the outward-directed surface element on a sphere of large radius. Here $P^0 = -P_0 = M$ gives the mass of the spacetime (assuming that the three- and four-dimensional masses are equal – see section 4.4), and $P^i = P_i$ gives its momentum, while $Q_A, \bar{Q}_{A'}$ give its supercharge. The generators $P_{AA'}, Q_A$, and $Q_{A'}$ obey a global supersymmetry algebra [Teitelboim 1977b].

3.3 Quantum representation

We now look for a quantum representation based on the classical theory of supergravity with dynamical variables $e^{AA'}{}_i, p_{AA'}{}^i, \psi^A{}_i$, and $\bar{\psi}^{A'}{}_i$, which have the Dirac brackets (3.2.28)–(3.2.34) and Hamiltonian (3.2.47). As in subsection 2.9.2, Dirac brackets $[\ ,\ ]^*$ must be replaced by commutators $[\ ,\ ]_-$ or anticommutators $[\ ,\ ]_+$ according to [Dirac 1965, Casalbuoni 1976].

$$[E_1, E_2]_- = i\hbar\,[E_1, E_2]^*,$$
$$[O, E]_- = i\hbar\,[O, E]^*, \tag{3.3.1}$$
$$[O_1, O_2]_+ = i\hbar\,[O_1, O_2]^*,$$

where E denotes an even element and O an odd element.

In the canonical quantization of metric gravity one is led most simply to the metric representation, in which the quantum state of the gravitational field is described by a complex-valued wave functional $f\left(h_{ij}(x)\right)$ of the intrinsic spatial metric h_{ij} [DeWitt 1967]. Similarly, it will be simplest here if the quantum state is described by a wave functional f, of which one argument is the Hermitian quantity $e^{AA'}{}_i(x)$ which gives the spatial metric $h_{ij} = -e_{AA'i}e^{AA'}{}_j$. One would like to treat the fermionic variables $\psi^A{}_i, \bar{\psi}^{A'}{}_i$ on the same footing as the tetrad variables $e^{AA'}{}_i$. However, the relation (3.2.32) shows that one cannot have a simultaneous eigenstate of $\psi^A{}_i$ and $\bar{\psi}^{A'}{}_i$. Rather, one must choose a basis of eigenstates of one or the other. Suppose we choose $\psi^A{}_i$, so that we have a wave functional $f\left(e^{AA'}{}_i(x), \psi^A{}_i(x)\right)$, and the operator $\psi^A{}_i(x)$ is given by multiplication on the left by $\psi^A{}_i(x)$. Then, in order to satisfy the anticommutation relation following from Eq. (3.2.32), $\bar{\psi}^{A'}{}_i$ will be given by the momentumlike operator

$$\bar{\psi}^{A'}{}_i(x) = -i\hbar D^{AA'}{}_{ji}(x)\frac{\delta}{\delta\psi^A{}_j(x)}\,, \tag{3.3.2}$$

where as before $\psi^A{}_i$ must be brought to the left in any expression before the functional derivative is applied. Note that this operator obeys $\left[\bar{\psi}^{A'}{}_i(x), \bar{\psi}^{B'}{}_j(x')\right]_+ = 0$ as required. If we had instead chosen to use eigenstates of $\bar{\psi}^{A'}{}_j$, then $\psi^A{}_i$ would be represented by an operator proportional to $\delta/\delta\bar{\psi}$. The use of two-component spinors is essential in this description.

An alternative approach, following [Deser et al. 1977a], would have been to take the weighted tetrad components $\phi^A{}_a = h^{1/4}e_a{}^\mu\psi^A{}_\mu$ and $\bar{\phi}^{A'}{}_a$ as the basic fermionic variables in the classical Hamiltonian treatment; this leads to simplifications in the Dirac brackets as compared to Eqs. (3.2.28)–(3.2.34) [Pilati 1978]. In this approach $e^{AA'}{}_i$ and $\phi^A{}_a$ would not be completely independent variables in the wave functional, which would be subject to an additional constraint of being independent of $\psi^A{}_0$. We prefer to treat the bosonic and fermionic variables as symmetrically as possible, using, e.g. , $e^{AA'}{}_i$ and $\psi^A{}_i$ which have equal numbers of components since $e^{AA'}{}_i$ is being taken here to be Hermitian.

A quantum state is given by a wave functional $f\left(e^{AA'}{}_i(x), \psi^A{}_i(x)\right)$ taking values in the Grassmann algebra formed from complex linear combinations of products of the odd elements $\psi^A{}_i(x)$. Such a wave functional could, for example, describe a state containing one pair of incoming gravitinos. Sometimes it will be necessary to consider 'states' given by functions $f(e, \psi)$ which also depend on extraneous odd variables. A state in which $\psi^A{}_i(x)$ takes a definite value $\Psi^A{}_i(x)$ on a spacelike hypersurface is of this type, with wave functional proportional to

$$\prod_x \left[\psi^A{}_i(x) - \Psi^A{}_i(x)\right].$$

Wave functionals of odd variables have been treated by [Berezin 1966] and by [Faddeev & Slavnov 1980], and much of the treatment in this section follows the discussion of the holomorphic representation for fermions given in the latter.

We also need a representation of the operator $p_{AA'}{}^i(x)$ acting on wave functionals $f(e,\psi)$, such that the commutation relations following from Eqs. (3.2.29),(3.2.30), and (3.2.34) and their conjugates hold. This can be achieved by taking

$$\begin{aligned} p_{AA'}{}^i(x) &= -i\hbar\frac{\delta}{\delta e^{AA'}{}_i(x)} + \frac{1}{2}\epsilon^{ijk}\psi_{Aj}(x)\bar{\psi}_{A'k}(x) \\ &= -i\hbar\frac{\delta}{\delta e^{AA'}{}_i(x)} \\ &\quad -\frac{1}{2}i\hbar\epsilon^{ijk}\psi_{Aj}(x)D^B{}_{A'lk}(x)\frac{\delta}{\delta\psi^B{}_l(x)} \end{aligned} \tag{3.3.3}$$

This choice gives the quantum version of Eq. (3.2.30) with the factor ordering given there. We have chosen the factor ordering in $p_{AA'}{}^i(x)$ with derivative operators on the right, but could equally well have taken an ordering with $\psi\bar{\psi}$ replaced by $-\bar{\psi}\psi$, or any convex linear combination of these. The particular choice made will determine the formal inner product with respect to which $p_{AA'}{}^i(x)$ is Hermitian.

The formal inner product between two wave functionals

$$f\left(e^{AA'}{}_i(x), \psi^A{}_i(x)\right) \text{ and } g\left(e^{AA'}{}_i(x)\psi^A{}_i(x)\right)$$

is defined to be

$$\begin{aligned} \langle f,g\rangle = \int \bar{f}(e,\psi)g(e,\psi)\exp\left[\frac{-i}{\hbar}\int C_{AA'}{}^{ij}(x')\psi^A{}_i(x')\bar{\psi}^{A'}{}_j(x')d^3x'\right] \\ D^{-1}(e)\mathscr{D}e^{AA'}{}_i(x)\mathscr{D}\psi^A{}_i(x)\mathscr{D}\bar{\psi}^{A'}{}_i(x)\,. \end{aligned} \tag{3.3.4}$$

Here $\bar{f}(e,\psi)$ is the Hermitian conjugate of $f(e,\psi)$, which includes conjugation of extra fermionic variables present in f. We define

$$C_{AA'}{}^{ij}(x) = -\epsilon^{ijk}e_{AA'k}(x), \tag{3.3.5}$$

which obeys

$$C_{AA'}{}^{ji}D^{AB'}{}_{jk} = \epsilon_{A'}{}^{B'}\delta^i{}_k, \tag{3.3.6}$$

and we will usually denote the term

$$\exp\left[\frac{-i}{\hbar}\int C_{AA'}{}^{ij}(x')\psi^A{}_i(x')\bar{\psi}^{A'}{}_j(x')d^3x'\right]$$

by

$$\exp\left[\frac{-i}{\hbar}C\psi\bar{\psi}\right].$$

We also define

$$D(e) = \prod_x \det\left[\frac{-i}{\hbar}C_{AA'}{}^{ij}(x)\right]. \tag{3.3.7}$$

The integral $\int(\ )\mathscr{D}e\mathscr{D}\psi\mathscr{D}\bar{\psi}$ denotes a functional integration over all values of $e^{AA'}{}_i(x), \psi^A{}_i(x)$, and $\bar{\psi}^{A'}{}_i(x)$ at all points x. Berezin integration [Berezin 1966] is used for the variables $\psi, \bar{\psi}$ where (for example) for a single fermionic variable θ

$$\int 1 d\theta = 0, \tag{3.3.8}$$

$$\int \theta d\theta = 1. \tag{3.3.9}$$

The inner product between two states given by f, g is complex-valued, although the inner product between two 'states' involving extraneous odd variables will itself involve the extra variables. It obeys the condition

$$\langle \overline{f, g}\rangle = \langle g, f\rangle. \tag{3.3.10}$$

The term

$$\exp\left[\frac{-i}{\hbar}C\psi\bar{\psi}\right]$$

in the inner product ensures that the operators $\psi^A{}_i(x)$ and $\bar{\psi}^{A'}{}_i(x)$ are Hermitian adjoints, so that

$$\langle f, \bar{\psi}^{A'}{}_i(x) g\rangle = \left\langle \psi^A{}_i(x) f, g\right\rangle. \tag{3.3.11}$$

This may easily be checked, using the representation (3.3.2) of $\bar{\psi}^{A'}{}_i(x)$, integrating by parts and using Eq. (3.3.6). Note that the Berezin rule (3.3.8) implies that the integral of any total fermionic derivative is automatically zero. One can also show [D'Eath 1984] that the operator $p_{AA'}{}^i(x)$ is Hermitian:

$$\left\langle f, p_{AA'}{}^i(x) g\right\rangle = \left\langle p_{AA'}{}^i(x) f, g\right\rangle, \tag{3.3.12}$$

provided that the $D^{-1}(e)$ determinant factor is included in the definition of $\langle\ ,\ \rangle$. Had one instead used the opposite factor ordering in $p_{AA'}{}^i$ to that of Eq. (3.3.3), then the $D^{-1}(e)$ factor would not have been needed.

One can move from a representation using eigenstates of $\psi^A{}_i$ to a representation using eigenstates of $\bar{\psi}^{A'}{}_i$, where a wave functional

$$f\left(e^{AA'}{}_i(x), \psi^A{}_i(x)\right)$$

gives a wave functional

$$\tilde{f}\left(e^{AA'}{}_i(x), \bar{\psi}^{A'}{}_i(x)\right)$$

through the functional Fourier transform

$$\tilde{f}(e, \bar{\psi}) = D^{-1}(e) \int f(e, \psi) \exp\left[\frac{-i}{\hbar} C\psi\bar{\psi}\right] \mathscr{D}\psi. \tag{3.3.13}$$

The operators $\psi^A{}_i(x)$ and $\bar{\psi}^{A'}{}_i(x)$ acting on f become the operators

$$\begin{aligned} & -i\hbar D^{AA}{}_{ij}(x) \frac{\delta}{\delta \bar{\psi}^{A'}{}_j(x)} \tilde{f}(e, \bar{\psi}) \\ & = D^{-1}(e) \int \psi^A{}_i(x) f(e, \psi) \exp\left[\frac{-i}{\hbar} C\psi\bar{\psi}\right] \mathscr{D}\psi, \end{aligned} \tag{3.3.14}$$

$$\begin{aligned} & \bar{\psi}^{A'}{}_i(x) \tilde{f}(e, \bar{\psi}) = D^{-1}(e) \\ & \times \int \left[-i\,\hbar D^{AA'}{}_{ji}(x) \frac{\delta}{\delta \psi^A{}_j(x)} f(e, \psi)\right] \exp\left[\frac{-i}{\hbar} C\psi\bar{\psi}\right] \mathscr{D}\psi, \end{aligned} \tag{3.3.15}$$

acting on $\tilde{f}(e, \bar{\psi})$. One can also check [D'Eath 1984] that the operator $p_{AA'}{}^i(x)$ acting on $f(e, \psi)$ corresponds to the operator

$$\begin{aligned} & \left[-i\hbar \frac{\delta}{\delta e^{AA'}{}_i(x)} + \frac{1}{2} i\hbar \epsilon^{ijk} D_A{}^{B'}{}_{jl}(x) \frac{\delta}{\delta \bar{\psi}^{B'}{}_l(x)} \bar{\psi}_{A'k}(x)\right] \tilde{f}(e, \bar{\psi}) \\ & = D^{-1}(e) \int \left[p_{AA'}{}^i(x) f(e, \psi)\right] \exp\left[\frac{-i}{\hbar} C\psi\bar{\psi}\right] \mathscr{D}\psi \end{aligned} \tag{3.3.16}$$

acting on $\tilde{f}(e, \bar{\psi})$.

The wave functional $f(e, \psi)$ can be reconstructed from $\tilde{f}(e, \bar{\psi})$ by the inverse transform

$$f(e, \psi) = \int \tilde{f}(e, \bar{\psi}) \exp\left[\frac{i}{\hbar} C\psi\bar{\psi}\right] \mathscr{D}\bar{\psi}. \tag{3.3.17}$$

Note that the determinant factor $D(e)$ appears asymmetrically in the pair of equations (3.3.13) and (3.3.17). The way in which $D(e)$ appears here is again determined ultimately by our factor-ordering choice for $p_{AA'}{}^i(x)$. We shall see in section 3.4 that our choice for $p_{AA'}{}^i(x)$ enforces a particular factor ordering in the constraint operator $\bar{S}_{A'}$, in order that the quantum constraint $\bar{S}_{A'} f = 0$ should correctly describe the transformation properties of $f(e, \psi)$ under left-handed supersymmetry transformations acting on e, ψ. In order that the quantum constraint $S_A f = 0$, involving the Hermitian adjoint operator $S_{A'}$, should then correctly describe the transformation of $\tilde{f}(e, \bar{\psi})$ under right-handed supersymmetry transformations acting on $e, \bar{\psi}$, one needs the property (3.3.16) of $p_{AA'}{}^i$, which only holds with the definition (3.3.13).

The inner product (3.3.4) can then be rewritten in the equivalent forms

$$\begin{aligned}\langle f,g\rangle &= \int f(\bar{e,\psi})\tilde{g}\,(e,\bar{\psi})\,\mathscr{D}e\mathscr{D}\bar{\psi} \\ &= \int \tilde{f}(\bar{e,\psi})g(e,\psi)\mathscr{D}e\mathscr{D}\psi \\ &= \int \tilde{f}\,(\bar{e,\bar{\psi}})\tilde{g}(e,\psi)\exp\left[\frac{i}{\hbar}C\psi\bar{\psi}\right]\mathscr{D}e\mathscr{D}\psi\mathscr{D}\bar{\psi},\end{aligned} \qquad (3.3.18)$$

The formal inner product $\langle f,f\rangle$ is not positive-definite for an arbitrary wave functional $f(e,\psi)$; we only expect it to be positive when defined with the help of gauge fixing on physically allowed wave functionals f which obey the quantum constraints (section 3.4). To see this, consider the expression

$$-\frac{i}{\hbar}C_{AA'}{}^{ij}\psi^{A}{}_{i}\bar{\psi}^{A'}{}_{j}$$

involved in the exponential in Eq. (3.3.4) defining $\langle f,g\rangle$. Instead of $\psi^{A}{}_{i}$, use the spinor version

$$\chi^{A}{}_{BB'} = h^{1/4}e_{BB'}{}^{i}\psi^{A}{}_{i}\,, \qquad (3.3.19)$$

which obeys $n^{BB'}\chi^{A}{}_{BB'} = 0$. On using the relation

$$\epsilon_{abcd} \leftrightarrow i\left(\epsilon_{AD}\epsilon_{BC}\epsilon_{A'C'}\epsilon_{B'D'} - \epsilon_{AC}\epsilon_{BD}\epsilon_{A'D'}\epsilon_{B'C'}\right) \qquad (3.3.20)$$

between the tetrad alternating symbol ϵ_{abcd} and its spinor version, one finds

$$-\frac{i}{\hbar}C_{AA'}{}^{ij}\psi^{A}{}_{i}\bar{\psi}^{A'}{}_{j} = \hbar^{-1}\left(n_{AB'}\chi^{A}{}_{BA'}\bar{\chi}^{A'B'B} - n_{BA'}\chi^{A}{}_{AB'}\bar{\chi}^{A'B'B}\right). \qquad (3.3.21)$$

To split this into its positive and negative parts, make the decomposition

$$\chi_{ABB'} = \rho_{ABB'} + \frac{2}{3}\left(n_{A}{}^{A'}\bar{\beta}_{A'}n_{BB'} + n_{B}{}^{A'}\bar{\beta}_{A'}n_{AB'}\right) + \epsilon_{AB}\bar{\beta}_{B'}, \qquad (3.3.22)$$

where

$$\bar{\beta}_{A'} = \frac{1}{2}\chi_{A}{}^{A}{}_{A'}. \qquad (3.3.23)$$

Then $\rho_{ABB'}$, so defined, obeys

$$\rho_{ABB'} = \rho_{(AB)B'},\ n^{BB'}\rho_{ABB'} = 0, \qquad (3.3.24)$$

so that $\rho_{ABB'}$ corresponds to $\rho_{Ai} = -e^{BB'}{}_{i}\rho_{ABB'}$. Then one obtains

$$-\frac{i}{\hbar}C_{AA'}{}^{ij}\psi^{A}{}_{i}\bar{\psi}^{A'}{}_{j} = \hbar^{-1}\left(-n_{AA'}\rho^{A}{}_{i}\bar{\rho}^{A'i} + \frac{16}{3}n_{AA'}\bar{\beta}^{A'}\beta^{A}\right), \qquad (3.3.25)$$

a difference of quantities with definite signs. The finite-dimensional inner product between two functions $f\,(\psi^A{}_i)$, $g\,(\psi^A{}_i)$, given by

$$\langle f, g\rangle_{FD} = \int f(\bar{\psi})g(\psi)\exp\left[-\frac{i}{\hbar}C_{AA'}{}^{ij}\psi^A{}_i\bar{\psi}^{A'}{}_j\right]\times\prod_{A,i}\left(d\psi^A{}_i d\bar{\psi}^{A'}{}_i\right) \quad (3.3.26)$$

will only be positive-definite on functions $f(\psi^A{}_i)$ which depend only on $\rho^A{}_i$ but not on $\bar{\beta}_{A'}$. A similar statement holds for the inner product (3.3.4).

Now the formal inner product $\langle\, ,\,\rangle$ can only be made precise when defined by means of gauge fixing or gauge averaging. The quantity $\bar{\beta}_{A'}$ is gauge dependent, since it changes under an infinitesimal primed supersymmetry transformation $\delta e^{AA'}{}_i = -i\kappa\bar{\epsilon}^{A'}\psi^A{}_i$, $\delta\psi^A{}_i = 0$ applied to the variables $e^{AA'}{}_i, \psi^A{}_i$. And a physical state must obey the quantum constraint $\bar{S}_{A'}f = 0$ (section 3.4) which describes the transformation of f under such a supersymmetry transformation. Here we allow $e^{AA'}{}_i$ to become non-Hermitian and extend the definition of $f(e^{AA'}{}_i, \psi^A{}_i)$ accordingly. Thus for a physical state we can pick a gauge in which $\bar{\beta}_{A'} = 0$ by using the left-handed supersymmetry freedom, and can evaluate the inner product $\langle f, g\rangle$ in this gauge, provided we include a factor analogous to a Faddeev–Popov determinant [Itzykson & Zuber 1980] to allow for the integration over the supersymmetry parameter $\bar{\epsilon}^{A'}$. A similar procedure must be applied for the other quantum constraints, corresponding to $\mathscr{H}_{AA'}$, S_A, J_{AB}, and $\bar{J}_{A'B'}$, so that the integration in $\langle f, g\rangle$ is effectively only over the remaining physical degrees of freedom. Although the rigorous definition of the inner product in a system with constraints is apparently a major unsolved problem, we do expect the result to give a positive-definite expression in this case.

3.4 The quantum constraints

Once a representation has been found for the operators

$$e^{AA'}{}_i,\; p_{AA'}{}^i,\; \psi^A{}_i,\; \bar{\psi}^{A'}{}_i$$

acting on states described by wave functionals, one can examine the properties of physically allowed states. In the quantum theory the constraints (3.2.20) and (3.2.49) associated with the invariances (3.2.13)–(3.2.15) of the volume part of the action become conditions on the wave functions [Dirac 1965], so that any physically allowed state must obey the quantum constraints $J_{AB}f = 0$, $\bar{J}_{A'B'}f = 0$, $\mathscr{H}_{AA'}f = 0, S_Af = 0$, and $\bar{S}_{A'}f = 0$.

The conditions $J_{AB}f = 0$, $\bar{J}_{A'B'}f = 0$ simply describe the invariance of $f(e^{AA'}{}_i, \psi^A{}_i)$ under local Lorentz transformations. Choosing the factor ordering for $J_{AB}, \bar{J}_{A'B'}$ with momentum operators on the right, as in Eq. (3.2.21), one has

$$J_{AB} = -\frac{1}{2}i\hbar\left[e_B{}^{A'}{}_i\frac{\delta}{\delta e^{AA'}{}_i} + e_A{}^{A'}{}_i\frac{\delta}{\delta e^{BA'}{}_i} + \psi_{Bi}\frac{\delta}{\delta\psi^A{}_i} + \psi_{Ai}\frac{\delta}{\delta\psi^B{}_i}\right], \tag{3.4.1}$$

$$\bar{J}_{A'B'} = -\frac{1}{2}i\hbar\left[e^A{}_{B'i}\frac{\delta}{\delta e^{AA'}{}_i} + e^A{}_{A'i}\frac{\delta}{\delta e^{AB'}{}_i}\right]. \tag{3.4.2}$$

Then $J_{AB}f = 0$ gives the invariance of f under an infinitesimal unprimed Lorentz transformation

$$\begin{aligned}\delta e^{AA'}{}_i &= M^A{}_B e^{BA'}{}_i,\\ \delta\psi^A{}_i &= M^A{}_B\psi^B{}_i,\end{aligned} \tag{3.4.3}$$

with $M^{AB} = M^{(AB)}$, while $\bar{J}_{A'B'}f = 0$ gives the invariance of f under the primed transformation

$$\begin{aligned}\delta e^{AA'}{}_i &= N^{A'}{}_{B'}e^{AB'}{}_i,\\ \delta\psi^A{}_i &= 0,\end{aligned} \tag{3.4.4}$$

with $N^{A'B'} = N^{(A'B')}$. Here, as remarked previously, we allow the argument $e^{AA'}{}_i(x)$ to become non-Hermitian and correspondingly extend the definition of $f(e^{AA'}{}_i, \psi^A{}_i)$. Thus we have the freedom to make independent unprimed and primed Lorentz transformations.

We next consider the constraint $\bar{S}_{A'}f = 0$. In order to be able to write out this equation using the representations of operators given in section 3.3, one must start with a form of the classical quantity $\bar{S}_{A'}$ in which the dependence on the variables $e, p, \psi, \bar{\psi}$ is manifest. The form of $\bar{S}_{A'}$ given by conjugating Eq. (3.2.48) has extra $\psi, \bar{\psi}$ dependence involved through torsion in the derivative ${}^3D_j\psi^A{}_k$. When this is expanded out in terms of the torsion-free derivative ${}^{3s}D_j\psi^A{}_k$ based on the connection ${}^{3s}\omega^{AB}{}_i$ of Eq. (3.2.43), one finds that all the $\bar{\psi}\psi\psi$ terms cancel to give classically

$$\bar{S}_{A'} = \epsilon^{ijk}e_{AA'i}\,{}^{3s}D_j\psi^A{}_k + \frac{1}{2}i\kappa^2\psi^A{}_i p_{AA'}{}^i. \tag{3.4.5}$$

Quantum-mechanically one must also choose this factor ordering for the second term in order that the condition $\bar{S}_{A'}f = 0$ should correctly describe the transformation of f under a left-handed supersymmetry transformation. Note that when $\psi^A{}_i p_{AA'}{}^i$ is written out using Eq. (3.3.3) for the quantum operator $p_{AA'}{}^i$, the term $\frac{1}{2}\psi^A{}_i\epsilon^{ijk}\psi_{Aj}\bar{\psi}_{A'k}$ gives zero because of the anticommutation of $\psi^A{}_i$ with $\psi^B{}_j$. Hence in the quantum theory $\bar{S}_{A'}$ is given by the first-order operator

$$\bar{S}_{A'} = \epsilon^{ijk} e_{AA'i} \, {}^{3s}D_j \psi^A{}_k + \frac{1}{2}\hbar\kappa^2 \psi^A{}_i \frac{\delta}{\delta e^{AA'}{}_i}. \tag{3.4.6}$$

The quantum constraint $\bar{S}_{A'} f = 0$ shows that under a primed supersymmetry transformation

$$\begin{aligned} \delta e^{AA'}{}_i &= i\kappa \tilde{\epsilon}^{A'} \psi^A{}_i, \\ \delta \psi^A{}_i &= 0, \end{aligned} \tag{3.4.7}$$

f changes by

$$\delta f = \frac{-2if}{\hbar\kappa} \int d^3x \epsilon^{ijk} e_{AA'i} \left({}^{3s}D_j \psi^A{}_k \right) \tilde{\epsilon}^{A'}. \tag{3.4.8}$$

One can verify this in a different way by examining the transformation of the Feynman path integral under supersymmetry (section 3.5).

Although a primed supersymmetry transformation (3.4.7) acts simply on our variables $e^{AA'}{}_i, \psi^A{}_i$, the same is not true of an unprimed transformation. In the classical theory (see Eqs. (4.1.19),(4.1.20) below), the unprimed supersymmetry variation of $e^{AA'}{}_i, \psi^A{}_i$ involves also the variables $p_{AA'}{}^i$ and $\bar{\psi}^{A'}{}_i$, which are not known exactly in a quantum state given by $f(e, \psi)$. In this respect, viewed in terms of the coordinate variables $e^{AA'}{}_i, \psi^A{}_i$, an unprimed supersymmetry transformation is similar to a normal deformation of the spacelike surface on which the fields are defined. This is to be expected, since the anticommutator of a primed and unprimed supersymmetry transformation gives a coordinate transformation (plus a Lorentz transformation and a further supersymmetry transformation) [van Nieuwenhuizen 1981]. Correspondingly, the operator

$$S_A = \epsilon^{ijk} e_{AA'i} \, {}^{3s}D_j \bar{\psi}^{A'}{}_k - \frac{1}{2} i\kappa^2 p_{AA'}{}^i \bar{\psi}^{A'}{}_i, \tag{3.4.9}$$

given by the Hermitian adjoint of Eq. (3.4.5), is of second order in terms of the variables e, ψ:

$$S_A = i\hbar \, {}^{3s}D_i \left[\frac{\delta}{\delta \psi^A{}_i} \right] + \frac{1}{2} i\hbar^2\kappa^2 \frac{\delta}{\delta e^{AA'}{}_i} \left[D^{BA'}{}_{ji} \frac{\delta}{\delta \psi^B{}_j} \right]. \tag{3.4.10}$$

The constraint $S_A f = 0$, which appears complicated in terms of $f(e, \psi)$, becomes a simple first-order equation in terms of the transformed wave functional $\tilde{f}(e, \bar{\psi})$. The action of $p_{AA'}{}^i$ on $\tilde{f}$ given by Eq. (3.3.10), together with the factor-ordering of S_A given in Eq. (3.4.9), show that this constraint simply describes the transformation of $\tilde{f}$ under an unprimed supersymmetry transformation on $e^{AA'}{}_i, \bar{\psi}^{A'}{}_i$ by equations analogous to Eqs. (3.4.7) and (3.4.8). As remarked at the end of section 3.3, our choice of the $D^{-1}(e)$ determinant factor in Eq. (3.3.13), giving $\tilde{f}$ in terms of f, was enforced by the requirement that $p_{AA'}{}^{i}$ should act on $\tilde{f}$ as in Eq. (3.3.16), so that the constraint $S_A f = 0$ should correctly describe the transformation of a wave functional $\tilde{f}(e^{AA'}{}_{i}, \bar{\psi}^{A'}{}_{i})$ under unprimed supersymmetry.

It is instructive to compare the constraints $\bar{S}_{A'} f = 0$, $S_A f = 0$ in supergravity with those in the free spin-3/2 theory. There a quantum state can be described by a wave functional $F(\psi^{A}{}_{i}(x))$, and the constraints $\bar{S}_{A'} F = 0, S_A F = 0$ read

$$\epsilon^{ijk} e_{AA'i} \left(\partial_j \psi^{A}{}_{k} \right) F(\psi) = 0, \tag{3.4.11}$$

$$\partial_i \left[\frac{\delta F(\psi)}{\delta \psi^{A}{}_{i}} \right] = 0. \tag{3.4.12}$$

Thus, $\bar{S}_{A'} F = 0$ implies that F only resides on solutions of the classical constraint $\epsilon^{ijk} e_{AA'i} \partial_j \psi^{A}{}_{k} = 0$, while $S_A F = 0$ describes the invariance of F under unprimed gauge transformations $\psi^{A}{}_{i} \to \psi^{A}{}_{i} + \partial_i \epsilon^{A}$. In supergravity, the supersymmetric coupling of the spin-3/2 field to gravity involves the ψp term in $\bar{S}_{A'}$, which may allow the wave functional f to have support over the whole space of $e^{AA'}{}_{i}(x), \psi^{A}{}_{i}(x)$ (once one includes auxiliary fields – see sections 4.3, 8.1), and which makes the constraints give a simple description of primed rather than unprimed transformations on e, ψ.

Now that we have determined the factor ordering of the operators $\bar{S}_{A'}, S_A$ from consideration of the supersymmetry properties of quantum states, we can evaluate their anticommutators,

$$[S_A(x), S_B(x')]_+ = 0, \tag{3.4.13}$$

$$[\bar{S}_{A'}(x), \bar{S}_{B'}(x')]_+ = 0, \tag{3.4.14}$$

$$[S_A(x), \bar{S}_{A'}(x')]_+ = -\frac{1}{2} \hbar \kappa^2 \, {}_2\mathscr{H}_{AA'}(x) \delta(x, x'), \tag{3.4.15}$$

where

$$\begin{aligned} {}_2\mathscr{H}_{AA'}(x) = {} & \frac{1}{2} i \hbar^2 \kappa^2 \psi^{B}{}_{i} \frac{\delta}{\delta e^{AB'}{}_{j}} \left[\epsilon^{ilm} D_{B}{}^{B'}{}_{mj} D^{C}{}_{A'kl} \frac{\delta}{\delta \psi^{C}{}_{k}} \right] \\ & - \frac{1}{2} i \hbar^2 \kappa^2 \frac{\delta}{\delta e^{AB'}{}_{j}} \left[D^{BB'}{}_{ij} \frac{\delta}{\delta e^{BA'}{}_{i}} \right] \\ & + \frac{1}{2} \epsilon^{ijk} \left[\left({}^{3s}D_j \psi_{Ak} \right) \bar{\psi}_{A'i} + \psi_{Ai} \left({}^{3s}D_j \bar{\psi}_{A'k} \right) \right] \\ & + {}^{3s}D_i p_{AA'}{}^{i} + \frac{1}{2\kappa^2} h^{1/2} \, {}^{3s}R \, n_{AA'} \end{aligned} \tag{3.4.16}$$

and ${}^{3s}R$ is the three-dimensional curvature scalar formed with the torsion-free connection ${}^{3s}\omega^{ab}{}_{i}$. Thus a state in which f and $\tilde{f}$ have the transformation properties under supersymmetry implied by $\bar{S}_{A'} f = 0$, $S_A f = 0$ also obeys the quantum constraint

$$_2\mathscr{H}_{AA'}f = 0. \tag{3.4.17}$$

Now, as in Eq. (3.2.50), $\mathscr{H}_{AA'}$ differs from $_2\mathscr{H}_{AA'}$ by terms proportional to J and $\bar{J}$. By making the appropriate field-dependent change from the Lagrange multipliers $-\omega_{AB0}, -\bar{\omega}_{A'B'0}$ for $J^{AB}, \bar{J}^{A'B'}$ in Eq. (3.2.47), to new multipliers $M_{AB}, \bar{M}_{A'B'}$, one can replace $\mathscr{H}_\perp, \mathscr{H}_i$ in Eq. (3.2.47) by $_2\mathscr{H}_\perp, {}_2\mathscr{H}_i$. Thus one can regard $_2\mathscr{H}_\perp, {}_2\mathscr{H}_i, S_A, \bar{S}_{A'}, J_{AB}$ and $\bar{J}_{A'B'}$ as the classical constraints to be solved. When the system is quantized, one sees from Eq. (3.4.17) that it is sufficient that a physical state should obey $J_{AB}f = 0, \bar{J}_{A'B'}f = 0, S_Af = 0, \bar{S}_{A'}f = 0$, describing its behaviour under Lorentz and supersymmetry transformations. The remaining quantum constraint $_2\mathscr{H}_{AA'}f = 0$ will then be automatic.

The tangential projection $e^{AA'}{}_i\ {}_2\mathscr{H}_{AA'}f = 0$, modulo factor-ordering terms [Wulf 1995] should imply that $f\left(e^{AA'}{}_i(x), \psi^A{}_i(x)\right)$ is invariant under spatial coordinate transformations applied to $e^{AA'}{}_i, \psi^A{}_i$, while the normal projection $n^{AA'}\ {}_2\mathscr{H}_{AA'}f = 0$ gives a many-time evolution equation for f, analogous to the Wheeler–DeWitt equation of canonical metric quantum gravity.

3.5 The path integral

As in any quantum theory, one can find the most general physically allowed state by superposing propagators or elementary amplitudes to go from one field configuration to another. For example, in quantum gravity one can consider the amplitude to go from an initial to a final three-metric h_{ij}, given by a Lorentzian Feynman path integral of $\exp(iS/\hbar)$ over all four-geometries $g_{\mu\nu}$ which fill in between the two bounding three-surfaces, inducing the correct spatial metrics on them, where S is a suitably defined gravitational action. (For the Euclidean path integral, see chapter 4. The treatment of the present section can, if desired, be rephrased in terms of the Euclidean path integral.) The corresponding amplitude in a theory containing fermions has been discussed by [Faddeev & Slavnov 1980], and following their treatment we are led to consider the amplitude to go from a prescribed field configuration $(e^{AA'}{}_i(x), \tilde{\psi}^{A'}{}_i(x))_I$ on an initial surface to another configuration $(e^{AA'}{}_i(x), \psi^A{}_i(x))_F$ on a final surface. This is given formally by a path integral

$$K\left(e_F, \psi_F; e_I, \tilde{\psi}_I\right) = \int \exp(iS/\hbar)\mathscr{D}e\mathscr{D}\psi\mathscr{D}\tilde{\psi} \tag{3.5.1}$$

over all infilling fields $e^{AA'}{}_\mu, \psi^A{}_\mu, \tilde{\psi}^{A'}{}_\mu$ where $(e^{AA'}{}_\mu, \tilde{\psi}^{A'}{}_\mu)$ restricted to the initial surface give the correct initial data $(e^{AA'}{}_i, \tilde{\psi}^{A'}{}_i)_I$, and similarly $(e^{AA'}{}_\mu, \psi^A{}_\mu)$ give the correct final data $(e^{AA'}{}_i, \psi^A{}_i)_F$. Here $\psi^A{}_\mu$ and $\tilde{\psi}^{A'}{}_\mu$ are to be regarded as independent quantities, with $\tilde{\psi}^{A'}{}_\mu$ no longer the Hermitian conjugate of $\psi^A{}_\mu$. Further, the path integral can be regarded as a contour integral over $e^{AA'}{}_\mu(x)$, since the action S is an analytic functional of $e^{AA'}{}_\mu(x)$, so that the contour can be moved away from the space of Hermitian $e^{AA'}{}_\mu(x)$, and the end values $e^{AA'}{}_i(x)_{I,F}$ can be taken to be non-Hermitian if desired. More strictly, in order to evaluate K one should include in the action S auxiliary field, gauge-fixing, and ghost terms [van Nieuwenhuizen 1981] (with corresponding extra functional integrations) in addition to the terms given in Eq. (3.2.12), with $\bar{\psi}^{A'}{}_\mu$ replaced by $\tilde{\psi}^{A'}{}_\mu$. S will also contain boundary terms at the initial and final surfaces, and terms at spatial infinity in the asymptotically flat case.

The boundary terms in S on the initial and final surfaces are necessary in order that, when one composes the amplitude to go from an initial state I to an intermediate state J with the amplitude to go from J to a final state F by summing over J, one correctly recovers the amplitude to go from I to F. They must have the property that variation of S with our prescribed boundary data leads to the classical solution with those data [Faddeev & Slavnov 1980]. If for simplicity of exposition we continue to omit auxiliary fields, then S has the form

$$S = S_2 + S_{3/2} + S_{2B} + S_{3/2B} + S_\infty \,, \tag{3.5.2}$$

where S_2 and $S_{3/2}$ are the volume terms in Eq. (3.2.12), S_{2B} and $S_{3/2B}$ are corresponding surface terms on the bounding hypersurfaces, and S_∞ is the contribution at spatial infinity. Here S_{2B} has the same form as in pure gravity [Eq. (2.3.68)]:

$$S_{2B} = \kappa^{-2} \int_F d^3x h^{1/2} \mathrm{tr} K - \kappa^{-2} \int_I d^3x h^{1/2} \mathrm{tr} K \,, \tag{3.5.3}$$

where the integrals are taken over the final and initial hypersurfaces. One can check that this term is already allowed for in the Hamiltonian reduction of the action:

$$\begin{aligned} S_2 + S_{3/2} + S_{2B} = \int_{t_I}^{t_F} dt \int d^3x \Big(& \dot{e}^{AA'}{}_i p_{AA'}{}^i + \dot{\psi}^A{}_i \pi_A{}^i + \dot{\tilde{\psi}}^{A'}{}_i \tilde{\pi}_{A'}{}^i \\ & - N\mathscr{H}_\perp - N^i \mathscr{H}_i - \psi^A{}_0 S_A - \tilde{S}_{A'} \tilde{\psi}^{A'}{}_0 \\ & - M_{AB} J^{AB} - \tilde{M}_{A'B'} \tilde{J}^{A'B'} \Big). \end{aligned} \tag{3.5.4}$$

The term $S_{3/2B}$ corresponding to boundary conditions with $\tilde{\psi}^{A'}{}_i$ fixed initially and $\psi^A{}_i$ fixed finally is

$$\begin{aligned} S_{3/2B} &= \int_I d^3x\psi^A{}_i\pi_A{}^i - \int_F d^3x\tilde{\psi}^{A'}{}_i\tilde{\pi}_{A'}{}^i \\ &= \left[\int_I + \int_F\right] d^3x\epsilon^{ijk}\psi^A{}_ie_{AA'j}\tilde{\psi}^{A'}{}_k. \end{aligned} \tag{3.5.5}$$

With these boundary terms the variational condition $\delta S = 0$ gives the classical field equations.

In the asymptotically flat case, where one must specify also the relation of the two surfaces at infinity, one finds, following Eqs. (3.2.51) and (3.2.52), the contribution

$$S_\infty = -MT + P^iX_i - \phi^AQ_A - \tilde{Q}_{A'}\tilde{\phi}^{A'}. \tag{3.5.6}$$

Here the two surfaces are separated by a proper time interval

$$T = -\int_{t_I}^{t_F} \alpha^0 dt \tag{3.5.7}$$

at infinity, with a relative spatial translation

$$X^i = -\int_{t_I}^{t_F} \alpha^i dt, \tag{3.5.8}$$

and we allow a supersymmetry transformation at infinity, measured by

$$\begin{aligned} \phi^A &= \int_{t_I}^{t_F} \psi^A{}_0(\infty)dt, \\ \tilde{\phi}^{A'} &= \int_{t_I}^{t_F} \tilde{\psi}^{A'}{}_0(\infty)dt. \end{aligned} \tag{3.5.9}$$

The resulting amplitude $K(e_F, \psi_F; e_I, \tilde{\psi}_I; T, X^i, \phi^A, \tilde{\phi}^{A'})$ then depends on the additional information at infinity. Since global supersymmetry transformations do not commute, the parameters $(\phi^A, \tilde{\phi}^{A'})$ must refer to a particular representation of global supersymmetry, in fact the vector or symmetric representation [Roček 1981].

By standard arguments $K(e, \psi; e_I, \tilde{\psi}_I; \ldots)$, regarded as a functional of the final data $e^{AA'}{}_i, \psi^A{}_i$, will obey the quantum constraints

$$\begin{aligned} &J_{AB}K = 0, \quad \bar{J}_{A'B'}K = 0, \\ &S_AK = 0, \quad \bar{S}_{A'}K = 0, \quad {}_2\mathscr{H}_{AA'}K = 0, \end{aligned} \tag{3.5.10}$$

with analogous equations holding at the initial surface. K will also obey the Schrödinger evolution equation: for example if we only allow the Lagrange multiplier N to be non-zero, with $N \to 1$ as $r \to \infty$, we find

$$i\hbar\frac{\partial K}{\partial T} = \left[P^0 + \int d^3xN\,{}_2\mathscr{H}_\perp\right]K, \tag{3.5.11}$$

using the normal projection ${}_2\mathscr{H}_\perp = -n^{AA'}\ {}_2\mathscr{H}_{AA'}$ of the operator given in Eq. (3.4.16). Because of the anticommutation relations between $Q_A, \tilde{Q}_{A'}$ and $S_A, \tilde{S}_{A'}$, this amounts to evolving K by considering the commutator between

$$\left[\psi^A{}_0(\infty)Q_A + \int d^3x\psi^A{}_0 S_A\right]$$

and

$$\left[\tilde{Q}_{A'}\tilde{\psi}^{A'}{}_0(\infty) + \int d^3x\tilde{S}_{A'}\tilde{\psi}^{A'}{}_0\right],$$

applied to K for suitable choices of $\psi^A{}_0(\infty)$ and $\tilde{\psi}^{A'}{}_0(\infty)$. Note that the volume term does not actually contribute to $\partial K/\partial T$, because of the constraints (3.5.10).

The path-integral representation (3.5.1) of K makes evident the transformation properties of K under local symmetry transformations, and hence, those of a general wave functional f given by a convolution of initial data with the propagator K. It is immediate that K is invariant under local Lorentz and spatial coordinate transformations applied to the data e_F, ψ_F, and $e_I, \tilde{\psi}_I$. Now consider the effect of a primed supersymmetry transformation, which is non-zero at the final surface, but zero at the initial surface. Take a typical infilling field $e^{AA'}{}_\mu, \psi^A{}_\mu, \tilde{\psi}^{A'}{}_\mu$ appearing in the path integral (3.5.1) and evaluate the change in its action S. This consists only of surface terms, which can be found by applying a supersymmetry transformation generated by the term $\tilde{S}_{A'}\tilde{\psi}^{A'}{}_0$ in the Hamiltonian over a short parameter time δt. Allowing for the extra contribution $S_{3/2B}$ [Eq. (3.5.5)] to be added to the action (3.5.4), one finds

$$\begin{aligned}
\delta S &= \int d^3x \left(\delta e^{AA'}{}_i p_{AA'}{}^i + \delta\psi^A{}_i \pi_A{}^i - \tilde{\psi}^{A'}{}_i \delta\tilde{\pi}_{A'}{}^i - \tilde{S}_{A'}\tilde{\psi}^{A'}{}_0 \delta t\right) \\
&= \int d^3x \left[\delta e^{AA'}{}_i \left(p_{AA'}{}^i - \frac{1}{2}\epsilon^{ijk}\psi_{Aj}\tilde{\psi}_{A'k}\right)\right. \\
&\quad \left. - \delta\psi^A{}_i \epsilon^{ijk}\tilde{\psi}^{A'}{}_j e_{AA'k} - \tilde{S}_{A'}\tilde{\psi}^{A'}{}_0 \delta t\right],
\end{aligned} \tag{3.5.12}$$

where all variables are evaluated at the final surface. The supersymmetry variations of $e^{AA'}{}_i, \psi^A{}_i$ are

$$\begin{aligned}
\delta e^{AA'}{}_i &= \frac{1}{2} i\kappa^2 \psi^A{}_i \tilde{\psi}^{A'}{}_0 \delta t, \\
\delta\psi^A{}_i &= 0,
\end{aligned} \tag{3.5.13}$$

and the resulting change in S is then

$$\delta S = -\int d^3x \epsilon^{ijk} e_{AA'i} \left({}^{3s}D_j \psi^A{}_k\right) \tilde{\psi}^{A'}{}_0 \delta t. \tag{3.5.14}$$

Thus under the primed supersymmetry transformation (3.5.14) with $\tilde{\epsilon}^{A'} = \frac{1}{2}\kappa\tilde{\psi}^{A'}{}_0\delta t$, the action S of any infilling $e^{AA'}{}_\mu, \psi^A{}_\mu, \tilde{\psi}^{A'}{}_\mu$ with prescribed final data $e^{AA'}{}_i, \psi^A{}_i$ changes by the same amount δS, which can be calculated from the final data and $\tilde{\epsilon}^{A'}$. Adding all these variations in the path integral (3.5.1), we find that K changes according to

$$\delta(\ln K) = \frac{-2i}{\hbar\kappa}\int d^3x\epsilon^{ijk}e_{AA'i}\left({}^{3s}D_j\psi^A{}_k\right)\tilde{\epsilon}^{A'} \qquad (3.5.15)$$

under a supersymmetry transformation parametrized by $\tilde{\epsilon}^{A'}$, in agreement with Eq. (3.4.8). This provides an alternative derivation of the left-handed supersymmetry transformation property [(3.4.7,8)] of a wave functional $f(e,\psi)$, associated with the constraint $\tilde{S}_{A'}f = 0$.

The path integral yields no such simple description of the change of K under an unprimed supersymmetry transformation at the final surface; under an infinitesimal unprimed transformation the change δS in the action of any infilling $e^{AA'}{}_\mu, \psi^A{}_\mu, \tilde{\psi}^{A'}{}_\mu$ depends not only on the final data $e^{AA'}{}_i, \psi^A{}_i$, but also on $p_{AA'}{}^i$ and $\tilde{\psi}^{A'}{}_i$. Corresponding statements apply to the variation of K under supersymmetry transformations at the initial surface.

4
The quantum amplitude

4.1 Semi-classical expansion of the quantum amplitude

As was discussed in sections 2.4 and 2.6, it is most natural, in studying a path integral in quantum field theory, to work in the Riemannian ('Euclidean') rather than the Lorentzian context. The classical Lorentzian boundary-value problem is already badly posed, so that one cannot even find a semi-classical Lorentzian approximation $A\exp(iS/\hbar)$ to the amplitude. Rather, one should work with a Euclidean time separation τ at spatial infinity, and study the Euclidean path integral. The elliptic boundary-value problem will be well posed, at least for weak fields, and one can approximate the Euclidean path integral as $A\exp(-I_{\text{class}}/\hbar)$, where I_{class} is the classical action and A is a prefactor. Once the path integral is known analytically as a function of τ, one may rotate back towards real time $t = -i\tau$. This corresponds to the $+i\epsilon$ prescription for Feynman propagators in Feynman-diagram theory [Itzykson & Zuber 1980].

In the case of asymptotically flat initial and final data $e^{AA'}{}_{iI}$, $e^{AA'}{}_{iF}$, where the three-metric $h_{ij} = -(e^{AA'}{}_i e_{AA'j})_{I,F}$ tends to flatness rapidly (say) for distances $r \to \infty$, one can study the Euclidean path integral, with Euclidean time τ at spatial infinity between the two surfaces and other data at spatial infinity as in Eqs. (3.5.8),(3.5.9). This is given formally by

$$K(e_F, \psi_F; e_I, \tilde{\psi}_I; \tau, \ldots) = \int \exp(-I/\hbar)\mathscr{D}e\mathscr{D}\psi\mathscr{D}\tilde{\psi}, \qquad (4.1.1)$$

where $I = -iS$ is the Euclidean action, which should include boundary contributions as in Eqs. (3.5.2) ff. As in Eq. (3.5.1), one should include in the action I auxiliary fields, gauge-fixing and ghost terms [van Nieuwenhuizen 1981]. Auxiliary fields will be studied further in section 4.3. Again, as for the Lorentzian path integral (3.5.1), the Euclidean path integral

(4.1.1) is a contour integral over the infilling $e^{AA'}{}_{\mu}$; this integral will in general not be Hermitian, because of the contribution to the action from the fermions, whose boundary data are not restricted by any hermiticity conditions. Note, following section 2.6, that if, say, one has Hermitian boundary data $(e^{AA'}{}_{i})_{I,F}$, and one takes an infilling four-geometry $g_{\mu\nu}$ in the path integral which is real and Riemannian, corresponding to Euclidean time-separation τ at spatial infinity, then if one rotates τ back to real Lorentzian time t, there will in general be no real Lorentzian section of the four-metric $g_{\mu\nu}$. Thus one cannot simply convert the Euclidean path integral into the Lorentzian path integral by a change of integration contour. This indicates that the Lorentzian path integral (3.5.1) and Euclidean path integral (4.1.1) are not obviously related, at least for strong-field geometries. From now on, the Euclidean path integral will be considered, since this appears to be better behaved.

Consider now the classical boundary-value problem in supergravity, in the Euclidean context, where (say) asymptotically flat initial data $(e^{AA'}{}_{i}, \tilde{\psi}^{A'}{}_{i})_{I}$ are specified on an initial surface, asymptotically flat final data $(e^{AA'}{}_{i}, \psi^{A}{}_{i})_{F}$ are specified on a final surface, and a Euclidean time separation τ at spatial infinity is specified between the surfaces (together with the parameters at infinity X^{i}, ϕ^{A} and $\tilde{\phi}^{A'}$ of Eqs. (3.5.8),(3.5.9)). As is well known [Bao *et al.* 1985], the classical field equations of $N = 1$ supergravity form a hierarchy in powers of fermions. First, one must solve the classical vacuum Einstein equations for $e^{AA'}{}_{\mu}$ or $g_{\mu\nu}$ subject to the given boundary data $(e^{AA'}{}_{i})_{I,F}$. With Euclidean time separation τ at infinity specified, this gives a non-linear elliptic boundary-value problem. Very little is known about this classical problem in general, although it has been shown [Reula 1987] that there is a unique (up to gauge) nonlinear solution in the case of boundary data close to flat space. This problem is potentially of great interest to those working in classical and in quantum gravity, and to mathematicians. One would like to know whether the uniqueness property continues to hold for general bounding three-geometries, or whether one may have no classical Riemannian solutions or many such solutions for strongly deformed gravitational boundary data, or whether complex conjugate pairs of solutions may appear. Here, only Reula's result will be assumed, and only weak gravitational boundary data near flat space will be considered, with classical bosonic action $I_{B}(e^{AA'}{}_{iF}; e^{AA'}{}_{iI}; \tau, \ldots)$. The classical bosonic part of the momentum, $p_{AA'}{}^{i}{}_{B}$, evaluated (say) at the final surface, is given by

$$p_{AA'}{}^{i}{}_{BF} = i\frac{\delta I_{B}}{\delta e^{AA'}{}_{iF}}. \tag{4.1.2}$$

The boundary data obey the bosonic parts of the ${}_{1}\mathscr{H}_{\perp} = 0$ and ${}_{1}\mathscr{H}_{i} = 0$

constraint equations (3.2.38), (3.2.39):

$$-\frac{1}{2\kappa^2}h^{1/2}\,{}^3R + 2\kappa^2 h^{-\frac{1}{2}}[\pi_{ij}\pi^{ij} - 1/2(tr\pi)^2] = 0, \tag{4.1.3}$$

$$h_{ij}\pi^{jk}{}_{\|k} = 0, \tag{4.1.4}$$

where π^{ij} is given in terms of $p_{AA'}{}^i$ by Eqs. (3.2.22),(3.2.24).

The dependence of the classical action I_B on the Euclidean time coordinate τ is given, following Eq. (3.5.6), by

$$\frac{\partial I_B}{\partial \tau} = M_B, \tag{4.1.5}$$

where $M_B = M_B(e^{AA'}{}_{iF}; e^{AA'}{}_{iI}; \tau, \ldots)$ is the bosonic part of the mass of the four-dimensional spacetime, given by surface integrals at spatial infinity (see section 4.4) [D'Eath 1981], which includes an extra term beyond that of Eq. (3.2.52). Even if the three-dimensional mass of the boundary data $e^{AA'}{}_{iI}$, $e^{AA'}{}_{iF}$ is chosen to be zero, by requiring that the three-metric h_{ij} should approach flatness at a rate faster than r^{-1}, there will in general be a non-zero bosonic mass M_B of the four-dimensional classical solution spacetime, since the three-surfaces of constant τ may not be embedded near infinity in the usual way. An example is provided in [Wheeler 1968], where spacelike hypersurfaces of zero three-dimensional mass are described in the Lorentzian Schwarzschild geometry. These ideas will be treated further in section 4.4.

At next order in the hierarchy for the classical solution expanded in powers of fermions, one must study the order linear in fermions. One solves the linearized (Rarita–Schwinger) gravitino field equations about the classical gravitational background [Bao *et al.* 1985]. The spatial components of these field equations are first-order evolution equations. The zero component gives the classical supersymmetry constraints $S_A = 0$ and $\tilde{S}_{A'} = 0$. For example, at the final surface, the field $\psi^A{}_{iF}$ must obey $\tilde{S}_{A'} = 0$, namely, from Eq. (3.4.5):

$$\epsilon^{ijk} e_{AA'i}\,{}^{3s}D_j \psi^A_k + \tfrac{1}{2} i\kappa^2 \psi^A{}_i p_{AA'}{}^i = 0, \tag{4.1.6}$$

where the bosonic part $p_{AA'}{}^i{}_{BF}$ of $p_{AA'}{}^i$ is given in terms of the boundary data $(e^{AA'}{}_{iF}; e^{AA'}{}_{iI}; \tau, \ldots)$ by $p_{AA'}{}^i{}_{BF} = i\delta I_B/\delta e^{AA'}{}_{iF}$ [Eq. (4.1.2)]. Similarly, the initial data $\tilde{\psi}^{A'}{}_{iI}$ must obey $S_A = 0$ at the initial surface. Conversely, if the boundary data $(e^{AA'}{}_{iF}, \psi^A{}_{iF}; e^{AA'}{}_{iI}, \tilde{\psi}^{A'}{}_{iI}; \tau, \ldots)$ are such that the constraints (4.1.6) at the final surface, and the corresponding equation at the initial surface hold, then the results of [Bao *et al.* 1985] show that the linearized gravitino field equations may be solved to give infilling classical linear-order fields $(\psi^A{}_\mu, \tilde{\psi}^{A'}{}_\mu)$ up to the local supersymmetries (3.2.13–15). In general, arbitrary boundary data will not admit a classical solution. Note that this was wrongly stated in Section VI of [D'Eath 1984], where it was stated that all boundary data admit classical solutions. The supersymmetry constraints provide perhaps an unfamiliar situation in mathematical physics. Usually, any bosonic system of spins 0,1,2 admits classical solutions (possibly complex) for an elliptic boundary-value problem. The same holds when one adds in spin-1/2 fields $(\chi^A_{(I)}, \tilde{\chi}^{A'}_{(I)})$, and treats $\chi^A_{(I)}$ as final boundary data, and $\tilde{\chi}^{A'}_{(I)}$ as initial boundary data. The difficulty only arises when one adds in a gravitino field, when local supersymmetry is present, with the associated constraints. Eq. (4.1.6) assumes that the auxiliary fields [Wess & Bagger 1992] have been set to zero. In section 4.3 it will be seen how one can solve the supersymmetry constraints more generally with the help of auxiliary fields.

At the next, quadratic, order in fermions, one evaluates the energy–momentum tensor $T_{\mu\nu}$, which contains (in general) non-Hermitian terms of the form $\psi(\partial\tilde{\psi})$ and $\tilde{\psi}(\partial\psi)$. Solving the linearized Einstein field equations for the quadratic-in-fermions correction to the bosonic $e^{AA'}{}_{\mu}$, with boundary data for the full problem given by $e^{AA'}{}_{iI}$, $e^{AA'}{}_{iF}$, one will obtain a non-Hermitian (in general) correction to $e^{AA'}{}_{\mu}$. This process can be iterated to all orders, giving a classical solution $e^{AA'}{}_{\mu}$ which depends on the boundary data $\tilde{\psi}^{A'}{}_{iI}$, $\psi^A{}_{iF}$ through an infinite series. Note that the number of powers of $\tilde{\psi}^{A'}{}_{iI}$ in the classical $e^{AA'}{}_{\mu}$ equals the number of powers of $\psi^A{}_{iF}$. There are analogous infinite series for $\psi^A{}_{\mu}$, $\tilde{\psi}^{A'}{}_{\mu}$ in powers of the boundary data $\tilde{\psi}^{A'}{}_{iI}$, $\psi^A{}_{iF}$. These higher-order expressions in fermions follow for $(e^{AA'}{}_{\mu}, \psi^A{}_{\mu}, \tilde{\psi}^{A'}{}_{\mu})$, provided that one imposes the full constraint (4.1.6) on $\psi^A{}_{iF}$, and correspondingly for $\tilde{\psi}^{A'}{}_{iI}$, where $p_{AA'}{}^i$ is expressed as a power series in fermions. (In fact, the boundary data $\psi^A{}_{iF}$ and $\tilde{\psi}^{A'}{}_{iI}$, just like the four-dimensional versions $\psi^A{}_{\mu}$ and $\tilde{\psi}^{A'}{}_{\mu}$, must also be given by a power series in odd Grassmann quantities, in order to satisfy the exact constraint (4.1.6) and its conjugate. The lowest-order terms in $\psi^A{}_{iF}$ obey the constraint (4.1.6) with $p_{AA'}{}^i{}_{BF} = i\delta I_B/\delta e^{AA'}{}_{iF}$. Solving the supersymmetry constraints with the help of supergravity auxiliary fields, as in section 4.3 below, simplifies this picture.) The full action $I_{\text{class}}(e^{AA'}{}_{iF}, \psi^A{}_{iF}; e^{AA'}{}_{iI}, \tilde{\psi}^{A'}{}_{iI}; \tau, \ldots)$ will correspondingly be an infinite series in the fermionic boundary data (with equal powers of $\tilde{\psi}$ and ψ). The bosonic part of I, obtained by setting the fermionic fields to zero, equals $I_B(e^{AA'}{}_{iF}; e^{AA'}{}_{iI}; \tau, \ldots)$ above.

Note that the classical boundary-value problem becomes simpler if one takes (say) the initial data to have zero fermionic part: $\tilde{\psi}^{A'}{}_{iI} = 0$. In this case, $\psi^A{}_{\mu}$ in the interior is linear in the boundary data $\psi^A{}_{iF}$, up to the local symmetries (3.2.13–15), while $\tilde{\psi}^{A'}{}_{\mu}$ will be zero, again up to the local symmetries. In the simplest gauge $\tilde{\psi}^{A'}{}_{\mu} = 0$, the quadratic-order $\tilde{\psi}\psi$ contribution to the gravitational solution is zero, and the classical gravitational solution is given only by the bosonic solution for $e^{AA'}{}_{\mu}$, up to the local symmetries. Similarly, the classical action reduces to the bosonic action $I_B(e^{AA'}{}_{iF}; e^{AA'}{}_{iI}; \tau, \ldots)$ (see below). These properties will be used in section 4.3 below, where the constraints are solved with the help of auxiliary fields, in studying a quantum state which may correspond to the classical boundary-value problem specified by data $(e^{AA'}{}_{iF}, \psi^A{}_{iF}; e^{AA'}{}_{iI}, 0; \tau, \ldots)$.

In the case above with $\tilde{\psi}^{A'}{}_{iI} = 0$, the necessary and sufficient condition for there to exist a classical solution $(e^{AA'}{}_{\mu}, \psi^A{}_{\mu}, \tilde{\psi}^{A'}{}_{\mu})$ is that $\psi^A{}_{iF}$ should obey the constraint (4.1.6) with $p_{AA'}{}^{i}{}_{BF} = i\delta I_B/\delta e^{AA'}{}_{iF}$. Here the property above that $I_{\text{class}} = I_B$ is being used, i.e. that the full action is the bosonic action $I_B(e^{AA'}{}_{iF}; e^{AA'}{}_{iI}; \tau, \ldots)$. In the gauge with $\tilde{\psi}^{A'}{}_{\mu} = 0$, this is obvious. Now consider an infinitesimal primed supersymmetry transformation (3.2.15) applied to $\tilde{\psi}^{A'}{}_{\mu}$. If the parameter $\tilde{\epsilon}^{A'}$ were non-zero at the final boundary, then $e^{AA'}{}_{i}$ would be changed by $-i\kappa\tilde{\epsilon}^{A'}\psi^A{}_{i}$, which is not permitted, since $e^{AA'}{}_{i}$ is fixed at the final boundary. Hence one must take $\tilde{\epsilon}^{A'} = 0$ at the final boundary. Hence, from Eq. (3.5.15) with $\tilde{\epsilon}^{A'} = \frac{1}{2}\kappa\tilde{\psi}^{A'}{}_{0}\delta t$, one has $\delta I = 0$ under infinitesimal local supersymmetry. The same conclusion will hold for a finite supersymmetry transformation. Hence $I = I_B$ in any gauge, for these boundary conditions with $\tilde{\psi}^{A'}{}_{iI} = 0$.

Solutions $\psi^A{}_{i}$ on the final surface, obeying the $\tilde{S}_{A'} = 0$ classical constraint (4.1.6) with $p_{AA'}{}^{i}{}_{BF} = i\delta I_B/\delta e^{AA'}{}_{iF}$ [Eq. (4.1.2)], will be denoted by ${}^{s}\psi^A{}_{iF}$. Similarly, solutions $\tilde{\psi}^{A'}{}_{i}$ of $S_A = 0$ on the initial surface are denoted by ${}^{s}\tilde{\psi}^{A'}{}_{iI}$. Given one solution ${}^{s}\psi^A{}_{iF}$ (say), in the case $\tilde{\psi}^{A'}{}_{\mu} = 0$ discussed above, one can obtain other solutions by making a local supersymmetry transformation (written in three-dimensional form)

$$\delta e^{AA'}{}_{i} = 0, \tag{4.1.7}$$

$$\delta\psi^A{}_{i} = 2\kappa^{-1}\,{}^{3s}D_i\epsilon^A - i\kappa\epsilon^B D^{AA'}{}_{ij}\, p_{BA'}{}^{j}{}_{B}. \tag{4.1.8}$$

This gauge transformation is parametrized by two degrees of freedom ϵ^A per space point.

In the case of flat boundary surfaces in flat Euclidean four-space, the general solution of Eq. (4.1.6) for ${}^{s}\psi^A{}_{iF}$ is [D'Eath 1986a,b]

$$\psi^{\text{lin}A}{}_{i} = \psi^{\text{TT}A}{}_{i} + \alpha^A{}_{,i}, \tag{4.1.9}$$

where

$$e_{AA'i}\psi^{\text{TT}A}{}_{i} = 0, \; \psi^{\text{TT}A}{}_{i,i} = 0. \tag{4.1.10}$$

The part of the solution given by α^A corresponds to the two gauge degrees of freedom per space point of Eq. (4.1.8). The part $\psi^{\text{TT}A}{}_{i}$ describes the freely specifiable gravitino wave data, also with two degrees of freedom per space point. These data can be written using a spatial Fourier decomposition. One might call $\psi^{\mathrm{TT}A}{}_i$ 'physical gravitino data' in the flat case.

One would like to generalise these results to the curved case studied here, with boundaries at finite Euclidean times τ. The result of [D'Eath 1995] shows that solutions ${}^s\psi^A{}_i$ of the supersymmetry constraint (4.1.6) on a single boundary exist and are in one-to-one correspondence with the flat-space solutions $\psi^{\mathrm{lin}A}{}_i$, provided that the deviation $(e^{AA'}{}_i - \sigma^{AA'}{}_i)$ of the spatial tetrad from its flat-space value $\sigma^{AA'}{}_i$, its first derivatives and the field $p_{AA'}{}^i$ are sufficiently small. One sets up a Hölder function space [Ladyzhenskaya & Uraltseva 1968, Choquet-Bruhat & Deser 1973, D'Eath 1976] containing $(e^{AA'}{}_i - \sigma^{AA'}{}_i)$ and $p_{AA'}{}^i$, with suitable Hölder differentiability and fall-off which allow for the mass contribution present in $(e^{AA'}{}_i - \sigma^{AA'}{}_i) = O(r^{-1})$, $p_{AA'}{}^i = O(r^{-3/2})$ (see section 4.4 for a discussion of this mass-induced fall-off). The implicit function theorem for Banach spaces [Dieudonné 1960] gives the existence result. One would also like to combine this result with that of [Reula 1987] on the existence of weak-field solutions of the Riemannian Einstein equations, given suitable weak-field data $e^{AA'}{}_{iI}$ and $e^{AA'}{}_{iF}$. Because [Reula 1987] used weighted Sobolev spaces [Choquet-Bruhat & Christodoulou 1981], while [D'Eath 1995] uses weighted Hölder spaces, it is hard to match up the asymptotic fall-off. If one could match these mathematical arguments together, one would have the first two steps in the hierarchy in powers of fermions [Bao *et al.* 1985] in a proof that suitable supergravity boundary data, whose gravitational part $e^{AA'}{}_{iF}$ and $e^{AA'}{}_{iI}$ is weak-field, admit a unique classical solution (up to gauge).

One might ask what might happen with solutions ${}^s\psi^A{}_i$ of Eq. (4.1.6), in the event that the result of [D'Eath 1995] is no longer applicable for strong gravitational fields. One expects in the strong-field case that there might be bound states of Eq. (4.1.6). Recall from Eqs. (3.3.19),(3.3.22) (but not using the weighted variable $\chi^A{}_{BB'}$ of Eq. (3.3.19)), that the general $\psi^A{}_i$ can be written in terms of

$$\begin{aligned}\psi^A{}_{BB'} &= e_{BB'}{}^i\psi^A{}_i \\ &= \gamma^A{}_{BB'} + \frac{2}{3}(n^{AA'}\tilde{\delta}_{A'}n_{BB'} + n_B{}^{A'}\tilde{\delta}_{A'}n^A{}_{B'}) + \epsilon^A{}_B\tilde{\delta}_{B'},\end{aligned} \tag{4.1.11}$$

where

$$\gamma_{ABC} = \gamma_{ABB'}n_C{}^{B'} = \gamma_{(ABC)} \tag{4.1.12}$$

is totally symmetric, giving the spin-3/2 part of the data. A bound state occurs when there is an asymptotically flat solution to Eq. (4.1.6) of the form (4.1.11) given by $\tilde{\delta}_{A'}$ with $\gamma_{ABC} = 0$. The constraint then reduces to the spatial Dirac equation with the four-dimensional connection, on the final surface. A bound state will give a solution to the constraint, but is not part of the one-to-one correspondence with flat solutions, which involves the spin-3/2 field γ_{ABC}.

In specifying boundary data for the supergravity boundary-value problem without the help of auxiliary fields (section 4.3), one must, as above, take solutions ${}^s\psi^A{}_i$ and ${}^s\tilde{\psi}^{A'}{}_i$ of the classical supersymmetry constraints. This is analogous to working with (local) physical gravitino data in the flat case.

The Euclidean path integral (4.1.1), corresponding to boundary data $(e^{AA'}{}_{iF}, {}^s\psi^A{}_{iF}; e^{AA'}{}_{iI}, {}^s\tilde{\psi}^{A'}{}_{iI}; \tau, \ldots)$, will have a semi-classical expansion of the form

$$K(e_F, {}^s\psi_F; e_I, {}^s\tilde{\psi}_I; \tau, \ldots) \sim (A + \hbar A_1 + \hbar^2 A_2 + \cdots)\exp(-I_{\text{class}}/\hbar), \quad (4.1.13)$$

where $I_{\text{class}}(e_F, {}^s\psi_F; e_I, {}^s\tilde{\psi}_I; \tau, \ldots)$ is the action of the classical solution of the supergravity field equations, which, as described above, involves all powers of ${}^s\psi^A{}_{iF}$ and ${}^s\tilde{\psi}^{A'}{}_{iI}$. Similarly the one-loop, two-loop,... factors $A, A_1, \ldots$ are functionals of the boundary data. The semi-classical expansion of the path integral has been written in Eq. (4.1.13) for boundary data giving classical solutions, involving ${}^s\psi^A{}_{iF}$ and ${}^s\tilde{\psi}^{A'}{}_{iI}$. One can of course consider the path integral in the case that the fermionic boundary data do not give a solution of the classical supersymmetry constraints. Because of the general form of the path integral, one expects that this depends smoothly on the boundary data. But it is hard to estimate the form of the path integral (4.1.1) in the non-classical case, unless one takes suitable auxiliary fields in the boundary data (see section 4.3).

Since the amplitude K obeys the quantum constraints (3.5.10), the functions $I_{\text{class}}, A, A_1, \ldots$ will also obey constraint equations resulting from the substitution of Eq. (4.1.13) into Eqs. (3.5.10), and setting the coefficient of each power of $\hbar$ to zero. The leading terms in each equation as $\hbar \to 0$ simply give back the classical constraint equations (3.2.20) and (3.2.49) at the final surface, where one uses the variational derivatives

$$\frac{\delta I_{\text{class}}}{\delta e^{AA'}{}_i} = -i(p_{AA'}{}^i - \tfrac{1}{2}\epsilon^{ijk}\psi_{Aj}\tilde{\psi}_{A'k}), \quad (4.1.14)$$

$$\frac{\delta I_{\text{class}}}{\delta \psi^A{}_i} = i\epsilon^{ijk}\tilde{\psi}^{A'}{}_j e_{AA'k}, \quad (4.1.15)$$

of I_{class} with respect to the final data $e^{AA'}{}_i, \psi^A{}_i$, which are found as in Eq. (3.5.13). Corresponding equations hold at the initial surface. It is understood that, as in Eq. (4.1.13), the fermionic boundary data are of the form ${}^s\psi^A{}_{iF}, {}^s\tilde{\psi}^{A'}{}_{iI}$, from which (e.g.) the final $\tilde{\psi}^{A'}{}_i$ will also be of the form ${}^s\tilde{\psi}^{A'}{}_{iF}$, since the classical constraints are preserved under time evolution.

The time evolution of

$$I_{\text{class}}(e^{AA'}{}_{iF}, {}^s\psi^A{}_{iF}; e^{AA'}{}_{iI}, {}^s\tilde{\psi}^{A'}{}_{iI}; \tau, \ldots)$$

is given by

$$\frac{\partial I_{\text{class}}}{\partial \tau} = M, \tag{4.1.16}$$

where

$$M(e^{AA'}{}_{iF}, {}^{s}\psi^{A}{}_{iF}; e^{AA'}{}_{iI}, {}^{s}\tilde{\psi}^{A'}{}_{iI}; \tau, \dots)$$

is the mass (including fermionic contributions) of the four-dimensional classical solution spacetime. As in Eq. (4.1.5), the four-dimensional mass will in general differ from the three-dimensional mass on the initial and final surfaces – see section 4.4.

The leading constraint in the semi-classical expansion of the quantum constraint $\bar{S}^{A'}\Psi = 0$ is the classical constraint (4.1.6) with $\psi^{A}{}_{i} = {}^{s}\psi^{A}{}_{iF}$, and $p_{AA'}{}^{i} = i\delta I_{\text{class}}/\delta e^{AA'}{}_{i}$. (Note that the $\psi\tilde{\psi}$ term on the right-hand side of Eq. (4.1.14) gives zero when substituted into the term $-\frac{1}{2}\kappa^{2}\psi^{A}{}_{i}\delta I_{\text{class}}/\delta e^{AA'}{}_{i}$ of the leading constraint.) In full, the $\bar{S}_{A'}K = 0$ constraint, as given by Eqs. (3.4.6–8), shows that under the primed supersymmetry transformation

$$\delta e^{AA'}{}_{i} = -i\kappa\tilde{\epsilon}^{A'}\psi^{A}{}_{i}, \ \delta\psi^{A}{}_{i} = 0, \tag{4.1.17}$$

at the final surface, the variations of $I_{\text{class}}, A, A_1, \dots$ are

$$\delta I_{\text{class}} = 2i\kappa^{-1}\int d^{3}x\epsilon^{ijk}e_{AA'i}({}^{3s}D_{j}\psi^{A}{}_{k})\tilde{\epsilon}^{A'}, \tag{4.1.18}$$

$$\delta A = \delta A_1 = \delta A_2 = \dots = 0. \tag{4.1.19}$$

Thus I_{class} changes according to Eqs. (3.5.14,15), while the one- and higher-loop terms $A, A_1, A_2, \dots$ are exactly invariant under a primed supersymmetry transformation applied to the final data $e^{AA'}{}_{iF}, \psi^{A}{}_{iF}$. This invariance property will be used in the following section 4.2 to constrain the boundary divergences or counterterms in the loop expansion (4.1.13).

The S_A quantum constraint can also be studied for the case of the semi-classical expansion (4.1.13) of the quantum amplitude

$$K(e^{AA'}{}_{iF}, {}^{s}\psi^{A}{}_{iF}; e^{AA'}{}_{iI}, {}^{s}\tilde{\psi}^{A'}{}_{iI}; \tau, \dots).$$

The leading term in powers of $\hbar$ gives the classical equation $S_A = 0$, which is satisfied because (as described above) $\tilde{\psi}^{A'}{}_{iF}$ is of the form ${}^{s}\tilde{\psi}^{A'}{}_{iF}$. The next term gives a first-order equation for A :

$$\begin{aligned} &{}^{3s}D_{i}\Big(\frac{\delta A}{\delta\psi^{A}{}_{i}}\Big) + \tfrac{1}{2}i\kappa^{2}\tilde{\psi}^{A'}{}_{i}\frac{\delta A}{\delta e^{AA'}{}_{i}} \\ &+ \tfrac{1}{2}i\kappa^{2}D^{BA'}{}_{ji}(p_{AA'}{}^{i} - \tfrac{1}{2}\epsilon^{ikl}\psi_{Ak}\tilde{\psi}_{A'l})\frac{\delta A}{\delta\psi^{B}{}_{j}} \\ &+ \tfrac{1}{2}i\kappa^{2}\frac{\delta}{\delta e^{AA'}{}_{i}}\left(-iD^{BA'}{}_{ji}\frac{\delta I_{\text{class}}}{\delta\psi^{B}_{j}}\right)A = 0, \end{aligned} \tag{4.1.20}$$

where Eqs. (4.1.14,15) have been used. This formally describes how A changes under an unprimed supersymmetry transformation (as generated by the S_A term in the Hamiltonian)

$$\delta e^{AA'}{}_i = -i\kappa\epsilon^A \tilde{\psi}^{A'}{}_i, \tag{4.1.21}$$

$$\delta\psi^A{}_i = 2\kappa^{-1}\,{}^{3s}D_i\epsilon^A - i\kappa\epsilon^B D^{AA'}{}_{ij} p_{BA'}{}^j + \tfrac{1}{2} i\kappa\epsilon^B \epsilon^{jkl} D^{AA'}{}_{ij}\psi_{Bk}\tilde{\psi}_{A'l}, \tag{4.1.22}$$

applied to the final data in the classical solution. This equation is to be regarded at present only as formal, since it involves a second functional derivative of I_{class} at a point x, or equivalently the derivative of $-iD^{BA'}{}_{ji}\delta I_{\text{class}}/\delta\psi^B{}_j = \tilde{\psi}^{A'}{}_i(x)$ at the final surface produced by varying $e^{AA'}{}_i(x)$ there while holding $\psi^A{}_i(x)$ fixed at the initial surface and fixing the initial data (but see below). Equations analogous to Eq. (4.1.20) hold for the supersymmetry transformation of $A_1, A_2, \ldots$.

Note that the final $\tilde{\psi}^{A'}{}_i(x) = -iD^{BA'}{}_{ji}\delta I_{\text{class}}/\delta\psi^B{}_j$ is of the form ${}^s\tilde{\psi}^{A'}{}_i(x)$. Further, the functional derivative $\delta/\delta e^{AA'}{}_i$ in the final term of Eq. (4.1.20) involves comparing ${}^s\tilde{\psi}^{A'}{}_{iF}(x)$ for different final $e^{AA'}{}_{iF}(x)$. Hence the final term of Eq. (4.1.20) can be computed by working solely within the space of I_{class} and A evaluated as functionals of boundary data of the form $(e^{AA'}{}_{iF}, {}^s\psi^A{}_{iF}; e^{AA'}{}_{iI}, {}^s\tilde{\psi}^{A'}{}_{iI}; \tau, \ldots)$. A similar comment applies to the first three terms in Eq. (4.1.20); they involve the variation of A under unprimed supersymmetry, which preserves the property that $\psi^A{}_i = {}^s\psi^A{}_i$ and $\tilde{\psi}^{A'}{}_i = {}^s\tilde{\psi}^{A'}{}_i$. Hence Eq. (4.1.20) refers only to fermionic fields of the form ${}^s\psi^A{}_i$, ${}^s\tilde{\psi}^{A'}{}_i$ as arguments of A and I_{class}. One need not consider (e.g.) the variation of A in the direction of those $\psi^A{}_i$ which do not obey the classical constraints.

In fact the one-loop result of the following section 4.2 (with our trivial topology) shows that A is a finite functional of the boundary data, and hence that $\delta(-iD^{BA'}{}_{ji}\delta I_{\text{class}}/\delta\psi^B_j)/\delta e^{AA'}{}_i$ in Eq. (4.1.20) is a finite quantity. Alternatively [D'Eath & Wulf 1995] one can note that ${}^s\tilde{\psi}^{A'}{}_{iF} = -iD^{BA'}{}_{ji}\delta I_{\text{class}}/\delta\psi^B_j$ on the final surface is given by the convolution of a propagator, depending on $e^{AA'}{}_{iI}$ and $e^{AA'}{}_{iF}$, with the initial ${}^s\tilde{\psi}^{A'}{}_{iI}$. This will give a finite expression for the second derivatives, even though the derivatives $\delta/\delta e^{AA'}{}_i$ and $\delta/\delta\psi^A{}_i$ act at the same spatial point. One can build up a series for the one-loop prefactor A near flat space (or more precisely, for technical reasons, near flat Euclidean four-space with large three-sphere boundaries), in powers of $\tilde{\psi}^{A'}{}_{iI}$, $\psi^A{}_{iF}$, $(e^{AA'}{}_{iI} - \sigma^{AA'}{}_i)$ and $(e^{AA'}{}_{iF} - \sigma^{AA'}{}_i)$, where $\sigma^{AA'}{}_i$ is the flat-space spatial tetrad, using Eqs. (4.1.17)–(4.1.22) and their analogues at the initial surface.

As mentioned above, there will be equations analogous to Eq. (4.1.20) at higher loop order, giving (e.g.) the unprimed supersymmetry variation of the two-loop term A_1 in terms of A and I_{class}. Again, these equations will involve second functional derivative terms at the same space point. But these should again lead to a finite supersymmetry variation of A_1 since (section 4.2) $N = 1$ supergravity is also finite at two loops, having no on-shell two-loop counterterms, even in the presence of boundaries. This will also be interesting to check directly, through evaluation of the second-derivative terms. More generally, a study of the finiteness of $N = 1$ supergravity, with the present local boundary conditions, could be made by considering the form of the second-derivative terms in the loop expansion of the quantum $S_A K = 0$ constraint, subject also to $\bar{S}_{A'} K = 0$ as given by Eqs. (4.1.17,19).

4.2 Two-loop finiteness of supergravity with boundaries

The one-loop term A in Eq. (4.1.13) can be found from the super-determinant of the matrix of second variations of the action I about the classical path [van Nieuwenhuizen 1981]. More precisely, let us write the action of a path close to the classical path $({e_{\text{class}}}^{AA'}{}_{\mu}, {\psi_{\text{class}}}^{A}{}_{\mu}, {\tilde{\psi}_{\text{class}}}{}^{A'}{}_{\mu})$ joining initial to final data as

$$\begin{aligned} I[e, \psi, \tilde{\psi}] = {} & I[{e_{\text{class}}}^{AA'}{}_{\mu}, {\psi_{\text{class}}}^{A}{}_{\mu}, {\tilde{\psi}_{\text{class}}}{}^{A'}{}_{\mu}] \\ & + I_2[{e_{\text{class}}}^{AA'}{}_{\mu}, {\psi_{\text{class}}}^{A}{}_{\mu}, {\tilde{\psi}_{\text{class}}}{}^{A'}{}_{\mu}; f^{AA'}{}_{\mu}, \phi^{A}{}_{\mu}, \tilde{\phi}^{A'}{}_{\mu}] \\ & + \text{higher-order terms}. \end{aligned} \tag{4.2.1}$$

Here $I[e_{\text{class}}, \psi_{\text{class}}, {\tilde{\psi}_{\text{class}}}{}^{A'}{}_{\mu}]$ will be written as I_{class}, and

$$\begin{aligned} e^{AA'}{}_{\mu} &= {e_{\text{class}}}^{AA'}{}_{\mu} + f^{AA'}{}_{\mu}, \\ \psi^{A}{}_{\mu} &= {\psi_{\text{class}}}^{A}{}_{\mu} + \phi^{A}{}_{\mu}, \tilde{\psi}^{A'}{}_{\mu} = {\tilde{\psi}_{\text{class}}}{}^{A'}{}_{\mu} + \tilde{\phi}^{A'}{}_{\mu}. \end{aligned} \tag{4.2.2}$$

with I_2 being quadratic in the fluctuations $f^{AA'}{}_{\mu}, \phi^{A}{}_{\mu}, \tilde{\phi}^{A'}{}_{\mu}$. In supergravity, I_2 splits into a sum of a bosonic and fermionic part

$$I_2 = I_{2B} + I_{2F}, \tag{4.2.3}$$

where I_{2B} does not involve the fermionic perturbations $(\phi^{A}{}_{\mu}, \tilde{\phi}^{A'}{}_{\mu})$ whereas I_{2F} is bilinear in $(\phi^{A}{}_{\mu}, \tilde{\phi}^{A'}{}_{\mu})$. The one-loop semi-classical expansion for the amplitude is based on the Euclidean path integral (4.1.1), with I replaced by $I_{\text{class}} + I_{2B} + I_{2F}$, giving

$$\begin{aligned} K_{\text{one-loop}} = {} & \exp(-I_{\text{class}}/\hbar) \left[\int \mathscr{D} f \exp(-I_{2B}/\hbar) \right] \\ & \times \left[\int \mathscr{D}\phi \mathscr{D}\tilde{\phi} \exp(-I_{2F}/\hbar) \right]. \end{aligned} \tag{4.2.4}$$

As in the full theory, so in this linearized version one should strictly include gauge-fixing, Faddeev–Popov and auxiliary field terms in Eq. (4.2.4). The general features of the one-loop approximation can, however, be seen from the schematic version of Eq. (4.2.4).

The quadratic actions I_{2B} and I_{2F} can be written symbolically in the form

$$I_{2B}[\text{background}; f^{AA'}{}_{\mu}] = \tfrac{1}{2}\int (fBf)(g_{\text{class}})^{1/2} d^4x, \tag{4.2.5}$$

$$I_{2F}[\text{background}; \phi^{A}{}_{\mu}, \tilde{\phi}^{A'}{}_{\mu}] = \tfrac{1}{2}\int (\phi C\tilde{\phi})(g_{\text{class}})^{1/2} d^4x, \tag{4.2.6}$$

where B and C are operators which depend on the background tetrad $e_{\text{class}}{}^{AA'}{}_{\mu}$ (and possibly also on the background fermion field ($\psi_{\text{class}}{}^{A}{}_{\mu}$, $\tilde{\psi}^{A'}_{\text{class}\,\mu}$)). Here B is the second-order wave operator [Misner *et al.* 1973] which governs linear perturbations of the vacuum gravitational field around a background. And C is the first-order Rarita–Schwinger operator [see below – Eq. (4.2.10)], which is obtained by varying the supergravity action (3.2.12) with respect to the fermion fields. The boundary conditions are that $f^{AA'}{}_{i} = 0$ at the boundaries (Dirichlet), and that $\phi^{A}{}_{i} = 0$ on the final surface, while $\tilde{\phi}^{A'}{}_{i} = 0$ on the initial surface.

Now one makes an eigenfunction decomposition of the fluctuations [Hawking 1979]. Consider first the bosonic fluctuations. Let λ_n be the eigenvalues and f_n the eigenfunction of B, with $f^{AA'}{}_{i(n)} = 0$ on the boundaries. The f_n can be taken to be normalized by

$$\int f_{(m)} \cdot f_{(n)}(g_{\text{class}})^{1/2} d^4x = \delta_{mn}. \tag{4.2.7}$$

An arbitrary fluctuation $f^{AA'}{}_{\mu}$ with $f^{AA'}{}_{i} = 0$ on the boundaries can be expanded as

$$f = \Sigma_n y_n f_n. \tag{4.2.8a}$$

One can also write the measure on the space of all fluctuations f as

$$\mathscr{D}f = \Pi_n \mu dy_n, \tag{4.2.8b}$$

where the normalization factor μ has dimensions of mass. The bosonic one-loop factor in Eq. (4.2.4) can then be expressed as

$$\begin{aligned}\int \mathscr{D}f \exp(-I_{2B}/\hbar) &= \Pi_n \int \mu dy_n \exp(-\tfrac{1}{2}\lambda_n {y_n}^2) \\ &= \Pi_n (2\pi\mu^2 {\lambda_n}^{-1})^{1/2} \\ &= \left[\det\left(\tfrac{1}{2}\pi^{-1}\mu^{-2}B\right)\right]^{-1/2}.\end{aligned} \tag{4.2.9}$$

In the fermionic case, one solves the coupled equation based on the Rarita–Schwinger operator for the eigenfunctions:

$$\begin{aligned}\tfrac{1}{2}\epsilon^{\mu\nu\rho\sigma}e_{AA'\nu}D_\rho\phi_{(n)}{}^{A}{}_{\sigma} &= \lambda_n\tilde{\phi}_{(n)A'}{}^{\mu},\\ \tfrac{1}{2}\epsilon^{\mu\nu\rho\sigma}e_{AA'\nu}D_\rho\tilde{\phi}_{(n)}{}^{A'}{}_{\sigma} &= \lambda_n\phi_{(n)A}{}^{\mu},\end{aligned} \tag{4.2.10}$$

subject to the boundary conditions that $\phi_{(n)}{}^{A}{}_{i} = 0$ on the final surface and $\tilde{\phi}^{A'}{}_{i(n)} = 0$ on the initial surface. Because of the Berezin rules (Eqs. (3.3.8–9)), the fermionic one-loop factor in Eq. (4.2.4) is given by

$$\int \mathscr{D}\phi\mathscr{D}\tilde{\phi}\,\exp(-I_{2F}/\hbar) = [\det(\tfrac{1}{2}\pi^{-1}\mu^{-2}C^{\dagger}C)]^{1/2}. \tag{4.2.11}$$

The ratio of the determinants $\det(\frac{1}{2}\pi^{-1}\mu^{-2}B)/\det(\frac{1}{2}\pi^{-1}\mu^{-2}C^{\dagger}C)$, appearing in Eqs. (4.2.9,11) is the super-determinant referred to in the first sentence of this section 4.2. Analogous formal expressions can be found when one includes gauge-fixing and Faddeev–Popov terms for the gravitational fluctuations in pure gravity [Hawking 1979] and these terms plus auxiliary fields for supergravity.

A determinant such as that in Eq. (4.2.9) diverges badly, since it is formally given by

$$\det\left(\tfrac{1}{2}\pi^{-1}\mu^{-2}B\right) = \Pi_n\left(\tfrac{1}{2}\pi^{-1}\mu^{-2}\lambda_n\right), \tag{4.2.12}$$

and $\lambda_n \to \infty$ as $n \to \infty$. A suitable regularization [Hawking 1977], known as zeta-function regularization, involves studying

$$\zeta(s) = \sum_{n=0}^{\infty}(\lambda_n)^{-s}, \tag{4.2.13}$$

where the λ_n are the eigenvalues for a second-order operator such as B (or $C^{\dagger}C$). Then the series in Eq. (4.2.13) will converge for $\mathrm{Re}(s) > 2$, and its analytic continuation defines a meromorphic function of s, analytic at $s = 0$. Formally one has

$$\log(\det B) = -\zeta'(0). \tag{4.2.14}$$

The expression (4.2.12), following from the path integral, suggests the definition

$$\det\left(\tfrac{1}{2}\pi^{-1}\mu^{-2}B\right) = (2\pi\mu^2)^{\zeta(0)/2}\exp\left[\tfrac{1}{2}\zeta'(0)\right] \tag{4.2.15}$$

as the zeta-function regularization of Eq. (4.2.12).

Because of the appearance of the unknown mass parameter μ on the right-hand side of Eq. (4.2.15), the one-loop amplitude will be ambiguous unless $\zeta(0) = 0$. When one includes the fermionic fluctuations with second-order operator $C^{\dagger}C$, it is the difference in the $\zeta(0)$ terms for the two operators which should vanish. (When gauge-fixing, Faddeev–Popov terms and auxiliary fields are included, there is a similar condition.) Now [Gibbons 1979], $\zeta(0)$ for a given operator is known to have the form

$$4\pi\zeta(0) = \int_m B_4\, g^{1/2} d^4x + \int_{\partial m} C_4\, h^{1/2} d^3x, \tag{4.2.16}$$

where $B_4(x)$ is a local invariant formed from the

$$\text{(four-dimensional curvature)}^2$$

and $C_4(x)$ is a local invariant formed from the intrinsic and extrinsic curvature of the boundary, with the property that both the volume term and the surface term are dimensionless. In a Feynman-diagram type of expansion around the background, the total $\zeta(0)$ measures the one-loop divergence of the quantum amplitude. The general algorithms needed to evaluate $\zeta(0)$ for $N = 1$ supergravity with boundaries have been set up [Moss & Poletti 1994], but the explicit calculation has not yet been carried out. Nevertheless, we shall see that it is possible to show using local supersymmetry that the total $\zeta(0)$ is zero for our boundary conditions.

One works 'on shell', i.e. around a classical solution, as above. The possible divergent terms at one loop (i.e. possible contributions to $\zeta(0)$ in Eq. (4.2.16)) will be proportional to a basis of *counterterms*, i.e. dimensionless integrals contributing to Eq. (4.2.16). The possible one-loop counterterms on shell are

$$I_0 = \int d^4x g^{1/2} R_{\lambda\mu\nu\rho} R^{\lambda\mu\nu\rho}, \tag{4.2.17}$$

and

$$I_1 = \int d^3x h^{1/2}\, {}^3R_{ij} K^{ij}, \tag{4.2.18}$$

$$I_2 = \int d^3x h^{1/2} (\mathrm{tr} K)^3, \tag{4.2.19}$$

$$I_3 = \int d^3x h^{1/2} K_{ij} K^{ij} (\mathrm{tr} K), \tag{4.2.20}$$

$$I_4 = \int d^3x h^{1/2} K^i{}_j K^j{}_k K^k{}_i. \tag{4.2.21}$$

Here $R_{\lambda\mu\nu\rho}$ is the four-dimensional Riemann tensor, ${}^3R_{ij}$ is the three-dimensional Ricci tensor, and K_{ij} is the second fundamental form, all taken without torsion. Note that there is no need for a term of the form $\int d^3x (h^{1/2})^3 RK$, since by the scalar constraint equation at bosonic order,Eqs. (2.3.57,59), this can be replaced by a linear combination of Eqs. (4.2.19–20).

Now the volume term I_0 is associated with the topological Euler number of the classical spacetime, which in the presence of boundaries is given by a suitable linear combination of $I_0, \ldots, I_4$ [Eguchi *et al.* 1980]. In the case of simple topology $[0,1] \times \mathbb{R}^3$ considered in the present Hamiltonian context, the Euler number is zero, so that there is no volume divergence [van Nieuwenhuizen 1981, Duff 1982]. In a general classical spacetime with more complicated topology, the Euler number and associated one-loop counterterm will be non-zero.

In quantizing Einstein gravity [Moss & Poletti 1994, Esposito *et al.* 1995], one finds contributions of the form $I_1, \ldots, I_4$ above to $\zeta(0)$, showing that quantum Einstein gravity is already divergent at one loop, in the presence of boundaries. (Non-zero volume counterterms in quantum Einstein gravity first appear at two-loop order [Goroff & Sagnotti 1985].) However, in supergravity the local supersymmetry invariance (4.1.17–19) under a primed supersymmetry transformation at the final surface, and the analogous invariance at the initial surface, imply that the pure surface integrals (4.2.18–21) are forbidden as the bosonic parts of one-loop counterterms [D'Eath 1986a,b]. Nor can one have one-loop counterterms without a bosonic part. Only the Euler-number counterterm will be present, when the topology of the spacetime is non-trivial.

To see this, first note that there are general restrictions on the form of local surface counterterms in supergravity [D'Eath 1986a,b]. In addition to obeying the invariance (4.1.17),(4.1.19), they should be given by surface integrals of quantities invariant under local Lorentz and spatial coordinate transformations. The integrands should be formed only from the basic Hamiltonian data $e^{AA'}{}_i, p_{AA'}{}^i, \psi^A{}_i, \tilde{\psi}^{A'}{}_i$ and their spatial derivatives. If this is not so, and the integrand depends also (say) on $\tilde{\psi}^{A'}{}_0$, then the counterterm will not be invariant under local primed supersymmetry transformations parametrized by $\tilde{\epsilon}^{A'}$ with the property that $\tilde{\epsilon}^{A'} = 0$ but $\partial_0 \tilde{\epsilon}^{A'} \neq 0$ at the surface. Similarly, if the integrand depends on $e^{AA'}{}_0$, then the counterterm will not be invariant under a class of spacetime coordinate transformations which reduce to the identity at the boundary.

Surface counterterms must involve equal numbers of powers of $\psi^A{}_i$ (or its spatial derivatives) and of $\tilde{\psi}^{A'}{}_i$ (or derivatives), as in the case of volume counterterms. This is because counterterms can in principle be evaluated by studying perturbations of flat spacetime, and obey the rigid chiral invariance

$$\begin{gathered} e^{AA'}{}_\mu \to e^{AA'}{}_\mu, \\ \psi^A{}_\mu \to e^{i\theta} \psi^A{}_\mu, \; \tilde{\psi}^{A'}{}_\mu \to e^{-i\theta} \tilde{\psi}^{A'}{}_\mu, \end{gathered} \tag{4.2.22}$$

where $\theta =$ const. Parity also provides a considerable restriction on the form of surface counterterms. In the case of simple topology $[0,1] \times \mathbb{R}^3$, the amplitude $K(e^{AA'}{}_{iF}, \psi^A{}_{iF}; e^{AA'}{}_{iI}, \tilde{\psi}^{A'}{}_{iI}; \tau, \ldots)$ will be invariant under the parity transformation P which maps

$$\begin{aligned} e^{AA'}{}_i(x) &\to e^{AA'}{}_i(-x), \\ \psi^A(x) &\to -i\sqrt{2}n^A{}_{A'}(-x)\tilde{\psi}^{A'}{}_i(-x), \\ \tilde{\psi}^{A'}{}_i(x) &\to i\sqrt{2}n_A{}^{A'}(-x)\psi^A{}_i(-x), \end{aligned} \tag{4.2.23}$$

applied to the boundary data. Thus

$$K(e_F, \psi_F; e_I, \tilde{\psi}_I; \tau, \ldots) = K(Pe_F, P\psi_F; Pe_I, P\tilde{\psi}_I; \tau, \ldots). \tag{4.2.24}$$

Note that parity changes the chirality of the fermionic boundary data, and that it is necessary in establishing (4.2.24) to check that the action I is mapped correctly to the action appropriate to fixing fermionic fields of the opposite chirality at the boundaries. Because of this chirality change, Eq. (4.2.24) alone is not sufficient to constrain the form of possible surface counterterms, and it must be combined with the property

$$\bar{K}(e_F, \psi_F; e_I, \tilde{\psi}_I; \tau, \ldots) = K(\bar{e}_F, \bar{\psi}_F; \bar{e}_I, \bar{\tilde{\psi}}_I; \tau, \ldots), \tag{4.2.25}$$

where a bar denotes Hermitian conjugation.

A great deal can be learnt simply by considering possible on-shell surface counterterms (i.e. counterterms evaluated at a classical solution) at lowest order in weak fields, and applying the condition that they be invariant under rigid primed supersymmetry transformations with $\tilde{\epsilon}^{A'} =$ constant at the final surface, and similarly under rigid unprimed transformations at the initial surface. One consequence of local supersymmetry is useful in this approach: since the primed variation of $\tilde{\psi}^{A'}{}_i$ involves a spatial derivative of $\tilde{\epsilon}^{A'}$ through

$$\delta\tilde{\psi}^{A'}{}_i = 2\kappa^{-1}\partial_i\tilde{\epsilon}^{A'} + \cdots, \tag{4.2.26}$$

$\tilde{\psi}^{A'}{}_i$ can only appear at lowest order in on-shell surface counterterms at the final surface through the field strength

$$\tilde{F}^{\text{lin}\,A'}{}_{ij} = \partial_i\tilde{\psi}^{\text{lin}\,A'}{}_j - \partial_j\tilde{\psi}^{\text{lin}\,A'}{}_i, \tag{4.2.27}$$

where $\tilde{\psi}^{\text{lin}\,A'}{}_i$ is the linearized $\tilde{\psi}^{A'}{}_i$. Similarly the way in which $\psi^A{}_i$ appears at the initial surface is simplified.

At linear order, the bosonic ingredients for surface counterterms are constructed from the intrinsic metric $h_{ij} = \delta_{ij} + h^{\text{lin}}_{ij}$ and second fundamental form $K_{ij} = K^{\text{lin}}_{ij}$ on a surface $x^0 =$ const. These are subject on-shell to the linearized classical constraint equations

$${}^3R^{\text{lin}} = 0, \tag{4.2.28}$$

$$K^{\text{lin}}_{ij,j} - K^{\text{lin}}_{jj,i} = 0. \tag{4.2.29}$$

The general solution of Eqs. (4.2.28,29) is

$$h_{ij}^{\rm lin} = h_{ij}^{\rm TT} + \xi_{i,j} + \xi_{j,i}\,, \tag{4.2.30}$$

$$K_{ij}^{\rm lin} = K_{ij}^{\rm TT} + \lambda_{,ij}\,, \tag{4.2.31}$$

where $h_{ij}^{\rm TT}$ and $K_{ij}^{\rm TT}$ are symmetric transverse-traceless tensors obeying

$$h_{ij,j}^{\rm TT} = 0,\ h_{ii}^{\rm TT} = 0, \tag{4.2.32}$$

and similarly for $K_{ij}^{\rm TT}$. The initial data for the wave modes of the gravitational field are given by $h_{ij}^{\rm TT}$ and $K_{ij}^{\rm TT}$, while ξ_i and λ describe gauge degrees of freedom, corresponding to infinitesimal coordinate transformations within the hypersurface and infinitesimal deformations in the normal direction.

Spacetime curvature is carried in the initial data through the electric and magnetic curvature tensors

$$E_{ij} = {}^3R_{ij}^{\rm lin} \quad = -\tfrac{1}{2} h_{ij,kk}^{\rm TT}, \tag{4.2.33}$$

$$B_{ij} = \epsilon_{ikl} K_{jk,l}^{\rm lin} = \epsilon_{ikl} K_{jk,l}^{\rm TT}. \tag{4.2.34}$$

Both E_{ij} and B_{ij} are symmetric, transverse and traceless. Knowledge of E_{ij} and B_{ij} is equivalent to knowledge of the linearized Weyl tensor through $(E_{ij} + iB_{ij}) \leftrightarrow C_{ABCD}$, $(E_{ij} - iB_{ij}) \leftrightarrow \tilde{C}_{A'B'C'D'}$ in the spinor language of [Penrose & Rindler 1984], and hence on shell to knowledge of the four-dimensional linearized Riemann tensor. Useful potentials for $(E_{ij} \pm iB_{ij})$ are provided by $(L_{ij} \pm iK_{ij}^{\rm TT})$, where

$$L_{ij} = \tfrac{1}{2} \epsilon_{ikl} h_{jk,l}^{\rm TT} \tag{4.2.35}$$

is symmetric, transverse and traceless. These potentials obey

$$\epsilon_{ikl} \partial_l (L_{jk} \pm iK_{jk}^{\rm TT}) = E_{ij} \pm iB_{ij}. \tag{4.2.36}$$

The fermionic surface data $\psi^A{}_i$ and $\tilde{\psi}^{A'}{}_i$ obey the supersymmetry constraint equations, given at linear order by

$$\epsilon_{ijk} e_{AA'i} \psi^{{\rm lin}A}{}_{j,k} = 0, \tag{4.2.37}$$

and similarly for $\tilde{\psi}^{{\rm lin}\,A'}{}_i$, where the $e_{AA'i}$ take their flat values. As in Eqs. (4.1.9,10), this has the general solution

$$\psi^{{\rm lin}A}{}_i = \psi^{{\rm TT}A}{}_i + \alpha^A{}_{,i}\,, \tag{4.2.38}$$

where $\psi^{{\rm TT}A}{}_i$ obeys the conditions

$$e_{AA'i} \psi^{{\rm TT}A}{}_i = 0,\ \psi^{{\rm TT}A}{}_{i,i} = 0. \tag{4.2.39}$$

As described in section 4.1, the physical degrees of freedom of the weak spin-3/2 field are carried by $\psi^{{\rm TT}A}{}_i$ and $\tilde{\psi}^{{\rm lin}\,TTA'}{}_i$, while α^A and $\tilde{\alpha}^{A'}$ represent the gauge freedom, corresponding to infinitesimal local supersymmetry transformations. Knowledge of the linearized spatial field strengths $F^{\mathrm{lin}A}{}_{ij}$ and $\tilde{F}^{\mathrm{lin}\,A'}{}_{ij}$ is equivalent to knowledge of the on-shell linearized four-dimensional field strengths through $\tilde{F}^{\mathrm{lin}\,A'}{}_{ij} \leftrightarrow F_{ABC}$, $\tilde{F}^{\mathrm{lin}A'}_{ij} \leftrightarrow \tilde{F}_{A'B'C'}$, in the spinor language of [Penrose & Rindler 1984].

The transformation laws of these field strengths can be studied under rigid supersymmetry, in which the parameters ϵ^A and $\tilde{\epsilon}^{A'}$ defining an infinitesimal supersymmetry transformation are constant in space. It will be sufficient to consider rigid supersymmetry in order to rule out one- and two-loop boundary counterterms in supergravity. The relevant transformations can be found with the help of the language of superspace (in the sense of supersymmetry) [Wess & Bagger 1992]. One uses an odd Grassmann coordinate θ^D which is a fermionic counterpart of the bosonic spacetime coordinates x^μ. The field strengths are contained in the on-shell chiral multiplet $W_{ABC} \sim \kappa F_{ABC} + \theta^D C_{ABCD}$ (in a chiral representation) and antichiral multiplet $\tilde{W}_{A'B'C'}$. For a rigid primed supersymmetry transformation, one obtains

$$\delta(E_{ij} - iB_{ij}) = 0, \tag{4.2.40}$$

$$\delta(E_{ij} + iB_{ij}) = -i\kappa\tilde{\epsilon}^{A'}(F^{\mathrm{lin}A}{}_{ik,j}e_{AA'k} + F^{\mathrm{lin}A}{}_{ki,k}e_{AA'j}) + (i \leftrightarrow j), \tag{4.2.41}$$

$$\delta(F^{\mathrm{lin}A}{}_{ij}) = 0, \tag{4.2.42}$$

$$\delta(\tilde{F}^{\mathrm{lin}\,A'}{}_{ij}) = -2i\kappa^{-1}\epsilon_{jkl}n_A{}^{A'}e^{AB'}{}_l(E_{ki} - iB_{ki})\tilde{\epsilon}_{B'} - (i \leftrightarrow j). \tag{4.2.43}$$

Since $K^{\mathrm{lin}}_{ij} = K^{\mathrm{TT}}_{ij} + \lambda_{,ij}$ will also be an ingredient in surface counterterms, we need its on-shell transformation laws, summarized by

$$\delta(L_{ij} - iK^{\mathrm{TT}}_{ij}) = 0, \tag{4.2.44}$$

$$\delta(L_{ij} + iK^{\mathrm{TT}}_{ij}) = i\kappa\epsilon_{kil}e_{AA'l}\tilde{\epsilon}^{A'}\psi^{\mathrm{TT}A}{}_{j,k} + (i \leftrightarrow j), \tag{4.2.45}$$

$$\delta\lambda = i\kappa n_{AA'}\tilde{\epsilon}^{A'}\alpha^A. \tag{4.2.46}$$

The bare $\psi^{\mathrm{lin}A}{}_i$ can also appear in surface counterterms, but is of course invariant under primed supersymmetry. As described above, $\tilde{\psi}^{\mathrm{lin}\,A'}{}_i$ must appear through $\tilde{F}^{\mathrm{lin}\,A'}{}_{ij}$ at the final surface, so that the law (4.2.43) is sufficient. All on-shell surface counterterms at lowest order in weak fields must be formed from the quantities described here and their spatial derivatives; thus sufficient information has now been provided for computing their variations under rigid supersymmetry.

As an example, one can show that the simplest possible one-loop on-shell surface counterterm $I_1 = \int d^3x h^{1/2}\, {}^3R_{ij}K^{ij}$ of Eq. (4.2.18) for pure quantum gravity cannot be extended to a surface counterterm in supergravity, invariant under primed supersymmetry transformations at the final surface. Working at linear order on shell, one considers

$$\begin{aligned} I &= \int d^3x E_{ij} K_{ij}^{\text{lin}} \\ &= \int d^3x E_{ij} K_{ij}^{\text{TT}} \\ &= -\frac{1}{4} i \int d^3x (E_{ij} + iB_{ij})(L_{ij} + iK_{ij}^{\text{TT}}) \\ &\quad + \frac{1}{4} i \int d^3x (E_{ij} - iB_{ij})(L_{ij} - iK_{ij}^{\text{TT}}). \end{aligned} \tag{4.2.47}$$

Under a rigid primed supersymmetry transformation, Eqs. (4.2.40,41,44,45) show that

$$\delta I = \kappa \int d^3x \epsilon_{kil} e_{AA'l} \tilde{\epsilon}^{A'} F^{\text{lin}A}{}_{kj} (E_{ij} + iB_{ij}), \tag{4.2.48}$$

which is non-zero for generic choice of initial data for the gravitational and spin-3/2 field. Suppose that one could find a fermionic partner J such that $\delta(I + J) = 0$ under rigid primed supersymmetry. As described above, J must involve equal numbers of powers of $\psi^{\text{lin}A}{}_i$ and $\tilde{\psi}^{\text{lin}A'}{}_i$, and must involve $\tilde{\psi}^{\text{lin}A'}{}_i$ through $\tilde{F}^{\text{lin}A'}{}_{ij}$. Hence J must be symbolically of the form

$$J = \int \tilde{F} \partial \psi. \tag{4.2.49}$$

But then δJ is produced by the variation $\delta \tilde{F}^{\text{lin}A'}{}_{ij}$, which by Eq. (4.2.43) involves $(E_{ij} - iB_{ij})$. However, δI in Eq. (4.2.48) involves the combination $(E_{ij} + iB_{ij})$, which can be chosen independently of $(E_{ij} - iB_{ij})$ in arranging initial-value data for the gravitational field. Hence δJ can never balance δI, so that I_1 cannot be extended to a surface counterterm, with our boundary conditions, in supergravity.

Now I_1 is the only bosonic surface expression at one-loop order which is quadratic in weak fields. There are no further possibilities starting at quadratic order in weak fields, since the term $\int d^3x h^{1/2}\, {}^3R \operatorname{tr} K$ can be re-expressed in terms of higher-order quantities, using the classical constraint $\mathscr{H}_\perp = 0$ [Eq. (3.2.38)]. Any other allowed term must start at cubic order with a bosonic part of the form $\int d^3x KKK$ with some index coupling. The $\int d^3x \tilde{\epsilon} (\partial \psi) KK$ variation of $\int d^3x KKK$ under a transformation with $\tilde{\epsilon} = \text{const.}$ can only be compensated by a term $\int d^3x \psi \tilde{F} K$. But such a term will include in its variation a part $\int d^3x \psi \tilde{\epsilon} ({}^3R) K$. This cannot be compensated by the $\int d^3x \tilde{\epsilon} (\partial \psi) KK$ variation of $\int d^3x KKK$. Hence the on-shell surface counterterms I_2, I_3, I_4 of Eqs. (4.2.19–21) are forbidden by supersymmetry.

Similarly, one can show that it is impossible to construct one-loop surface terms in supergravity which have no bosonic part at lowest order in weak fields. Thus there are no on-shell purely surface counterterms at one loop in supergravity. Only the topological Euler invariant is allowed, which in the spacetimes of topology $[0,1] \times \mathbb{R}^3$ considered here gives zero.

At one loop there are only four possible independent surface terms. But at the next order, the possible terms proliferate, and one finds twenty-two possible on-shell two-loop bosonic surface terms [D'Eath 1986b]. A case-by-case analysis shows that all of these are forbidden by $\tilde{\epsilon}^{A'}$ supersymmetry invariance (and that one cannot make allowed two-loop surface counterterms with no bosonic part). Since there are no allowed volume counterterms at two loops in $N = 1$ supergravity, invariant under local supersymmetry [van Nieuwenhuizen 1981, Duff 1982], this shows that, with our local boundary conditions, supergravity is finite up to and including two loops in the presence of boundaries. Given the absence of purely surface counterterms at one and two loops, one might ask whether this absence of surface counterterms is a general property, holding at all loop orders. A general proof is lacking – it is not at present clear whether a superspace argument (in the sense of supersymmetry) can be devised. Any superspace treatment would be complicated by the fact that the anticommutator of two supersymmetry transformations applied to a surface quantity involves its normal derivatives, given on shell by a Hamiltonian evolution equation.

At three-loop order there is a possible volume counterterm [Deser *et al.* 1977b, van Nieuwenhuizen 1981, Duff 1982], including a bosonic volume term proportional to

$$J = \int d^4x g^{1/2} C_{ABCD} C^{ABCD} \tilde{C}_{A'B'C'D'} \tilde{C}^{A'B'C'D'}. \tag{4.2.50}$$

Here C_{ABCD} and $\tilde{C}_{A'B'C'D'}$ give the totally symmetric Weyl spinor, which on shell describes the curvature tensor through [Penrose & Rindler 1984]

$$R_{AA'BB'CC'DD'} = C_{ABCD}\epsilon_{A'B'}\epsilon_{C'D'} + \tilde{C}_{A'B'C'D'}\epsilon_{AB}\epsilon_{CD}. \tag{4.2.51}$$

Such a counterterm will vary under supersymmetry by surface terms at the initial and final boundaries, whereas the complete three-loop amplitude (and hence the counterterm) should by Eqs. (4.1.17,19) be invariant under a primed local supersymmetry transformation. Can suitable surface terms be found which, when added to the volume term starting with (4.2.50), yield an invariant under $\tilde{\epsilon}^{A'}$ at the final surface and ϵ^A at the initial surface? As discussed in [D'Eath 1986b], a suitable surface term can be found when one restricts attention in the weak-field limit to rigid supersymmetry transformations with $\tilde{\epsilon}^{A'} =$ const. It is not clear whether this work can be carried through to local supersymmetry transformations parametrized by $\tilde{\epsilon}^{A'}(x)$. That is, it is not clear whether one can form invariant counterterms for $N = 1$ supergravity at three loops and beyond in the presence of boundaries. Even if this is possible, one must ask whether the coefficients of the counterterms are zero or not. In later sections of this book, we shall see how it may be possible to sidestep Feynman-diagram methods, which lead to an immensely complicated (and never attempted) calculation to check whether the counterterm (4.2.50) is present. Rather, it may be possible using canonical methods to understand the divergences, or lack of them, directly.

4.3 Auxiliary fields

As described in section 4.1, general boundary data on the initial and final surfaces do not admit a classical solution of the Euclidean boundary-value problem. This difficulty is remedied by the use of the auxiliary fields of supergravity [Wess & Bagger 1992]. In the present context, these are independent complex scalar fields $M(x), \tilde{M}(x)$ and a complex Lorentz vector field $b_{AA'}(x)$. They contribute to the Euclidean action as (compare Eq. (3.5.2) and following)

$$I = I_2 + I_{3/2} + I_{2B} + I_{3/2\,B} + I_\infty + \int d^4x\, e[-6M\tilde{M} + 12b^{AA'}\bar{b}_{AA'}]. \quad (4.3.1)$$

Thus the classical field equations for the auxiliary fields are $M = \tilde{M} = 0$, $b_{AA'} = 0$. It will be helpful in the boundary-value problem to consider non-classical data for the auxiliary fields, in which $b_{AA'} \neq 0$.

Consider the classical boundary-value problem for supergravity, with final data $(e^{AA'}{}_i, \psi^A{}_i, b_{AA'})_F$, initial data $(e^{AA'}{}_i, \tilde{\psi}^{A'}{}_i = 0, b_{AA'})_I$ and Euclidean time separation τ at spatial infinity. Without auxiliary fields, the Rarita–Schwinger classical equation for the gravitino can be written as

$$-F_{CB'}{}^{C}{}_{A'B} + \epsilon_{B'A'}F_{AC'B}{}^{C'A} = 0. \quad (4.3.2)$$

Here $F_{AA'BB'C}$ is the spinor version of $F_{\mu\nu C}$, where

$$F_{\mu\nu C} = D_\mu \psi_{\nu C} - D_\nu \psi_{\mu C}. \quad (4.3.3)$$

With auxiliary fields, Eq. (4.3.2) is replaced by a linear combination of $\bar{\mathscr{D}}_{A'}G_{BB'}$ and $\epsilon_{A'B'}\mathscr{D}_B R$, where these quantities are described in chapter 17 of [Wess & Bagger 1992]. One has

$$\bar{\mathscr{D}}_{A'}G_{BB'} = \frac{1}{4}F^C{}_{A'CB'A} + \frac{1}{12}\epsilon_{A'B'}F_B{}^{C'C}{}_{C'C} + \frac{i}{6}\tilde{\psi}_{BB'A'}M$$
$$- \frac{i}{12}(\psi_{CB'}{}^{C}b_{BA'} + \psi_{CA'}{}^{C}b_{BB'} - \psi^C{}_{A'B}b_{CB'}), \quad (4.3.4)$$

$$\mathscr{D}_B R = -\frac{1}{3}(\sigma^{\mu\nu})_B{}^C F_{\mu\nu C} + \frac{i}{6}e_{BA'}{}^{\mu}\tilde{\psi}^{A'}_{\mu} M + \frac{i}{6}\psi_{AA'B}b^{AA'}. \quad (4.3.5)$$

Here

$$(\sigma^{\mu\nu})_B{}^C = \frac{1}{4}(e_{BB'}{}^{\mu}e^{CB'\nu} - e_{BB'}{}^{\nu}e^{CB'\mu}). \tag{4.3.6}$$

The $\tilde{S}_{A'} = 0$ constraint is obtained by contracting the Rarita–Schwinger equation (with auxiliary fields) with $n^{BB'}$.

Since $b_{AA'}$ has four independent complex components, and $\tilde{S}_{A'}$ has only two, then in general, for prescribed $(e^{AA'}{}_i, \psi^A{}_i)_F$ on the final surface, the constraint $\tilde{S}_{A'} = 0$ can be solved for $b_{AA'F}(x)$ (non-uniquely). In order that this should make sense with $\psi^A{}_{iF}(x)$ being a Grassmann variable, one has to assume that all the $\psi^A{}_{iF}(x)$ at different spatial points are proportional to the same odd Grassmann number. Then the solution $b_{AA'F}(x)$ will be bosonic, with no dependence on any odd parameter. For general choices of $\psi^A{}_{iF}(x)$, $b_{AA'}(x)$ will be non-singular. There will be special choices of $\psi^A{}_i$ for which $b_{AA'}$ cannot be computed at isolated spatial points x. This feature may be regarded as analogous to a feature discussed in section 2.7: in gravity with a Λ-term, there are three-geometries which, if deformed in one direction, allow a Euclidean classical solution, while if they are deformed in the other direction, they allow a Lorentzian classical solution. The $\psi^A{}_{iF}(x)$ for which $b_{AA'}$ cannot be found at isolated spatial points x may also correspond to boundaries in function space.

One can also solve the entire Rarita–Schwinger equation by choice of $b_{AA'}$ (as above) and of $\dot{\psi}^A_i$, where $\dot{}$ denotes $\partial/\partial\tau$. This defines $b_{AA'}$,$\psi^A{}_i$ (and $\psi^A{}_\mu$, by choice of gauge for $\psi^A{}_0$) in the infilling spacetime region, and hence on the initial three-surface. Hence $b_{AA'I}$ in the initial data above cannot be chosen freely, but is determined by $(e^{AA'}{}_i, \psi^A{}_i, b_{AA'})_F$ and $e^{AA'}{}_{iI}$: one uses up half the freedom of choosing $b_{AA'}$ in solving $S_A = 0$ on the initial surface.

Now consider the quantum constraints $\bar{S}_{A'}\Psi = 0$ and $S_A\Psi = 0$. Take the quantum state

$$\Psi = \exp(-I_B/\hbar), \tag{4.3.7}$$

where I_B is the classical gravitational action to go from the initial three-geometry $e^{AA'}{}_{iI}(x)$ to the final three-geometry $e^{AA'}{}_{iF}(x)$ in Euclidean time τ at spatial infinity (and with a spatial translation X^i between the surfaces, following Eqs. (3.5.6),(3.5.8)). Then Ψ obeys the quantum constraints at the final surface. First, it obeys

$$\bar{S}_{A'}\Psi = 0 \tag{4.3.8}$$

because the only derivative term in $\bar{S}_{A'}$ [Eq. (3.4.6)] involves $\hbar\delta/\delta e^{AA'}{}_i$, which yields $-(\delta I_B/\delta e^{AA'}{}_i)\exp(-I_B/\hbar) = -p_{AA'}{}^i{}_F\exp(-I_B/\hbar)$. Hence $\bar{S}_{A'}\Psi$ reduces to the classical constraint multiplied by $\exp(-I_B/\hbar)$, which vanishes for the above choice of auxiliary field $b_{AA'F}(x)$. Also, Ψ obeys

$$S_A\Psi = 0 \tag{4.3.9}$$

with $\bar{M} = 0$, because each term in S_A involves a derivative $\delta/\delta\psi^B{}_i$, which annihilates the bosonic wave function

$$\Psi = \exp(-I_B/\hbar)$$

of Eq. (4.3.7). By construction, $\Psi = \exp(-I_B/\hbar)$ obeys the Lorentz constraints

$$J_{AB}\Psi = 0, \bar{J}_{A'B'}\Psi = 0. \tag{4.3.10}$$

It also obeys the Hamiltonian constraint

$$_2\mathscr{H}_{AA'}\Psi = 0, \tag{4.3.11}$$

where, following Eq. (3.4.15), $_2\mathscr{H}_{AA'}(x)$ is defined in the presence of auxiliary fields by

$$[S_A(x), \bar{S}_{A'}(x')]_+ = -\tfrac{1}{2}\hbar\kappa^2\, {}_2\mathscr{H}_{AA'}(x)\delta(x, x'). \tag{4.3.12}$$

Hence $\Psi = \exp(-I_B/\hbar)$ obeys all the quantum constraint equations, provided $b_{AA'F}(x)$ is chosen, for each $\psi^A{}_{iF}(x)$ and for each three-geometry $e^{AA'}{}_{iF}(x)$, such that the classical equation $\tilde{S}_{A'} = 0$ holds. The above results suggest that $\exp(-I_B/\hbar)$ is a candidate for the quantum amplitude or propagator for the above boundary conditions with $\tilde{\psi}^{A'}{}_{iI} = 0$. In order to verify that $\exp(-I_B/\hbar)$ is the propagator, one should check that this wave function obeys the heat equation with respect to τ (the Riemannian version of the Schrödinger equation) with the correct initial data as $\tau \to 0_+$. These ideas will be discussed in the following section 4.4, where, as an example, the full propagator K will be studied in relation to the heat equation and boundary conditions.

4.4 Time evolution of the quantum amplitude

Consider, as in sections 4.1–3, the boundary-value amplitude denoted by $K(e^{AA'}{}_{iF}, \psi^A{}_{iF}, b_{AA'F}; e^{AA'}{}_{iI}, 0, b_{AA'I}; \tau, \ldots)$ for a Euclidean time separation $\tau > 0$. Its dependence on the variable τ is given by the heat equation

$$\hbar\frac{\partial K}{\partial\tau} + \hat{H}_E K = 0, \tag{4.4.1}$$

for $\tau > 0$. This is the quantum version of the classical equation (4.1.5). Here $\hat{H}_E$ is the quantum version of the classical Euclidean Hamiltonian

$$\begin{aligned} H_E = i\int d^3x(N\, {}_2\mathscr{H}_\perp + N_i\, {}_2\mathscr{H}^i + \psi^A{}_0 S_A + \tilde{S}_{A'}\tilde{\psi}^{A'}{}_0 \\ - \Omega_{AB}J^{AB} - \tilde{\Omega}_{A'B'}\tilde{J}^{A'B'}) \\ + i\lim_{r_0\to\infty}\int dS_i\left[\frac{N}{2\kappa^2}(h_{ij,j} - h_{jj,i}) + N_j\pi^{ij}\right], \end{aligned} \tag{4.4.2}$$

plus other terms of the form (3.2.51), (3.2.52) if relevant; here r_0 is the radius of a large sphere. This classical Hamiltonian is in general non-zero, even though the classical constraints ${}_2\mathscr{H}_\perp = 0$, ${}_2\mathscr{H}^i = 0$, $S_A = 0$, $\tilde{S}_{A'} = 0$, $J^{AB} = 0$, ; $\tilde{J}^{A'B'} = 0$ hold, because of the boundary contributions, which yield the mass M of the four-dimensional classical spacetime [D'Eath 1981]. The extra term $N_j\pi^{ij}$ in the final integrand of Eq. (4.4.2) provides a correction to the mass of the four-dimensional spacetime, showing that this mass may differ from the three-dimensional mass of the spacelike surface, given by Eq. (3.2.52) [DeWitt 1967, Regge & Teitelboim 1974]. The extra term arises as a boundary term from the bosonic part of Eq. (3.2.39). The Lagrange multipliers N, N_i are taken to be purely imaginary for a Riemannian spacetime. The heat equation (4.4.1) of course corresponds to the Schrödinger equation in the Lorentzian régime [Feynman & Hibbs 1965]. Note that the Schrödinger equation gives a unitary evolution, since the corresponding Lorentzian Hamiltonian operator $\hat{H} = -i\hat{H}_E$ is Hermitian with real N and N_i, with $\tilde{\psi}^{A'}{}_0 = \bar{\psi}_0^{A'}$, and with $\tilde{\Omega}_{A'B'} = \bar{\Omega}_{A'B'}$. Just as the classical H_E does not vanish, so the quantum $\hat{H}_E$ does not in general annihilate the propagator K, because of the boundary terms in $\hat{H}_E$, even though the constraint operators ${}_2\mathscr{H}_\perp$ etc. annihilate K.

For $\tau < 0$, one defines [Feynman & Hibbs 1965]

$$K = 0, \; (\tau < 0), \tag{4.4.3}$$

with the initial conditions given formally [D'Eath 1981] by

$$K \to \delta(e^{AA'}{}_{iF}, e^{AA'}{}_{iI}) \tag{4.4.4}$$

as $\tau \to 0_+$. These initial conditions are better described by making an asymptotic expansion of K as $\tau \to 0_+$ (see below), so that there is no ambiguity about a version of the delta functional in Eq. (4.4.4) which is invariant under spatial coordinate transformations.

Returning to the heat equation (4.4.1), consider the boundary terms in $\hat{H}_E$ at spatial infinity, acting on K. Suppose, for example, that the boundary data $e^{AA'}{}_{iF}$, $e^{AA'}{}_{iI}$ had been chosen to have rapid fall-off to flatness as the radius $r_0 \to \infty$, corresponding to zero three-dimensional mass. As remarked by [Wheeler 1968], a surface with zero mass in the Lorentzian Schwarzschild geometry

$$ds^2 = -\left(1 - \frac{2M}{r}\right) dt^2 + \left(1 - \frac{2M}{r}\right)^{-1} dr^2 + r^2(d\theta^2 + \sin^2\theta d\phi^2) \tag{4.4.5}$$

will be embedded near infinity with $t \sim t_0 + (8Mr)^{1/2}$. Hence the second fundamental form K_{ij} only falls off as $r^{-3/2}$, leading in particular to

$$\begin{aligned} \pi_{rr} &\sim -(2M)^{1/2}/8\pi r^{3/2}, \\ \mathrm{tr}\pi &\sim -3(2M)^{1/2}/16\pi r^{3/2}. \end{aligned} \tag{4.4.6}$$

One chooses the large timelike cylinder joining the initial and final surfaces to lie at a constant Schwarzschild r value; thus it will not intersect the surfaces orthogonally [D'Eath 1981]. This corresponds to the fall-off

$$N = 1 + O(r^{-2}),$$
$$N_r \sim -(2M/r)^{1/2}, \tag{4.4.7}$$

as $r \to \infty$, with other components of N_i decreasing more rapidly in magnitude. It may then be verified that the boundary term in the Lorentzian action S contributes $-MT$. The same is true if one allows for a non-zero three-dimensional mass. In the Euclidean régime, one finds that the action I acquires a boundary contribution $M\tau$, and the classical Euclidean Hamiltonian H_E equals the boundary contribution $M(e^{AA'}{}_{iF}; e^{AA'}{}_{iI}; \tau, \ldots)$. Here one must assume that the mass M is negative, in order that the Riemannian Schwarzschild metric

$$ds^2 = \left(1 - \frac{2M}{r}\right) d\tau^2 + \left(1 - \frac{2M}{r}\right)^{-1} dr^2 + r^2(d\theta^2 + \sin^2\theta d\phi^2) \tag{4.4.8}$$

should have no coordinate singularity in $0 < r < \infty$, and hence no periodicity with period $8\pi M$ in imaginary time τ [Hawking 1979]. A real period in τ would prevent one from considering the Euclidean evolution problem for all $\tau > 0$.

Now the operator $\hat{H}_E K$ will include a part

$$i \lim_{r_0 \to \infty} \int dS_i (N/2\kappa^2)(-h_{jj,i} + h_{ij,j})K$$

which contributes the three-dimensional mass M_{3D} times K. The other boundary term in $\hat{H}_E K$ contributes

$$-i \lim_{r_0 \to \infty} \int dS_i N_j (\delta K / \delta h_{ij}).$$

But in the far-field asymptotic region, the amplitude is free-field (cf. [Ashtekar 1987b]), and $-i\delta K/\delta h_{ij}$ is asymptotically $\pi_{ij} K$, where π_{ij} is the classical momentum. Hence the quantum operator $\hat{H}_E$ acting on K describes multiplication by $M(e^{AA'}{}_{iF}; e^{AA'}{}_{iI}; \tau, \ldots)$, and Eq. (4.4.1) gives the time evolution equation

$$\hbar \frac{\partial K}{\partial \tau} + MK = 0, \tag{4.4.9}$$

for $\tau > 0$, in the case with ${}^s\tilde{\psi}^{A'}{}_{iI} = 0$.

Now consider the initial or boundary conditions (4.4.4). As described in [D'Eath 1981], for $\tau \to 0_+$ with $e^{AA'}{}_{iF}$ and $e^{AA'}{}_{iI}$ fixed, one has a spacetime in which the extrinsic curvature is much stronger than the intrinsic curvature on the initial surface, in the sense

$$(K^{ij}K_{ij})^2 \gg R^{ij}R_{ij}. \tag{4.4.10}$$

Thus variations of the four-metric $g_{\mu\nu}$ with time must dominate over spatial variations. Hence the geometry must be described by a Kasner-type metric, of the type originally considered by [Lifschitz & Khalatnikov 1963] in order to approximate the behaviour of inhomogeneous cosmological models near their initial singularity. The exact Riemannian Kasner metric [Kasner 1921] is

$$ds^2 = d\tau^2 + \tau^{2p_1}dx^2 + \tau^{2p_2}dy^2 + \tau^{2p_3}dz^2, \tag{4.4.11}$$

where the indices p_1, p_2 and p_3 obey

$$p_1 + p_2 + p_3 = 1, \tag{4.4.12}$$

$$p_1^2 + p_2^2 + p_3^2 = 1. \tag{4.4.13}$$

From this one can construct metrics with spatial dependence which are approximate solutions of the Einstein equations at early times:

$$ds^2 = d\tau^2 + (\tau^{2p_1}l_il_j + \tau^{2p_2}m_im_j + \tau^{2p_3}n_in_j)dx^idx^j. \tag{4.4.14}$$

Here p_1, p_2, and p_3 are now functions of the spatial coordinates x^i, again subject to Eqs. (4.4.12,13), while l_i, m_i and n_i are also functions of x^j such that the limit of the $\mathscr{H}_i$ constraint is satisfied [Lifschitz & Khalatnikov 1963, D'Eath 1981]. Each of $p_1, p_2, p_3, l_i, m_i, n_i$ is allowed to vary by a fractional amount of $O(1)$ over the local spatial length scale l in the sense $|\nabla p_1|/|p_1| \sim l^{-1}$.

The second fundamental form K_{ij} is of $O(\tau^{-1})$ in magnitude. In the limit of a flat initial surface the corresponding mass can be evaluated from K_{ij} alone [D'Eath 1981]. In this case, one can solve the vector constraint equation (4.2.15) by Eq. (4.2.17):

$$K_{ij} = K_{ij}^{\mathrm{TT}} + \lambda_{,ij}. \tag{4.4.15}$$

The scalar constraint equation (4.1.3) reads

$$K_{ij}^{\mathrm{TT}}K_{ij}^{\mathrm{TT}} + 2\lambda_{,ij}K_{ij}^{\mathrm{TT}} + \lambda_{,ij}\lambda_{,ij} - (\lambda_{,kk})^2 = 0, \tag{4.4.16}$$

and can be regarded as an equation giving λ and hence the constrained part of the initial data in terms of the freely specifiable part K_{ij}^{TT}.

The asymptotic behaviour (4.4.6) of π_{ij} or equivalently K_{ij} as $r \to \infty$ implies that λ should obey

$$\lambda \sim -(8Mr)^{1/2} \tag{4.4.17}$$

as $r \to \infty$. Hence M can be read off from the asymptotic form of λ as

$$M = \frac{1}{2\kappa^2}\int dS_j(\lambda_{,j}\lambda_{,ii} - \lambda_{,i}\lambda_{,ij}), \tag{4.4.18}$$

where the integral is taken over a large sphere. Using Eq. (4.4.16), one obtains

$$
\begin{aligned}
M &= \frac{1}{2\kappa^2} \int d^3x (\lambda_{,jj}\lambda_{,ii} - \lambda_{,ij}\lambda_{,ij}) \\
&= \frac{1}{2\kappa^2} \int d^3x K_{ij}^{TT} K_{ij}^{TT} .
\end{aligned}
\tag{4.4.19}
$$

In this example, the mass evidently represents the kinetic energy present in the wave modes of the gravitational field, as described by K_{ij}^{TT} on the initial surface. Since K_{ij} or K_{ij}^{TT} will be pure imaginary for a Riemannian spacetime, the mass M is negative. Since K_{ij} scales as τ^{-1}, one has, in the case that h_{ijI} is flat,

$$
M(h_{ijF}; h_{ijI}; \tau, \ldots) \sim -\tau^{-2} v(h_{ijF}; h_{ijI}; \ldots), \tag{4.4.20}
$$

as $\tau \to 0_+$ for some functional v of the end-geometries and of any additional variables such as the spatial translation coordinate in Eq. (3.2.51). In the more general case, one has Eq. (4.4.20) for arbitrary h_{ijI} and h_{ijF}, because of the scaling of K_{ij}. The equation (4.1.5):

$$
\frac{\partial I_{\text{class}}}{\partial \tau} = M \tag{4.4.21}
$$

then shows that asymptotically

$$
I_{\text{class}}(h_{ijF}; h_{ijI}; \tau, \ldots) \sim \tau^{-1} v(h_{ijF}; h_{ijI}; \ldots) \tag{4.4.22}
$$

as $\tau \to 0_+$. This is in accordance with a (somewhat weakened) version of the positive action conjecture [Hawking 1979]. The positive action theorem is usually phrased for four-dimensional geometries which are asymptotically flat in all spacetime directions, showing that the Euclidean action I_{class} is positive provided ${}^4R = 0$ [Schoen & Yau 1979]. In our case, we have the stronger assumption ${}^4R_{\mu\nu} = 0$ (the classical Einstein vacuum equations), but different boundary conditions – two hypersurfaces and three-dimensional asymptotic flatness. A thorough treatment of this version of the positive action theorem would be of considerable interest. Related to this, one may also conjecture that the mass $M(e^{AA'}{}_{iF}; e^{AA'}{}_{iI}; \tau, \ldots)$ is negative for $\tau > 0$, as described in the discussion of Eq. (4.4.8). This would be a Riemannian (or Euclidean) version of the positive-mass theorem of Lorentzian general relativity [Witten 1981, Reula 1982].

In considering the behaviour as $\tau \to 0_+$ of the propagator

$$
K(e^{AA'}{}_{iF}, \psi^A{}_{iF}, b_{AA'F}; e^{AA'}{}_{iI}, 0, b_{AA'I}; \tau, \ldots),
$$

note that the classical action

$$
I_{\text{class}}(h_{ijF}; h_{ijI}; \tau, \ldots) \sim \tau^{-1} v(h_{ijF}; h_{ijI}; \ldots)
$$

becomes very large, for fixed end-geometries h_{ijF}, h_{ijI}. As in [Feynman & Hibbs 1965] for quantum mechanics, this implies that K is approximated for $\tau \to 0_+$ by its one-loop contribution $K \sim A \exp(-I_{\text{class}}/\hbar)$; higher-loop terms become asymptotically zero as $\tau \to 0_+$ when compared with $A \exp(-I_{\text{class}}/\hbar)$. Note also (see section 4.2) that the one-loop prefactor A for supergravity contains no divergences, even allowing for the presence of the boundaries.

Now the one-loop prefactor A must tend to 1 as $\tau \to 0_+$. This can be seen either from the quantum supersymmetry constraints (4.1.17), (4.1.19) for A, or by considering the eigenvalues involved in the one-loop path integral. In the latter case, the essential features can be seen by studying a model problem [D'Eath 1981] in one space and one time dimension, allowing for rapid time dependence of the metric on a short time-scale of $O(T)$ as $T \to 0$, with spatial dependence on length-scales of $O(1)$. The essential features are illustrated by the problem of finding eigenvalues λ for

$$a\left(\frac{t}{T}, x\right) \frac{\partial^2 F}{\partial t^2} + b\left(\frac{t}{T}, x\right) \frac{\partial^2 F}{\partial x^2} + \lambda F = 0, \tag{4.4.23}$$

subject to Dirichlet conditions $F = 0$ on boundaries $x = \pm X$, $t = 0$, $t = T\tau(x)$, in the limit $T \to 0$. The variable position $t = T\tau(x)$ of the final boundary corresponds to the variable proper time-separation between the boundary surfaces, over a spatial length scale of $O(1)$. The rapid time-dependence of the coeffients a, b of the elliptic operator mimics that of the bosonic or fermionic perturbation operator of the full theory of supergravity. There is a further boundary at a large distance $|x| = X$, to ensure a well-posed elliptic boundary-value problem. It is assumed that a, b and τ tend to 1 rapidly at large $|x|$. Fixing attention on a given eigenvalue in the limit $T \to 0$, one takes a solution of the form

$$F \sim \text{Re}\left[\exp\left(\frac{i\kappa}{T}\right) G\left(\frac{t}{T}, x\right)\right], \tag{4.4.24}$$

where $\kappa = \kappa(x)$ is a slowly-varying function. Then G must obey

$$\frac{a}{T^2} G_{11} - \frac{\kappa'^2 b}{T^2} G + \lambda G \simeq 0, \tag{4.4.25}$$

where a subscript 1 denotes a derivative with respect to (t/T). Thus $\kappa(x)$ must be chosen such that $G = 0$ at $(t/T) = 0, \tau(x)$, giving an eigenvalue problem for the ordinary differential equation (4.4.25) at each x. At large $|x|$, F must approach a solution of the form

$$\text{Re}[\text{const. } \exp(ikx) \sin(n\pi t/T)]$$

for some integer $n = 1, 2, \ldots$. Here k (which will depend on T) must be bounded as $T \to 0$, by consideration of the conditions $F = 0$ at $x = \pm X$, since λ is chosen to be the mth eigenvalue (say) for given n, independently of T. Thus $\kappa'(x) \to 0$ as $|x| \to \infty$, and

$$\lambda \sim n^2\pi^2/T^2 \tag{4.4.26}$$

as $T \to 0$. Hence the eigenvalues approach the eigenvalues of the flat-space wave operator as $T \to 0$, so that the determinant of the operator tends to 1. A similar argument will hold in the supergravity case; again the functional determinants of the gravitational or gravitino perturbation operators will approach the determinants of flat-space operators, giving $A \to 1$ in the Kasner limit $\tau \to 0_+$.

Alternatively, one can consider the one-loop approximation

$$A \exp(-I_B/\hbar)$$

to the quantum amplitude K, with the boundary conditions

$$(e^{AA'}{}_{iF}, \psi^A{}_{iF}, b_{AA'F}; e^{AA'}{}_{iI}, 0, b_{AA'I}; \tau, \ldots).$$

The classical action I_B depends only on the bosonic variables

$$(e^{AA'}{}_{iF}; e^{AA'}{}_{iI}; \tau, \ldots),$$

and with these boundary conditions, the one-loop factor A also depends only on $(e^{AA'}{}_{iF}; e^{AA'}{}_{iI}; \tau, \ldots)$. Then [Eqs. (4.1.17),(4.1.19)] A obeys

$$\psi^A{}_i(x) \frac{\delta A}{\delta e^{AA'}{}_i(x)} = 0. \tag{4.4.27}$$

Since $\psi^A{}_i(x)$ can be chosen arbitrarily, one has

$$\frac{\delta A}{\delta e^{AA'}{}_i} = 0 \Rightarrow A = \text{constant} \tag{4.4.28}$$

as a functional of $e^{AA'}{}_{iF}(x)$. Since A is constructed symmetrically from $e^{AA'}{}_{iF}(x)$ and $e^{AA'}{}_{iI}$, A is also independent of $e^{AA'}{}_{iI}(x)$, and could only depend on τ. Hence A equals the ratio of flat-space determinants for initial and final flat three-geometries, with separation τ, giving $A = 1$.

Hence the asymptotic behaviour as $\tau \to 0_+$ of the propagator

$$K(e^{AA'}{}_{iF}, \psi^A{}_{iF}, b_{AA'F}; e^{AA'}{}_{iI}, 0, b_{AA'I}; \tau, \ldots)$$

is

$$K \sim \exp[-\tau^{-1} v(h_{ijF}; h_{ijI}; \ldots)/\hbar], \tag{4.4.29}$$

where v appears in Eqs. (4.4.20,22). The detailed form of the functional $v(h_{ijF}; h_{ijI}; \ldots)$ is not known, but v is given implicitly by Eq. (4.4.19) for the mass M in terms of the second fundamental form K_{ij} in the case of a flat initial surface. The expression (4.4.29) is somewhat analogous to the fundamental solution

$$F(t,x) \sim \frac{\text{const.}}{t^{1/2}} \exp(-\text{const. } x^2/t) \tag{4.4.30}$$

of the one-dimensional heat equation

$$\frac{\partial F}{\partial t} + \text{const. } \frac{\partial^2 F}{\partial x^2} = 0, \tag{4.4.31}$$

subject to initial conditions

$$F(t,x) \to \delta(x) \text{ as } t \to 0_+. \tag{4.4.32}$$

In the supergravity case, the functional $v(h_{ijF}; h_{ijI}; \ldots)$ is analogous to x^2, the squared distance between the inital and final points.

The main results of this section are the initial condition or asymptotic data (4.4.29) for the amplitude K as $\tau \to 0_+$, together with the exact property $A = 1$ for the given boundary conditions. K obeys the initial condition (4.4.4) because it is given by the semi-classical expression (4.4.29) as $\tau \to 0_+$. Indeed, one should be able to verify by a classical calculation that K, given asymptotically by Eq. (4.4.29), obeys the initial condition (4.4.4). One expects that $v(h_{ijF}; h_{ijI}; \ldots)$ would be zero if h_{ijF} and h_{ijI} were taken to be the same three-metric, and that v would be small for slightly different three-geometries, but that v would be of $O(1)$ for three-geometries which differed considerably, so leading to the delta functional of Eq. (4.4.4).

In section 8.1 the results of this chapter will be assembled and used in a discussion to examine whether $N = 1$ supergravity, with the chosen local boundary conditions including $\tilde{\psi}^{A'}{}_{iI}$ and $\psi^{A}{}_{iF}$, might be a finite quantum field theory after all. The case of spectral boundary conditions for the Rarita–Schwinger operator, corresponding to the familiar fermionic boundary conditions involving positive and negative frequencies, also suggests the possibility of an eventual finiteness result (section 8.2). These ideas are also explored in the case of supermatter coupled to $N = 1$ supergravity [Wess & Bagger 1992] (section 8.3).

5
Supersymmetric mini-superspace models

5.1 Introduction

Having developed the general methods of chapters 3 and 4, one would like to understand how they work in simple examples of supersymmetric mini-superspace models. These are models in which the fields are assumed to be spatially homogeneous or isotropic. The fields are described by a finite-dimensional set of variables, leading to a quantum-mechanical model. We shall see in this chapter that the supersymmetry constraints $\bar{S}_{A'}\Psi = 0$, $S_A\Psi = 0$ lead to very restrictive quantum wave functions. Sometimes, the quantum degrees of freedom expected from the non-supersymmetric Hamiltonian constraints $\mathscr{H}_{AA'}\Psi = 0$ are missing. In the case of pure $N = 1$ supergravity, without a cosmological constant Λ or supermatter, the mini-superspace wave functions are proportional to $\exp(-I/\hbar)$ in the bosonic sector, where I is a certain Euclidean action for the gravitational field, and to $\exp(I/\hbar)$ in the filled fermionic sector. For homogeneous Bianchi models [Ryan & Shepley 1975], there are states at the intermediate fermionic levels [Csordás & Graham 1995] labelled by the solution of a Wheeler–DeWitt equation. The bosonic behaviour provides simple examples of the behaviour of $N = 1$ supergravity in the general case, without spacetime symmetries, described in chapter 8, where bosonic scattering amplitudes also take the form $\exp(-I_{\text{class}}/\hbar)$, with I_{class} being the classical bosonic action. When the simplest form of $N = 1$ supermatter, consisting of a complex scalar field and its fermionic spin-1/2 partner (sections 5.5, 5.6), is coupled to supergravity and quantized subject to a $k = +1$ Friedmann geometry, one again finds many fermionic physical quantum states, labelled by a solution of the Wheeler–DeWitt equation.

In 5.2–5.4, we describe the formulation and quantization of the $k = +1$ Friedmann model in supergravity [D'Eath & Hughes 1988,1992].

An Ansatz for the gravitational and spin-3/2 fields must be made such that the model is invariant under homogeneous local supersymmetry transformations, suitably modified by homogeneous coordinate and local Lorentz transformations. The gravitational field is described by the radius a of the three-sphere, and the gravitino field must be described by the spin-1/2 quantities $(\psi^A, \tilde{\psi}^{A'})$. The supersymmetry constraints are very simple, and one finds the general solution (5.4.7)

$$\Psi = C \exp\left(-3a^2/\hbar\right) + D \exp\left(3a^2/\hbar\right) \psi_A \psi^A, \tag{5.1.1}$$

where C and D are constants. The bosonic solution $C \exp\left(-3a^2/\hbar\right)$ gives the quantum wormhole state [Hawking 1988] (chapter 6), where one evaluates a path integral over all four-geometries with inner boundary at the three-sphere of radius a, subject to the condition that the four-geometry is asymptotically Euclidean at large distances. Here $3a^2$ is the action of flat Euclidean four-space outside the given three-sphere of radius a. Quantum wormhole states are discussed in more detail in chapter 6. Analogously, the filled fermionic solution $D \exp\left(3a^2/\hbar\right) \psi_A \psi^A$ gives the Hartle–Hawking state. Here the path integral is over all compact four-geometries with the three-sphere of radius a as their boundary; the action of flat Euclidean four-space (the classical solution) inside the three-sphere is $-3a^2$.

A much richer and more interesting class of models is given by coupling $N = 1$ supergravity to supermatter in a locally supersymmetric way [Wess & Bagger 1992]. These theories will be treated in the general case, without any spacetime symmetries, in section 8.4; there are indications that a certain class might give finite quantum theories of gravity coupled to matter. In sections 5.5 and 5.6 we shall consider the simplest such model, with a spin-1/2 field $(\chi^A, \tilde{\chi}^{A'})$ and complex scalar field $(\phi, \tilde{\phi})$ [Das *et al.* 1977]. The scalar field is taken to have the standard kinetic terms, proportional to $(\partial_\mu \phi)(\partial^\mu \tilde{\phi})$, corresponding to a flat Kähler metric [Wess & Bagger 1992]. In the Friedmann case, the Lorentz invariance implies that the wave function has the form (5.6.2)

$$\Psi = A + iB\psi^C\psi_C + iC\psi^C\chi_C + iD\chi^C\chi_C + E\psi^C\psi_C\chi^D\chi_D, \tag{5.1.2}$$

where A, B, C, D and E are functions of $\left(a, \phi, \tilde{\phi}\right)$. In the massless case, the supersymmetry constraints couple together only the terms with the same number of fermions. The equations for A and E in this case lead again to solutions proportional to $\exp(\mp 3a^2/\hbar)$. Again, E gives the Hartle–Hawking state; when one specifies $(a, \phi, \tilde{\phi})$ on an outer three-sphere boundary, the classical solution for $(\phi, \tilde{\phi})$ is constant, and the gravitational solution is flat Euclidean four-space. The quantities B, C, D are coupled together in the massless case by a system of first-order partial differential equations ('Dirac equations'). The consistency conditions are the wavelike Wheeler–DeWitt equations, which are decoupled. The general solution to the coupled system can be found (chapter 6) and used to find the ground wormhole state [Alty *et al.* 1992], which lies in the (B, C, D) sector. This can further be used (chapter 6) to show that for this massless system, the effective mass remains zero when wormhole effects are included.

Homogeneous but anisotropic Bianchi models are studied in sections 5.7, 5.8. The gravitational variables are now a symmetric matrix $h_{\alpha\beta}(t)$, where (5.7.1) [Ryan & Shepley 1975, Asano *et al.* 1993]

$$h_{ij} = h_{\alpha\beta}(t) E^{\alpha}{}_{i} E^{\beta}{}_{j}, \tag{5.1.3}$$

$(\alpha, \beta, \ldots = 1, 2, 3)$, and $E^{\alpha}{}_{i}$ is an invariant basis on the three-surface of homogeneity. The structure constants are $C^{\alpha}{}_{\beta\gamma} = C^{\alpha}{}_{[\beta\gamma]}$, which can be written as $C^{\alpha}{}_{\beta\gamma} = m^{\alpha\delta} \epsilon_{\delta\beta\gamma} + \delta^{\alpha}{}_{[\beta} a_{\gamma]}$, where $m^{\alpha\delta} = m^{(\alpha\delta)}$ and $a_{\gamma} = C^{\alpha}{}_{\alpha\gamma}$. The gravitino variables are now the triad components $\psi^{A}{}_{\alpha}$.

The spin-3/2 degrees of freedom are present in a Bianchi model. The wave function is, by Lorentz invariance, made from invariant polynomials in $\psi^{A}{}_{1}, \psi^{A}{}_{2}$ and $\psi^{A}{}_{3}$, of the form $\psi^{A}{}_{1}\psi_{A1}, \psi^{A}{}_{2}\psi_{A2}$, etc., and from $n_{AA'} e_{B}{}^{A'}{}_{s} \psi^{A}{}_{q} \psi^{B}{}_{r}$, multiplied by functions of $h_{\alpha\beta}$. The form of the fermionic wave function was taken to be unnecessarily restrictive in [D'Eath *et al.* 1993, D'Eath 1993, Asano *et al.* 1993], leading to the erroneous conclusion that there were no intermediate fermionic quantum states. The supersymmetry constraint $\bar{S}_{A'}\Psi = 0$ acting on the bosonic part of Ψ implies that $a_{\gamma} = 0$, so that the Bianchi model is of class A. It is straightforward then to solve the $\bar{S}_{A'}\Psi = 0$ constraint to find the bosonic part of Ψ. This was first done by [Hughes 1986], who found $\Psi_{\text{bosonic}} \propto \exp(-\text{const. } m^{\alpha\beta} h_{\alpha\beta}/\hbar)$. The corresponding result for the filled fermionic state is

$$\Psi_{\text{filled}} \propto \exp\left(\text{const. } m^{\alpha\beta} h_{\alpha\beta}/\hbar\right).$$

As stated above, the intermediate fermionic states are non-trivial, being based on a solution of the Wheeler–DeWitt equation. There is one apparent puzzle: in the Bianchi-IX case, corresponding to a deformed three-sphere, one can only find the ground wormhole state as above, but not the Hartle–Hawking state [D'Eath 1993]. If one uses a different set of homogeneous spinors $\psi^{A}{}_{i}$ [Graham & Luckock 1994], one can obtain the Hartle–Hawking state but not the ground wormhole state. This is explained if one recalls (section 2.9) that the Hartle–Hawking state in the presence of massless fermions can only be found if one specifies positive-frequency fermionic data on the boundary; the choice of negative-frequency data will lead to fermions which decay in the outward direction, and hence correspond to the ground wormhole state.

In the Friedmann case, a cosmological constant Λ is included in section 5.9. It is expected that inclusion of a Λ-term in a Bianchi model will be analogous. The claim in [D'Eath 1994, Cheng *et al.* 1994] that there are no quantum states in this case is wrong, based on the above erroneous assumption about the fermionic states.

5.2 Reduction of four-dimensional $N = 1$ supergravity to one dimension: Friedmann $k = +1$ case

As a first simplifying assumption, we choose the geometry to be that of a $k = +1$ Friedmann model, following [D'Eath & Hughes 1988,1992]. The tetrad of the four-dimensional theory is taken to be:

$$e_{a\mu} = \begin{pmatrix} N(\tau) & 0 \\ N^i a(\tau) E_{\hat{a}i} & a(\tau) E_{\hat{a}i} \end{pmatrix}, \; e^{a\mu} = \begin{pmatrix} N(\tau)^{-1} & -N^i N(\tau)^{-1} \\ 0 & a(\tau)^{-1} E^{\hat{a}i} \end{pmatrix}, \tag{5.2.1}$$

where $\hat{a}$ and i run from 1 to 3. The shift vector N^i is assumed to take the form $N^i = -N^{AA'}(\tau) e_{AA'}{}^i$, where $N^{AA'} n_{AA'} = 0$.

$E_{\hat{a}i}$ is a basis of left-invariant one-forms on the unit S^3 with volume $\sigma^2 = 2\pi^2$. The spatial tetrad $e^{AA'}{}_i$ satisfies the relation

$$\partial_i e^{AA'}{}_j - \partial_j e^{AA'}{}_i = 2a^2 \epsilon_{ijk} e^{AA'k} \tag{5.2.2}$$

as a consequence of the group structure of $SO(3)$.

This Ansatz reduces the number of degrees of freedom provided by $e_{AA'\mu}$. If supersymmetry invariance is to be retained, then we need an Ansatz for $\psi^A{}_\mu$ and $\tilde{\psi}^{A'}{}_\mu$ which reduces the number of fermionic degrees of freedom, so that there is equality between the number of bosonic and fermionic degrees of freedom. One is naturally led to take $\psi^A{}_0$ and $\tilde{\psi}^{A'}{}_0$ to be functions of time only. In the four-dimensional Hamiltonian theory, $\psi^A{}_0$ and $\tilde{\psi}^{A'}{}_0$ are Lagrange multipliers which may be freely specified. For this reason we do not allow $\psi^A{}_0$ and $\tilde{\psi}^{A'}{}_0$ to depend on ψ^i or $\tilde{\psi}^{A'}{}_i$ in our Ansatz. We further take

$$\begin{aligned} \psi^A{}_i &= e^{AA'}{}_i \tilde{\psi}_{A'}, \\ \tilde{\psi}^{A'}{}_i &= e^{AA'}{}_i \psi_A, \end{aligned} \tag{5.2.3}$$

where we introduce the new spinors ψ_A and $\tilde{\psi}_{A'}$, which are functions of time only. (It is possible to justify the Ansatz (5.2.3) by requiring that the form (5.2.1) of the tetrad be preserved under suitable homogeneous supersymmetry transformations, along the lines of the argument following (5.2.4) below.) When the auxiliary fields [Wess & Bagger 1992] are taken into account, one can extend the Ansatz to one in which the bosonic and fermionic degrees of freedom are equal.

Non-zero homogeneous spin-3/2 fields $\psi^A{}_\mu$ or $\tilde{\psi}^{A'}{}_\mu$ are of course not isotropic. However, it turns out that the constraints obeyed by classical solutions of the one-dimensional theory lead to a four-dimensional energy–momentum tensor which is isotropic, consistent with the assumption of a Friedmann geometry.

Ultimately, one would hope to understand some of the consequences of supersymmetry in the full theory by analogy with the simple mini-superspace model that we will construct. But there is no guarantee that the Ansatz we have chosen is invariant under any supersymmetry transformation in the full theory. If there is no supersymmetry transformation which preserves the form of (5.2.1) and (5.2.3), then one of the most important features of the full theory and its possible consequences would be missing from our model.

If one looks at the supersymmetry transformation (3.2.15), it is clear that this, on its own, will not preserve the form of the Ansatz. But suppose we consider a combination of a non-zero (spatially homogeneous) supersymmetry transformation (3.2.15) and possible local Lorentz and coordinate transformations, (3.2.13) and (3.2.14). If such a combination can be found to preserve the form of (5.2.1) and (5.2.3), then one should be more hopeful that a supersymmetric mini-superspace model exists.

We require that there exists a transformation of the four-dimensional theory (including a non-zero supersymmetry transformation) which will transform ${e^{AA'}}_i$ by the amount

$$\delta {e^{AA'}}_i = C(\tau) {e^{AA'}}_i, \tag{5.2.4}$$

$C(\tau)$ being spatially independent and possibly involving Grassmann elements, so that the form of (5.2.1) remains unchanged. We consider a combination of transformations of the form (3.2.13), (3.2.14) and (3.2.15) parametrised by $\xi^i = -\xi^{AA'}(\tau){e_{AA'}}^i$, $N^{AB}(\tau)$, $\tilde{N}^{A'B'}(\tau)$, $\epsilon^A(\tau)$ and $\tilde{\epsilon}^{A'}(\tau)$. ξ^0 is taken to be zero, since the transformations generated by taking $\xi^0(\tau)$ to be non-zero are translations in the time direction which preserve the Ansatz on their own. Consistent with this we take $\xi^{AA'} = i\xi^{AB}{n_B}^{A'}$, where ξ^{AB} depends only on τ and is symmetric in its two indices. Note that $\xi^{AA'}n_{AA'} = 0$, showing consistency with ξ^0 being zero. To simplify future results we also define

$$\tilde{\xi}^{A'B'} = 2\xi^{AB}{n_A}^{A'}{n_B}^{B'}. \tag{5.2.5}$$

Note that if $\tilde{\xi}^{A'B'} = \bar{\xi}^{A'B'}$ then ξ^i is real. Using (3.2.13), (3.2.14) and (3.2.15), the transformation of ${e^{AA'}}_i$ becomes

$$\begin{aligned}\delta {e^{AA'}}_i = {} & {N^A}_B {e^{BA'}}_i + {\tilde{N}^{A'}}_{B'} {e^{AB'}}_i \\ & - \xi^{BB'} {e_{BB'}}^j \partial_j {e^{AA'}}_i - {e^{AA'}}_j \partial_i \left(\xi^{BB'} {e_{BB'}}^j\right) \\ & - i\kappa\left(\epsilon^A {e^{BA'}}_i \psi_B + \tilde{\epsilon}^{A'} {e^{AB'}}_i \tilde{\psi}_{B'}\right).\end{aligned} \tag{5.2.6}$$

Using results given in subsection 2.9.1, the terms in this equation involving $\xi^{AA'}$ may be rearranged as follows:

$$
\begin{aligned}
&-\xi^{BB'} e_{BB'}{}^{j} \partial_j e^{AA'}{}_{i} - e^{AA'}{}_{j} \partial_i \left(\xi^{BB'} e_{BB'}{}^{j}\right) \\
&= -\xi^{BB'} e_{BB'}{}^{j} \left(\partial_j e^{AA'}{}_{i} - \partial_i e^{AA'}{}_{j}\right) \\
&= -\xi^{BB'} e_{BB'}{}^{j} \left(2a^2 \epsilon_{jik} e^{AA'k}\right) \\
&= -\xi^{BB'} a^2 \epsilon_{jik} \left(e_{BB'}{}^{j} e^{AA'k} - e_{BB'}{}^{k} e^{AA'j}\right) \\
&= -\xi^{BB'} a^2 \epsilon_{jik} \left(\epsilon_{B}{}^{A} e_{CB'}{}^{j} e^{CA'k} - \epsilon_{B'}{}^{A'} e_{BC'}{}^{k} e^{AC'j}\right) \\
&= -\xi^{BB'} a^{-1} \epsilon_{jik} \left(i\epsilon_{B}{}^{A} \epsilon^{jsk} n_{CB'} e^{CA'}{}_{s} - i\epsilon_{B'}{}^{A'} \epsilon^{jsk} n_{BC'} e^{AC'}{}_{s}\right) \\
&= -2ia^{-1} \xi^{AB'} n_{CB'} e^{CA'}{}_{i} + 2ia^{-1} \xi^{BA'} n_{BC'} e^{AC'}{}_{i}. \qquad (5.2.7)
\end{aligned}
$$

Returning to the transformation for $e^{AA'}{}_{i}$, we now have

$$
\begin{aligned}
\delta e^{AA'}{}_{i} = &\left(- N^{AB} + a^{-1} \xi^{AB} + i\kappa \epsilon^{(A} \psi^{B)}\right) e_{B}{}^{A'}{}_{i} \\
&+ \left(-\tilde{N}^{A'B'} + a^{-1} \tilde{\xi}^{A'B'} + i\kappa \tilde{\epsilon}^{(A'} \tilde{\psi}^{B')}\right) e^{A}{}_{B'i} \\
&+ \frac{i\kappa}{2} \left(\epsilon_C \psi^C + \tilde{\epsilon}_{C'} \tilde{\psi}^{C'}\right) e^{AA'}{}_{i}, \qquad (5.2.8)
\end{aligned}
$$

so that the Ansatz is preserved with

$$
\delta a = \frac{ia\kappa}{2} \left(\epsilon_C \psi^C + \tilde{\epsilon}_{C'} \tilde{\psi}^{C'}\right), \qquad (5.2.9)
$$

provided that the relations

$$
\begin{aligned}
N^{AB} - a^{-1} \xi^{AB} - i\kappa \epsilon^{(A} \psi^{B)} &= 0, \\
\tilde{N}^{A'B'} - a^{-1} \tilde{\xi}^{A'B'} - i\kappa \tilde{\epsilon}^{(A'} \tilde{\psi}^{B')} &= 0, \qquad (5.2.10)
\end{aligned}
$$

between the generators of Lorentz, coordinate and supersymmetry transformations are satisfied.

The Ansatz for the fields $\psi^{A}{}_{i}$ and $\tilde{\psi}^{A'}{}_{i}$ should also be preserved under the same combination of transformations. We concentrate only on $\psi^{A}{}_{i}$, since the two cases are similar. We have

$$
\begin{aligned}
\delta \psi^{A}{}_{i} &= N^{A}{}_{B} \psi^{B}{}_{i} + \xi^{j} \partial_j \psi^{A}{}_{i} + \psi^{A}{}_{j} \partial_i \xi^{j} + 2\kappa^{-1} D_i \epsilon^{A} \\
&= N^{A}{}_{B} \psi^{B}{}_{i} + \xi^{j} \left(\partial_j e^{AA'}{}_{i} - \partial_i e^{AA'}{}_{j}\right) \tilde{\psi}_{A'} + 2\kappa^{-1} D_i \epsilon^{A}. \qquad (5.2.11)
\end{aligned}
$$

The terms involving $\xi^{AA'}$ are simplified as before. The Ansatz for $e^{AA'}{}_{\mu}$ and $\psi^{A}{}_{\mu}$ is substituted into the expressions for the connection forms given in section 3.2. One finds that the spatial spinor versions of the torsion-free connection forms and the contorsion tensor given by ${}^3\omega^{AB}{}_{\mu}, \kappa^{AB}{}_{\mu}$ and their conjugates, with $\omega^{AB}{}_{\mu} = {}^3\omega^{AB}{}_{\mu} + \kappa^{AB}{}_{\mu}$, are

$$ {}^s\omega^{AB}{}_i = \left(\frac{\dot{a}}{aN} + \frac{i}{a}\right) n^A{}_{A'} e^{BA'}{}_i, \tag{5.2.12} $$

$$ \begin{aligned} \kappa^{AB}{}_i = &-\frac{i\kappa^2}{4N}(\psi_E \psi^E{}_0 + \tilde{\psi}_{E'0}\tilde{\psi}^{E'}) n^A{}_{A'} e^{BA'}{}_i \\ &+ \frac{i\kappa^2}{4} e^{(AE'}{}_i \epsilon^{B)E} \psi_E \tilde{\psi}_{E'}. \end{aligned} \tag{5.2.13} $$

Substituting the results into (5.2.11) and using (5.2.10), we obtain

$$ \begin{aligned} \delta\psi^A{}_i = &\frac{i\kappa}{4}\epsilon^A \psi^B \tilde{\psi}^{B'} e_{BB'i} + a^{-1}\tilde{\xi}^{A'B'} e^A{}_{B'i}\tilde{\psi}_{A'} \\ &+ \left[\frac{2}{\kappa}\left(\frac{\dot{a}}{aN} + \frac{i}{a}\right) - \frac{i\kappa}{2N}\left(\psi_F \psi^F{}_0 + \tilde{\psi}_{F'0}\tilde{\psi}^{F'}\right)\right] \\ &\times n_{BA'} e^{AA'}{}_i \epsilon^B. \end{aligned} \tag{5.2.14} $$

The Ansatz for $\psi^A{}_i$ is preserved if we impose the additional constraint

$$ \psi^B \tilde{\psi}^{B'} e_{BB'i} = 0 \tag{5.2.15} $$

and take $\tilde{\xi}^{A'B'} = 0$. Similarly, by considering the transformation of $\tilde{\psi}^{A'}{}_i$ we find the equivalent condition $\xi^{AB} = 0$.

By noting that (5.2.15) implies that $\psi^B \tilde{\psi}^{B'} \propto n^{BB'}$, we see that we can write (5.2.15) in the two equivalent forms:

$$ J_{AB} = \psi_{(A} \tilde{\psi}^{B'} n_{B)B'} = 0, \tag{5.2.16} $$

$$ \tilde{J}_{A'B'} = \tilde{\psi}_{(A'} \psi^B n_{BB')} = 0. \tag{5.2.17} $$

We will use the constraint in the form (5.2.16) and accept the resulting asymmetry in the appearance of some spinorial expressions. Note that the constraint $J_{AB} = 0$ has a natural interpretation as the reduced form of the Lorentz rotation constraint arising in the full theory (section 3.2). (Indeed, an alternative, and perhaps more natural way to proceed would have been to modify the Ansatz (5.2.1) for the tetrad by allowing a time-dependent rotation of $e^a{}_{\mu}$. In that case the constraint $J_{AB} = 0$ would have appeared, with its familiar interpretation, from the outset of the reduction from four dimensions to one.) Equation (5.2.16) can be solved classically, if desired, by taking $\psi_A \propto n_{AA'}\tilde{\psi}^{A'}$, since the right-hand side is then proportional to $n_{AA'}n_{BB'}\tilde{\psi}^{(A'}\tilde{\psi}^{B')}$ which vanishes because $\tilde{\psi}^{A'}$ is anticommuting (see also [Christodoulakis & Zanelli 1984a,b]). However, this is only possible provided that $\tilde{\psi}^{A'} \neq \bar{\psi}^{A'}$ (where a bar denotes Hermitian conjugation), as in the Euclidean regime which appears naturally in the interpretation of the quantum wave function.

By requiring that the constraint $J_{AB} = 0$ be preserved under the same combination of transformations as used above, one finds equations which could be regarded as restrictions on ϵ^A and $\tilde{\epsilon}^{A'}$, but which are satisfied provided the supersymmetry constraints $S_A = 0$, $\tilde{S}_{A'} = 0$ (see (5.3.13) below) hold. By further requiring that the supersymmetry constraints be preserved, one finds additionally that the Hamiltonian constraint $\mathscr{H} = 0$ of (5.3.13) should hold. All these constraints are part of the on-shell equations of the theory. Thus we have shown that a combination of transformations in the full theory preserves our Ansatz on shell. The fact that the Ansatz is only preserved on shell is understandable since we have neglected the auxiliary fields of the theory; it is well known that, once auxiliary fields have been eliminated, the supersymmetry algebra only closes on shell [van Nieuwenhuizen 1981].

We now restate the Ansatz in full, having shown that it is consistent with supersymmetry. The Ansatz for the tetrad is given by (5.2.1) while the Ansatz for the spin-3/2 field is given by (5.2.3). The supersymmetry transformation acts on a, N, ψ^A, $\tilde{\psi}^{A'}$, ${\psi^A}_0$ and ${\tilde{\psi}^{A'}}_0$ in the following way:

$$\begin{aligned}
\delta a &= \frac{ia\kappa}{2}\left(\epsilon_A \psi^A + \tilde{\epsilon}_{A'}\tilde{\psi}^{A'}\right),\\
\delta N &= -i\kappa n_{AA'}\left(\epsilon^A {\tilde{\psi}^{A'}}_0 + \tilde{\epsilon}^{A'}{\psi^A}_0\right),\\
\delta\left(a\psi^A\right) &= \left(\frac{2}{\kappa}\left(\frac{\dot{a}}{N} - i\right) - \frac{ia\kappa}{2N}\left(\psi_E {\psi^E}_0 + \tilde{\psi}_{E'0}\tilde{\psi}^{E'}\right)\right) {n^A}_{B'}\tilde{\epsilon}^{B'},\\
\delta\left(a\tilde{\psi}^{A'}\right) &= \left(\frac{2}{\kappa}\left(\frac{\dot{a}}{N} + i\right) - \frac{ia\kappa}{2N}\left(\psi_E {\psi^E}_0 + \tilde{\psi}_{E'0}\tilde{\psi}^{E'}\right)\right) {n_B}^{A'}\epsilon^B,\\
\delta {\psi^A}_0 &= \frac{2\dot{\epsilon}^A}{\kappa} - i\kappa\psi_B \epsilon^{(A}{\psi^{B)}}_0 - \frac{i\kappa}{2}N\epsilon^{E(A}n^{B)E'}\psi_E\tilde{\psi}_{E'}\epsilon_B,\\
\delta {\tilde{\psi}^{A'}}_0 &= \frac{2\dot{\tilde{\epsilon}}^{A'}}{\kappa} - i\kappa\tilde{\psi}_{B'}\tilde{\epsilon}^{(A'}{\tilde{\psi}^{B')}}_0\\
&\quad - \frac{i\kappa}{2}N\epsilon^{E'(A'}{n_E}^{B')}\psi^E\tilde{\psi}_{E'}\tilde{\epsilon}_{B'}.
\end{aligned} \tag{5.2.18}$$

The shift vector N^i has been left out here, since it drops out of the action completely when the Ansatz is substituted.

5.3 Reduced action for $N = 1$ supergravity model: Friedmann $k = +1$ case

The complete Ansatz can now be substituted into the four-dimensional action (3.2.12) to give a reduced one-dimensional action

$$S = \int d\tau \mathscr{L}_{sg}, \tag{5.3.1}$$

where the Lagrangian is given by

$$\begin{aligned}\mathscr{L}_{sg} = & -\frac{3a}{\kappa^2 N}\left(\dot{a} - \frac{ai\kappa^2}{4}\left(\tilde{\psi}_{E'0}\tilde{\psi}^{E'} - \psi^E\psi_{E0}\right)\right)^2 \\ & - \frac{3ia^3}{2}n_{AA'}\psi^A\dot{\tilde{\psi}}^{A'} - \frac{3ia^3}{2}n_{AA'}\tilde{\psi}^{A'}\dot{\psi}^A \\ & + \frac{3Na}{\kappa^2} + \frac{3a^2}{2}\left(\tilde{\psi}_{E'0}\tilde{\psi}^{E'} + \psi^E\psi_{E0}\right) - \frac{3}{2}Na^2n^{AA'}\psi_A\tilde{\psi}_{A'} \\ & + \frac{3\kappa^2a^3}{16}\left(n^E{}_{A'}\tilde{\psi}^{A'}{}_0\tilde{\psi}^{E'}\psi_E\tilde{\psi}_{E'} + n_A{}^{E'}\psi^E\psi^A{}_0\psi_E\tilde{\psi}_{E'}\right).\end{aligned} \tag{5.3.2}$$

In addition to Eqs. (5.2.12),(5.2.13), we have used

$$^s\omega^{AB}{}_0 = N^i\left(\frac{\dot{a}}{aN} - \frac{i}{a}\right)n^A{}_{A'}e^{BA'}{}_i, \tag{5.3.3}$$

$$\begin{aligned}\kappa^{AB}{}_0 = & -\frac{i\kappa^2}{4N}N^i(\psi_E\psi^E{}_0 + \tilde{\psi}_{E'0}\tilde{\psi}^{E'})n^A{}_{A'}e^{BA'}{}_i \\ & - \frac{i\kappa^2}{4}N^ie^{(AE'}{}_i\epsilon^{B)E}\psi_E\tilde{\psi}_{E'} \\ & + \frac{i\kappa^2}{2}\psi^{(A}\psi^{B)}{}_0 - \frac{i\kappa^2N}{4}\epsilon^{(AE}n^{B)E'}\psi_E\tilde{\psi}_{E'}.\end{aligned} \tag{5.3.4}$$

In order to simplify the form of future results it is useful, at this stage, to make slight redefinitions of the dynamical fields. We let $a \rightarrow \frac{\kappa}{\sigma}a$, $\psi^A \rightarrow \sqrt{\frac{2}{3}}\frac{\sigma^{1/2}}{(\kappa a)^{3/2}}\psi^A$ and $\tilde{\psi}^{A'} \rightarrow \sqrt{\frac{2}{3}}\frac{\sigma^{1/2}}{(\kappa a)^{3/2}}\tilde{\psi}^{A'}$, where, as before, $\sigma^2 = 2\pi^2$. The resulting changes to the transformation of these fields under supersymmetry are given in [D'Eath & Hughes 1988].

We include the constraint $J_{AB} = 0$ by adding $M^{AB}J_{AB}$ to the Lagrangian, where M^{AB} is a Lagrange multiplier. In order to achieve the simplest form of the generators (5.3.13) in the Hamiltonian and their Dirac brackets (5.3.11), we make the following redefinitions of the non-dynamical fields N, $\psi^A{}_0$, $\tilde{\psi}^{A'}{}_0$ and M^{AB}:

$$\begin{aligned}
\hat{N} &= \frac{\sigma N}{12\kappa}, \\
\rho^A &= \frac{i(\kappa\sigma)^{1/2}}{2\sqrt{6}a^{1/2}}\psi^A{}_0 + \frac{i\sigma N}{12\kappa a^2} n^{AA'}\tilde{\psi}_{A'}, \\
\tilde{\rho}^{A'} &= -\frac{i(\kappa\sigma)^{1/2}}{2\sqrt{6}a^{1/2}}\tilde{\psi}^{A'}{}_0 - \frac{i\sigma N}{12\kappa a^2} n^{AA'}\psi_A\,, \\
L^{AB} &= M^{AB} - \frac{(\kappa\sigma)^{1/2}}{3\sqrt{6}a^{3/2}}\psi^{(A}{}_0\psi^{B)} \\
&\quad - \frac{2(\kappa\sigma)^{1/2}}{3\sqrt{6}a^{3/2}}\tilde{\psi}^{(A'}{}_0\tilde{\psi}^{B')} n^A{}_{A'} n^B{}_{B'}.
\end{aligned} \tag{5.3.5}$$

The reduced action S_r is now given by:

$$\begin{aligned}
S_r &= \int d\tau \mathscr{L} \\
&= \int d\tau \Bigg[-\frac{a\dot{a}^2}{4\hat{N}} + 36a\hat{N} - in_{AA'}\psi^A\dot{\tilde{\psi}}^{A'} - in_{AA'}\tilde{\psi}^{A'}\dot{\psi}^A \\
&\quad - \frac{a}{4\hat{N}}\left(\rho_A\psi^A + \tilde{\psi}^{A'}\tilde{\rho}_{A'}\right)^2 + \rho_A\left(\frac{a\dot{a}}{2\hat{N}}\psi^A + 6ia\psi^A\right) \\
&\quad + \left(\frac{a\dot{a}}{2\hat{N}}\tilde{\psi}^{A'} - 6ia\tilde{\psi}^{A'}\right)\tilde{\rho}_{A'} \\
&\quad - L^{AB}\psi_{(A}\tilde{\psi}^{B'} n_{B)B'}\Bigg].
\end{aligned} \tag{5.3.6}$$

We now calculate the Hamiltonian of this system. The momentum π_a conjugate to a is given by

$$\pi_a = \frac{\partial \mathscr{L}}{\partial \dot{a}} = -\frac{a\dot{a}}{2\hat{N}} + \rho_A \frac{a}{2\hat{N}}\psi^A + \frac{a}{2\hat{N}}\tilde{\psi}^{A'}\tilde{\rho}_{A'}. \tag{5.3.7}$$

As usual with fermionic systems, calculating the momenta conjugate to the anticommuting dynamical variables gives rise to primary constraints which we denote here by $\mathscr{C}_A$ and $\mathscr{C}_{A'}$:

$$\begin{aligned}
\mathscr{C}_A &= \pi_{\psi A} - in_{AA'}\tilde{\psi}^{A'} = 0, \\
\mathscr{C}_{A'} &= \pi_{\tilde{\psi} A'} - in_{AA'}\psi^A = 0,
\end{aligned} \tag{5.3.8}$$

where $\pi_{\psi A} = \partial\mathscr{L}/\partial\dot{\psi}^A$ and $\pi_{\tilde{\psi}A'} = \partial\mathscr{L}/\partial\dot{\tilde{\psi}}^{A'}$, and left differentiation is being used [Wess & Bagger 1992]. The constraints $\mathscr{C}_A$ and $\mathscr{C}_{A'}$ are second-class and can be eliminated by the Dirac procedure [Dirac 1965, Hanson *et al.* 1976, Nelson & Teitelboim 1978]. We define the matrix $\mathscr{C}_{AA'}$ as the Poisson bracket of the constraints:

$$\mathscr{C}_{AA'} = [\mathscr{C}_A, \mathscr{C}_{A'}] = 2in_{AA'}\,, \tag{5.3.9}$$

where the conventions of [Casalbuoni 1976] have been used for brackets involving fermions. The Dirac bracket [,]* is then defined by

$$[A,B]^* = [A,B] - \left[A,\mathscr{C}^A\right]\mathscr{C}^{-1}_{AA'}\left[\mathscr{C}^{A'},B\right] - \left[A,\mathscr{C}^{A'}\right]\mathscr{C}^{-1}_{AA'}\left[\mathscr{C}^{A},B\right]. \quad (5.3.10)$$

The result of this procedure is the elimination of the momenta conjugate to the fermionic variables, leaving us with the following non-zero Dirac bracket relations:

$$[a,\pi_a]^* = 1\,, \qquad [\psi_A,\tilde{\psi}_{A'}]^* = in_{AA'}. \quad (5.3.11)$$

The Hamiltonian $H = \dot{a}\pi_a + \dot{\psi}^A\pi_{\psi A} + \dot{\tilde{\psi}}^{A'}\pi_{\tilde{\psi}A'} - \mathscr{L}$ is given by:

$$H = \hat{N}\mathscr{H} + \rho_A S^A + \tilde{S}^{A'}\tilde{\rho}_{A'} + L^{AB}J_{AB}, \quad (5.3.12)$$

with

$$S_A = \psi_A\pi_a - 6ia\psi_a\,, \qquad \mathscr{H} = -a^{-1}\left(\pi_a^2 + 36a^2\right),$$
$$\tilde{S}_{A'} = \tilde{\psi}_{A'}\pi_a + 6ia\tilde{\psi}_{A'}, \qquad J_{AB} = \psi_{(A}\tilde{\psi}^{B'}n_{B)B'}. \quad (5.3.13)$$

This is of the expected general form [Teitelboim 1977a, Deser *et al.* 1977a, Fradkin & Vasiliev 1977, Pilati 1978], with $\hat{N}$, ρ_A, $\tilde{\rho}_{A'}$ and L^{AB} being the Lagrange multipliers for the generator $\mathscr{H}$ of modified coordinate transformations in one dimension, the generators S^A and $\tilde{S}^{A'}$ of (modified) local supersymmetry, and the generator J_{AB} of local rotations. Now that the constraint $J_{AB} = 0$ has been included by means of a Lagrange multiplier, one may choose $\hat{N}$, ρ_A, $\tilde{\rho}_{A'}$ and L_{AB} freely as functions of τ during the classical evolution. The presence of the freely specifiable (odd) parameters ρ_A, $\tilde{\rho}_{A'}$ shows that this model has $N = 4$ local supersymmetry in one dimension, as might be expected from the reduction from four to one dimension of $N = 1$ supergravity.

The dynamical variables are a, π_a, ψ^A and $\tilde{\psi}^{A'}$. Classically $\mathscr{H} \approx 0$, $S^A \approx 0$, $\tilde{S}^{A'} \approx 0$ and $J^{AB} \approx 0$ form a set of first-class constraints, with the algebra:

$$\left[S_A,\tilde{S}_{A'}\right]^* = -ian_{AA'}\mathscr{H} - \frac{\tilde{\psi}_{A'}}{a}S_A - \frac{\psi_A}{a}\tilde{S}_{A'}, \quad (5.3.14a)$$

$$[S_A,\mathscr{H}]^* = -\frac{\psi_A}{a}\mathscr{H} - \frac{2}{a^2}\pi_a S_A, \quad (5.3.14b)$$

$$\left[\tilde{S}_{A'},\mathscr{H}\right]^* = -\frac{\tilde{\psi}_{A'}}{a}\mathscr{H} - \frac{2}{a^2}\pi_a\tilde{S}_{A'}, \quad (5.3.14c)$$

$$[J_{AB},S_C]^* = \frac{i}{2}S_{(A}\epsilon_{B)C}, \quad (5.3.14d)$$

$$\left[J_{AB},\tilde{S}_{C'}\right]^* = in^{B'}{}_{(B}n_{A)C'}\tilde{S}_{B'}, \quad (5.3.14e)$$

$$[J_{AB},J_{CD}]^* = \frac{i}{2}\left[J_{C(A}\epsilon_{B)D} + J_{D(A}\epsilon_{B)C}\right], \quad (5.3.14f)$$

all other brackets being zero.

The classical solutions for the reduced theory can now be investigated. The classical equation of motion for any dynamical variable v is given by $\dot{v} = [v, \mathscr{H}]^*$. Neglecting, for simplicity, Lorentz rotations generated by J_{AB}, this gives

$$\begin{aligned}
\dot{a} &= -\frac{2\hat{N}}{a}\pi_a + \rho_A \psi^A + \tilde{\psi}_{A'}\tilde{\rho}^{A'}, \\
\dot{\pi}_a &= -\frac{\hat{N}}{a^2}\pi_a^2 + 36\hat{N} + 6i\rho_A\psi^A - 6i\tilde{\psi}^{A'}\tilde{\rho}_{A'}, \\
\dot{\psi}^A &= in^{AA'}(\pi_a + 6ia)\,\tilde{\rho}_{A'}, \\
\dot{\tilde{\psi}}^{A'} &= in^{AA'}\rho_A(\pi_a - 6ia).
\end{aligned} \tag{5.3.15}$$

As boundary data, one can specify the radius $\frac{\kappa}{\sigma}a$ of a three-sphere and one of $\psi_A, \tilde{\psi}_{A'}$, the other being regarded as a momentum variable following (5.3.11). If, for example, we specify ψ_A, then the constraints (5.3.13) imply that $\pi_a = 6ia$ and $\tilde{\psi}_{A'} = 0$ on the boundary $a = a_0$; these equations continue to hold as a varies in the resulting Euclidean solution. One can, for example, find solutions with these boundary data in which, by suitable choice of the multiplier $\tilde{\rho}_{A'}(\tau)$, ψ_A is being reduced to zero as a is reduced from a_0 to a smaller value a_1, with $\psi_A = 0$ for $0 < a < a_1$, while $\tilde{\psi}_{A'} = 0$ for $0 < a < a_0$. The geometry is that of flat Euclidean four-space which has been subjected to a finite supersymmetry transformation, inside $a = a_0$. Analogous solutions can be found outside $a = a_0$ and also in the case that $\tilde{\psi}_{A'}$ is specified on $a = a_0$. It is straightforward to check that these one-dimensional solutions are also solutions of the four-dimensional Einstein and Rarita–Schwinger equations (see also [Isham & Nelson 1974, Christodoulakis & Zanelli 1984a,b, Christodoulakis & Papadopoulos 1988, Macías *et al.* 1987]).

Classically the solutions for pure supergravity are thus fairly trivial. However they illustrate an important feature which recurs in the models with supermatter, namely that the classical momenta are severely restricted by the constraints.

5.4 Quantization of supergravity model: $k = +1$ Friedmann case

Quantum-mechanically, one replaces Dirac brackets $[\ ,\ ]^*$ by anticommutators $\{\ ,\ \}$ if both arguments are odd, or commutators $[\ ,\ ]$ if either or both are even [Casalbuoni 1976]:

$$\begin{aligned}
[E_1, E_2] &= i\hbar\,[E_1, E_2]^*, \\
[O, E] &= i\hbar\,[O, E]^*, \\
\{O_1, O_2\} &= i\hbar\,[O_1, O_2]^*.
\end{aligned} \tag{5.4.1}$$

In particular, this gives

$$\left\{\psi^A, \bar{\psi}^{A'}\right\} = -n^{AA'}. \tag{5.4.2}$$

One can work with a representation in which, for example, a and ψ^A are 'position' operators, and study a wave function $\Psi(a, \psi^A)$ taking its value in a Grassmann algebra. The quantum version of (5.3.11) can be satisfied if we represent the 'momentum' operators π_a and $\bar{\psi}^{A'}$ by the differential operators $-i\hbar\frac{\partial}{\partial a}$ and $-\hbar n^{AA'}\frac{\partial}{\partial \psi^A}$ acting on $\Psi(a, \psi^A)$, where $\frac{\partial}{\partial \psi^A}$ denotes left differentiation. We have used the notation $\bar{\psi}^{A'}$ since quantum-mechanically $\bar{\psi}^{A'}$ is naturally regarded as the Hermitian adjoint of ψ^A with respect to the inner product

$$\langle \Psi, \Phi \rangle = \int \bar{\Psi}\Phi \exp\left[2\psi^A \bar{\psi}^{A'} n_{AA'}/\hbar\right] da\, d\psi^A d\bar{\psi}^{A'}, \tag{5.4.3}$$

appropriate to the holomorphic representation used here [D'Eath 1984, D'Eath & Halliwell 1987, Faddeev & Slavnov 1980], where Berezin integration is used for the fermionic variables.

The first-class classical constraints become conditions on physical wave functions:

$$\begin{aligned} S^A\Psi &= 0, & \mathscr{H}\Psi &= 0, \\ \bar{S}^{A'}\Psi &= 0, & J^{AB}\Psi &= 0. \end{aligned} \tag{5.4.4}$$

The operator J^{AB} is naturally ordered such that the constraint

$$J^{AB}\Psi = 0$$

describes the invariance of the wave function under Lorentz rotations:

$$J_{AB}\Psi = -\frac{\hbar}{2}\psi_{(A}\frac{\partial \Psi}{\partial \psi^{B)}} = 0. \tag{5.4.5}$$

In this example, as a result of the choice of Lagrange multipliers in (5.3.5), there is no operator-ordering ambiguity in defining the operator versions S^A and $\bar{S}^{A'}$ of (5.3.13). The ordering of $\mathscr{H}$ is then determined by consistency with the quantum version of (5.3.14a), and is found to be

$$\mathscr{H} = \hbar^2 \frac{\partial}{\partial a}\left(a^{-1}\frac{\partial}{\partial a}\right) - 36a, \tag{5.4.6}$$

which is Hermitian with the inner product (5.4.3) and suitable boundary conditions. It can further be verified that the quantum algebra (quantum version of (5.3.14)) is consistent.

In a more general example, one might attempt similarly to choose $S^A, \bar{S}^{A'}$ and $\mathscr{H}$ by requiring (1) that the equation $S^A\Psi = 0$, which is first-order in the $(a,\ \psi^A)$ representation, describes the transformation properties of the wave function under infinitesimal supersymmetry transformations parametrized by ρ_A [D'Eath 1984], (2) that $\bar{S}^{A'}\Psi = 0$ similarly describes transformation properties parametrized by $\bar{\rho}_{A'}$ in the $(a, \tilde{\psi}^{A'})$ representation, (3) that S^A and $\bar{S}^{A'}$ are Hermitian adjoints with respect to the inner product generalizing (5.4.3), and (4) that a Hermitian $\mathscr{H}$ is defined by consistency of the quantum algebra. A thorough treatment of this question would require consideration of the path-integral measure [Fradkin & Vasiliev 1977, Fradkin & Vilkovisky 1977], since the transformation properties of the measure are involved in the requirements (1) and (2).

In solving the quantum constraints (5.4.4) here, note that $J^{AB}\Psi = 0$ implies that Ψ can be written as $\Psi = A + B\psi_A\psi^A$, where A and B depend only on a and on other fermionic variables which may be introduced through boundary data. The $\mathscr{H}$ constraint will be automatically satisfied by virtue of the quantum algebra, provided that the constraints $S^A\Psi = 0$ and $\tilde{S}^{A'}\Psi = 0$ hold. These imply that

$$\Psi = C \exp\left(-3a^2/\hbar\right) + D \exp\left(3a^2/\hbar\right) \psi_A\psi^A, \tag{5.4.7}$$

where C and D are independent of a and ψ^A.

The exponential factors in (5.4.7) have a semi-classical interpretation as $\exp(-I/\hbar)$, where I is the Euclidean action for a classical solution outside or inside a three-sphere of radius $\frac{\kappa}{\sigma}a$ with a prescribed boundary value of ψ^A (with a further outer boundary introduced for finiteness in the former case). To see this, we consider the semi-classical approximation to the wave function. Consider, in particular, the wave function specified by the Hartle–Hawking or 'no-boundary' proposal, given by

$$\Psi = \int_C \mathscr{D}e^a{}_\mu \mathscr{D}\psi^A{}_\mu \mathscr{D}\tilde{\psi}^{A'}{}_\mu \exp(-I/\hbar), \tag{5.4.8}$$

where C is the class of compact 4-metrics and regular matter fields which match prescribed values on a given surface. The Euclidean action I is obtained by choosing the lapse function $\hat{N}$ to be negative imaginary in (5.3.6); I is then equal to $-iS_r$.

The Euclidean versions of the constraints (5.3.13) are:

$$\begin{aligned} S_A &= \psi_A\hat{\pi}_a - 6a\psi_A, & \mathscr{H} &= -a^{-1}\left(\hat{\pi}_a^2 - 36a^2\right), \\ \tilde{S}_{A'} &= \tilde{\psi}_{A'}\hat{\pi}_a + 6a\tilde{\psi}_{A'}, & J_{AB} &= \psi_{(A}\tilde{\psi}^{B'} n_{B)B'}. \end{aligned} \tag{5.4.9}$$

Classically these constraints must vanish for all time τ. $\mathscr{H} = 0$ implies that $\hat{\pi}_a = \pm 6a$. If $\hat{\pi}_a = +6a$ then $\tilde{S}_{A'} = 0$ implies that $\tilde{\psi}_{A'} = 0$ for all τ, while the case $\hat{\pi}_a = -6a$ similarly implies that $\psi_A = 0$ for all τ.

Consider the evolution from an initial surface on which we specify $a = a_{(i)}$ and $\psi^A = \psi^A{}_{(i)}$ to a final surface on which we specify $a = a_{(f)}$ and $\tilde{\psi}^{A'} = \tilde{\psi}^{A'}{}_{(f)}$. For the Hartle–Hawking state we look at Euclidean solutions where either the initial or final surface is shrunk to zero in a regular way. It is easy to see that the case $\hat{\pi}_a = -6a$ corresponds to the initial surface being shrunk to zero, and that $\hat{\pi}_a = +6a$ corresponds to the final surface being shrunk to zero. In both cases we choose the point $\tau = 0$ to be the point where the surface is shrunk to $a = 0$.

If the initial surface is shrunk to zero, then, because ψ_A is constrained to be zero for all τ, we are only free to specify $\tilde{\psi}_{A'}$ as well as a on the final surface. In this case, therefore, the wave function has the form $\Psi = \Psi(a, \tilde{\psi}_{A'})$. In the case where the final surface is shrunk to zero, we are only free to specify ψ_A and a at the initial surface; the wave function is therefore given by $\Psi = \Psi(a, \psi_A)$.

One can easily calculate the classical Euclidean action in each of the two cases. In the first case we get $I = -3a^2_{(f)}$ and in the second case $I = -3a^2_{(i)}$. Note that the fermionic boundary contribution [D'Eath 1984, D'Eath & Halliwell 1987] to I is zero for these classical solutions, since either $\psi_A = 0$ or $\tilde{\psi}_{A'} = 0$ at the boundary. A change of sign in the Euclidean action occurs because of the change of sign in $\hat{\pi}_a$. However, there is another change of sign, because one is now considering the final surface rather than the initial surface. Hence, the Euclidean action is the same in both cases.

Consider the semi-classical approximation to the Hartle–Hawking wave function, given by

$$\Psi_{sc} = P \exp\left(-I/\hbar\right) , \tag{5.4.10}$$

where I is the classical Euclidean action, and P is a prefactor. The prefactor can be determined by considering the $O(1)$ and $O(\hbar)$ terms in the quantum version of the constraints (5.3.13) or (5.4.9). In the case where the initial surface is shrunk to zero, the wave function depends on $\tilde{\psi}^{A'}$ and a, with ψ^A represented by a fermionic derivative. The prefactor P is found to be a constant for this case. When the final surface is shrunk to zero, the wave function depends on ψ^A and a, with $\tilde{\psi}^{A'}$ represented by a fermionic derivative. For this case, the prefactor is found to be $\psi_A \psi^A$, up to multiplication by a constant.

Hence the Hartle–Hawking wave function in the semi-classical approximation, regarded in terms of eigenstates of either ψ^A or $\tilde{\psi}^{A'}$, has the form

$$\begin{aligned} &\Psi\left(a, \psi^A\right) = \psi_A \psi^A \exp\left(3a^2/\hbar\right) \\ \text{or} \quad &\tilde{\Psi}\left(a, \tilde{\psi}^{A'}\right) = \exp\left(3a^2/\hbar\right), \end{aligned} \tag{5.4.11}$$

up to possible multiplication by a constant. These two different representations of the same wave function are related by a Fourier transform (section 3.3). Note that, for this particular model, the semi-classical approximation is an exact solution to the quantum constraints. Thus, the Hartle–Hawking state for this model is given by (5.4.7) with $C = 0$.

Similarly, the ground wormhole quantum state [Hawking & Page 1990] corresponding to a Riemannian path integral outside the three-sphere, with the gravitational field asymptotically Euclidean at infinity, is given by (5.4.7) with $D = 0$.

5.5 Locally supersymmetric model with scalar-spin-1/2 matter: $k = +1$ Friedmann case

As described in section 5.1, a much richer and more interesting class of one-dimensional mini-superspace models with local supersymmetry is found by coupling supermatter to $N = 1$ supergravity in four dimensions, then reducing the model to one dimension by making a suitable homogeneous Ansatz, analogous to that studied in section 5.2. The simplest such four-dimensional models contain a matter super-multiplet, consisting of a complex massive scalar field ϕ and massive spin-1/2 field $\chi^A, \tilde{\chi}^{A'}$ [Das *et al.* 1977]. In chapter 8, we shall discuss the most general locally supersymmetric model of $N = 1$ supergravity coupled to supermatter [Wess & Bagger 1992], of which this is a special case. It will be seen that the theory might be finite, with suitable local boundary conditions of the type studied in this chapter, when the analytic potential $P(\phi^I)$ vanishes. This is analogous to setting the mass M to zero in Eqs. (5.5.7–8). Clearly, for a one-dimensional model, these fields should be chosen to be spatially homogeneous, depending only on time. The Ansatz can then be substituted into the action to obtain a one-dimensional Lagrangian and Hamiltonian. This complicated procedure is simplified enormously by using the algebraic computing language REDUCE in the fermionic context [Hughes 1990a,b]. For the one-dimensional model resulting from this reduction, it can be verified that the constraints are first-class, and that their classical algebra has the form expected for supersymmetry. Since, as is well known [Dirac 1965, Hanson *et al.* 1976], first-class constraints generate gauge invariances of the theory, this shows that the one-dimensional theory is locally supersymmetric.

The main interest is in the quantization of this one-dimensional model, and for this purpose we only need the classical constraints and basic Dirac bracket relations. In the one-dimensional model, the gravitational field is again described by the radius a, while the spin-3/2 field is described by the odd variables ψ^A and $\tilde{\psi}^{A'}$ (there being also the usual Lagrange multipliers $N, N^i, \psi^A{}_0$ and $\tilde{\psi}^{A'}{}_0$). In addition to these variables, there is a complex scalar field ϕ and a spin-1/2 field described by the odd variables χ^A and $\tilde{\chi}^{A'}$. The complex scalar ϕ could be written as $\phi = \alpha + i\beta$, where α and β are independent real scalars. However, just as in the pure supergravity case, when specifying boundary data for the path integral or the corresponding classical boundary-value problem, it is natural to remove reality conditions, as implied by allowing ψ^A and $\tilde{\psi}^{A'}$, or χ^A and $\tilde{\chi}^{A'}$, to be independent quantities. In the case of the scalar field, this can be done formally by writing $\tilde{\phi} = \alpha - i\beta$, instead of using the complex conjugate $\bar{\phi}$, and allowing α and β in principle to be complex.

It is simplest to describe the theory using only (say) unprimed spinors, and, to this end, we define (for example) $\tilde{\psi}_A = 2n_A{}^{B'}\tilde{\psi}_{B'}$ and $\tilde{\chi}_A = 2n_A{}^{B'}\tilde{\chi}_{B'}$. The Dirac bracket relations are taken to be:

$$[\psi_A, \tilde{\psi}_B]^* = i\epsilon_{AB}, \tag{5.5.1}$$

$$[\chi_A, \tilde{\chi}_B]^* = -i\epsilon_{AB}, \tag{5.5.2}$$

$$[a, \pi_a]^* = 1, \tag{5.5.3}$$

$$[\phi, \pi_\phi]^* = 1, \tag{5.5.4}$$

$$\left[\tilde{\phi}, \pi_{\tilde{\phi}}\right]^* = 1. \tag{5.5.5}$$

The brackets (5.5.1) and (5.5.3) are as in (5.3.11), as written using the above definition of $\tilde{\psi}_A$. The relative minus sign between (5.5.1) and (5.5.2) can be seen to arise necessarily from the substitution of the Ansatz into the original action. Any other factors can be absorbed by rescaling.

We additionally assume the following generalisation of the J_{AB} constraint of pure supergravity:

$$J_{AB} = \psi_{(A}\tilde{\psi}_{B)} - \chi_{(A}\tilde{\chi}_{B)} \approx 0. \tag{5.5.6}$$

One can justify this by observing either that it arises from the corresponding constraint of the full theory, or that its quantum version describes the invariance of the wave function under Lorentz transformations.

The classical constraints are found to be

$$\begin{aligned} S_A = {} & \psi_A\left(a\pi_a - 6ia^2\right) + \chi_A\pi_\phi \\ & - 4i\psi_C\psi^C\tilde{\psi}_A - 6i\tilde{\phi}\psi_C\chi^C\tilde{\psi}_A + 3i\psi_C\chi^C\tilde{\chi}_A + 6i\tilde{\phi}\chi_C\chi^C\tilde{\chi}_A \\ & + a^3M\tilde{\phi}^2\exp\left(6\phi\tilde{\phi}\right)\tilde{\psi}_A \\ & + a^2\frac{M}{3}\tilde{\phi}\left[1 + 6\phi\tilde{\phi}\right]\exp\left(6\phi\tilde{\phi}\right)\tilde{\chi}_A, \end{aligned} \tag{5.5.7}$$

$$\begin{aligned}\tilde{S}_A = & \ \tilde{\psi}_A\left(a\pi_a + 6ia^2\right) + \tilde{\chi}_A\pi_{\tilde{\phi}} \\ & + 4i\tilde{\psi}_C\tilde{\psi}^C\psi_A + 6i\phi\tilde{\psi}_C\tilde{\chi}^C\psi_A - 3i\tilde{\psi}_C\tilde{\chi}^C\chi_A - 6i\phi\tilde{\chi}_C\tilde{\chi}^C\chi_A \\ & - 2a^3M\phi^2\exp\left(6\phi\tilde{\phi}\right)\psi_A \\ & - a^3\frac{2M}{3}\phi\left[1+6\phi\tilde{\phi}\right]\exp\left(6\phi\tilde{\phi}\right)\chi_A, \qquad (5.5.8)\end{aligned}$$

$$J_{AB} = \psi_{(A}\tilde{\psi}_{B)} - \chi_{(A}\tilde{\chi}_{B)}, \qquad (5.5.9)$$

where M is the mass in [Das *et al.* 1977]. These constraints obey the algebra

$$[S_{(A}, \tilde{S}_{B)}]^* \overset{*}{\approx} 0, \qquad (5.5.10)$$

$$[S_A, S_B]^* \overset{*}{\approx} 0, \qquad (5.5.11)$$

$$[\tilde{S}_A, \tilde{S}_B]^* \overset{*}{\approx} 0, \qquad (5.5.12)$$

where $\overset{*}{\approx}$ is used to denote equality provided that the constraints $S_A, \tilde{S}_A$ and J_{AB} are used after all Dirac brackets are evaluated. The remaining constraint $\mathscr{H}$ is now *defined* to be equal to the Dirac bracket $[S_A, \tilde{S}^A]^*$. The expansion of the constraint $\mathscr{H}$ for the case when $M = 0$ is given in [Hughes 1990b]. One finds that results such as $[\mathscr{H}, S_A]^* \approx 0$ (where ≈ 0 denotes weak equality in the usual sense, i.e. that the bracket vanishes provided that all the constraints $\mathscr{H} = 0$, $S_A = 0$, $\tilde{S}_A = 0$ and $J_{AB} = 0$ hold) can be established using the Jacobi identity. Note that

$$\begin{aligned}[\mathscr{H}, S_A]^* &= \left[\left[S_B, \tilde{S}^B\right]^*, S_A\right]^* \\ &= -\left[S^B, \left[\tilde{S}_B, S_A\right]^*\right]^* + \left[\tilde{S}^B, [S_A, S_B]^*\right]^* \\ &= -\left[S^B, \frac{\epsilon_{BA}}{2}\left[\tilde{S}_C, S^C\right]^* + \left[\tilde{S}_{(B}, S_{A)}\right]^*\right]^* \\ &\quad + \left[\tilde{S}^B, [S_A, S_B]^*\right]^* \\ \Rightarrow \tfrac{1}{2}[\mathscr{H}, S_A]^* &= -\left[S^B, \left[\tilde{S}_{(B}, S_{A)}\right]^*\right]^* \\ &\quad + \left[\tilde{S}^B, [S_A, S_B]^*\right]^* \approx 0, \qquad (5.5.13)\end{aligned}$$

since the brackets $[\tilde{S}_{(B}, S_{A)}]^*$ and $[S_A, S_B]^*$ can both be expanded in terms of the constraints $S_A, \tilde{S}_A$ and J_{AB} (as a consequence of (5.5.10) and (5.5.11)), and the bracket of S^B or $\tilde{S}^B$ with any combination of these constraints is already known to close.

5.6 Quantization of supersymmetric model with scalar-spin-1/2 matter: $k = +1$ Friedmann case

We replace the Dirac brackets by quantum brackets with the following representation for the 'momentum' operators:

$$\pi_a \to -i\frac{\partial}{\partial a}, \quad \pi_\phi \to -i\frac{\partial}{\partial \phi}, \quad \pi_{\bar{\phi}} \to -i\frac{\partial}{\partial \bar{\phi}},$$
$$\bar{\psi}_A \to \frac{\partial}{\partial \psi^A}, \quad \bar{\chi}_A \to -\frac{\partial}{\partial \chi^A}. \tag{5.6.1}$$

As in section 5.4, we return to a bar notation for conjugates of quantum operators. For simplicity, we set $\hbar = 1$.

If we order the derivative operators in J_{AB} on the right, then this constraint acting on the wavefunction Ψ has the natural interpretation as implying that Ψ is Lorentz invariant:

$$J_{AB}\Psi = \psi_{(A}\frac{\partial}{\partial \psi^{B)}}\Psi + \chi_{(A}\frac{\partial}{\partial \chi^{B)}}\Psi = 0$$

$$\Rightarrow \Psi = A + iB\psi^C\psi_C + iC\psi^C\chi_C + iD\chi^C\chi_C + E\psi^C\psi_C\chi^D\chi_D, \tag{5.6.2}$$

where A, B, C, D and E are functions of a, ϕ and $\bar{\phi}$ only. The factors of i are chosen to simplify future results.

When Ψ is written in the representation (5.6.2), S_A and $\bar{S}_A$ become differential operators. One can attempt to use the criteria (1)–(4) described after Eq. (5.4.6) to determine a suitable factor ordering. In fact, in the present model, not all of these criteria can be satisfied simultaneously. It is, however, possible to satisfy (e.g.) criteria (1), (2) and (4) together. In chapter 6 a different ordering will be used. With this (somewhat arbitrary) choice, fermionic derivatives are ordered on the right in S_A and on the left in $\bar{S}_A$. We obtain

$$\begin{aligned} S_A = &- ia\psi_A\frac{\partial}{\partial a} - 6ia^2\psi_A - i\chi_A\frac{\partial}{\partial \phi} \\ &- 4i\psi_C\psi^C\frac{\partial}{\partial \psi^A} - 6i\bar{\phi}\psi_C\chi^C\frac{\partial}{\partial \psi^A} \\ &- 3i\psi_C\chi^C\frac{\partial}{\partial \chi^A} - 6i\bar{\phi}\chi_C\chi^C\frac{\partial}{\partial \chi^A} \\ &+ a^3M\bar{\phi}^2\exp\left(6\phi\bar{\phi}\right)\frac{\partial}{\partial \psi^A} \\ &- a^3\frac{M}{3}\bar{\phi}\left[1 + 6\phi\bar{\phi}\right]\exp\left(6\phi\bar{\phi}\right)\frac{\partial}{\partial \chi^A}, \end{aligned} \tag{5.6.3}$$

$$\bar{S}_A = -ia\frac{\partial}{\partial\psi^A}\frac{\partial}{\partial a} + 6ia^2\frac{\partial}{\partial\psi^A} + i\frac{\partial}{\partial\chi^A}\frac{\partial}{\partial\bar{\phi}}$$
$$+ 4i\epsilon^{BC}\frac{\partial}{\partial\psi^B}\frac{\partial}{\partial\psi^C}\psi_A - 6i\epsilon^{BC}\phi\frac{\partial}{\partial\psi^B}\frac{\partial}{\partial\chi^C}\psi_A$$
$$+ 3i\epsilon^{BC}\frac{\partial}{\partial\psi^B}\frac{\partial}{\partial\chi^C}\chi_A$$
$$- 6i\phi\epsilon^{BC}\frac{\partial}{\partial\chi^B}\frac{\partial}{\partial\chi^C}\chi_A$$
$$- 2a^3M\phi^2\exp\left(6\phi\bar{\phi}\right)\psi_A$$
$$- a^3\frac{2M}{3}\phi\left[1+6\phi\bar{\phi}\right]\exp\left(6\phi\bar{\phi}\right)\chi_A. \quad (5.6.4)$$

The operator versions of these constraints, $S_A\Psi = 0$ and $\bar{S}_A\Psi = 0$, give a set of partial differential equations for A, B, C, D and E. The constraint $\mathscr{H}\Psi = 0$ will be automatically satisfied due to the closing of the quantum version of the constraint algebra (the consistency criterion (4)). Therefore, the only equations which need to be satisfied are those arising from the two supersymmetry constraints:

$$-a\frac{\partial A}{\partial a} - 6a^2A + 2a^3M\bar{\phi}^2\exp\left(6\phi\bar{\phi}\right)B$$
$$-\frac{1}{3}a^3M\bar{\phi}\left(1+6\phi\bar{\phi}\right)\exp\left(6\phi\bar{\phi}\right)C = 0, \quad (5.6.5a)$$

$$-\frac{\partial A}{\partial\phi} + a^3M\bar{\phi}^2\exp\left(6\phi\bar{\phi}\right)C$$
$$-\frac{2}{3}a^3M\bar{\phi}\left(1+6\phi\bar{\phi}\right)\exp\left(6\phi\bar{\phi}\right)D = 0, \quad (5.6.5b)$$

$$-\frac{\partial B}{\partial\phi} + \frac{a}{2}\frac{\partial C}{\partial a} + 3a^2C + \frac{5}{2}C - 6\bar{\phi}B$$
$$-\frac{1}{3}a^3M\bar{\phi}\left(1+6\phi\bar{\phi}\right)\exp\left(6\phi\bar{\phi}\right)E = 0, \quad (5.6.5c)$$

$$-a\frac{\partial D}{\partial a} - 6a^2D + \frac{1}{2}\frac{\partial C}{\partial\phi} - 3D + 3\bar{\phi}C$$
$$+ a^3M\bar{\phi}^2\exp\left(6\phi\bar{\phi}\right)E = 0, \quad (5.6.5d)$$

$$-a\frac{\partial B}{\partial a} + 6a^2B + \frac{1}{2}\frac{\partial C}{\partial\bar{\phi}} - 3B + 3\phi C$$
$$+ a^3M\phi^2\exp\left(6\phi\bar{\phi}\right)A = 0, \quad (5.6.6a)$$

$$-\frac{\partial D}{\partial \bar{\phi}} + \frac{a}{2}\frac{\partial C}{\partial a} - 3a^2 C + \frac{5}{2}C - 6\phi D$$
$$-\frac{1}{3}a^3 M\phi\left(1+6\phi\bar{\phi}\right)\exp\left(6\phi\bar{\phi}\right)A = 0, \tag{5.6.6b}$$
$$-\frac{\partial E}{\partial \bar{\phi}} + a^3 M\phi^2 \exp\left(6\phi\bar{\phi}\right)C$$
$$-\frac{2}{3}a^3 M\bar{\phi}\left(1+6\phi\bar{\phi}\right)\exp\left(6\phi\bar{\phi}\right)B = 0, \tag{5.6.6c}$$
$$-a\frac{\partial E}{\partial a} + 6a^2 E + 2a^3 M\phi^2 \exp\left(6\phi\bar{\phi}\right)D$$
$$-\frac{1}{3}a^3 M\phi\left(1+6\phi\bar{\phi}\right)\exp\left(6\phi\bar{\phi}\right)C = 0. \tag{5.6.6d}$$

The general solution to this coupled set of differential equations gives the general quantum wave function. An analytic solution to these equations in the massive case $M > 0$ may not be possible. In the massless case ($M = 0$), the general solution has been found as an integral expression [D'Eath *et al.* 1991] – see section 6.4. The supersymmetry constraints do provide a natural way of representing the second-order Wheeler–DeWitt equation as a system of first-order differential equations, but the need to solve second-order differential equations is not necessarily avoided. A numerical solution to these constraints is possible, but boundary conditions must be specified for this purpose. The natural boundary condition to choose is the one given by the Hartle–Hawking condition; the ground wormhole quantum state has also been studied – see section 6.4. However, there is a further difficulty. In the non-supersymmetric massive scalar field model [Hawking 1984, Hawking & Wu 1985] the Euclidean action is easily seen to tend to zero at the boundary of mini-superspace. This suggests that, at least in a first approximation, we can take the wave function to be constant on the boundary. But, in the supersymmetric model, the exponential term $\exp(6\phi\bar{\phi})$ dominates in the Euclidean action close to the boundary of mini-superspace, causing the action to tend to $+\infty$. The semi-classical approximation $\exp(-I)$ thus tends to zero at the boundary, so that an approximation similar to that made for the massive scalar field model would give a zero wave function on the boundary. A more careful treatment of the semi-classical wave function close to the superspace boundary is therefore needed. If the wave function can be approximated close to the boundary, then this approximation can be used as initial data for the computed solution. The computational difficulties associated with the exponentially large terms occurring in the equations have not thus far been overcome in such a way that accuracy of the computed solution is assured. However, the results that have been obtained suggest an oscillatory behaviour in each of the five components of the wave function, which shows similarity to the behaviour of the massive scalar field model [Hawking & Wu 1985].

The supersymmetry constraints provide the relationship between the various components of the wave function. Further, the components satisfy a set of hyperbolic partial differential equations (the equation $\mathscr{H}\Psi = 0$). Consider, for example, the simplest case, where the mass $M = 0$. In this case, the components A and E are unrelated to B, C and D, and A and E satisfy equations of the same type as occurred in the study of pure supergravity, with simple exponential solutions. The remaining components each satisfy a (decoupled) second-order differential equation, most simply described by defining

$$\tilde{B} = B \exp\left(6\phi\bar{\phi}\right),$$
$$\tilde{C} = C \exp\left(6\phi\bar{\phi}\right),$$
$$\tilde{D} = D \exp\left(6\phi\bar{\phi}\right). \tag{5.6.7}$$

Then

$$a\frac{\partial}{\partial a}\left(a\frac{\partial \tilde{B}}{\partial a}\right) + 8a\frac{\partial \tilde{B}}{\partial a} - \left(36a^4 + 24a^2 - 15\right)\tilde{B} - \frac{\partial^2 \tilde{B}}{\partial\phi\partial\bar{\phi}} = 0,$$
$$a\frac{\partial}{\partial a}\left(a\frac{\partial \tilde{C}}{\partial a}\right) + 8a\frac{\partial \tilde{C}}{\partial a} - \left(36a^4 - 15\right)\tilde{C} - \frac{\partial^2 \tilde{C}}{\partial\phi\partial\bar{\phi}} = 0,$$
$$a\frac{\partial}{\partial a}\left(a\frac{\partial \tilde{D}}{\partial a}\right) + 8a\frac{\partial \tilde{D}}{\partial a} - \left(36a^4 - 24a^2 - 15\right)\tilde{D} - \frac{\partial^2 \tilde{D}}{\partial\phi\partial\bar{\phi}} = 0. \tag{5.6.8}$$

5.7 Supersymmetric Bianchi models

As described in section 5.1, the bosonic and filled fermionic states for Bianchi models are unique, of the form $\exp(\pm I/\hbar)$ for a suitable action function I. The middle fermionic states, however, are arguably more interesting [Csordás & Graham 1995], being given in terms of a general solution of a suitable Wheeler–DeWitt equation.

The general Bianchi model [Ryan & Shepley 1975, Asano *et al.* 1993] has a spatial metric h_{ij} of the form

$$h_{ij} = h_{\alpha\beta}(t) E^{\alpha}{}_{i} E^{\beta}{}_{j}, \tag{5.7.1}$$

$(\alpha, \beta, \ldots = 1, 2, 3)$. The invariant basis $E^\alpha{}_i$ obeys

$$\partial_{[i} E^\alpha{}_{j]} = C^\alpha{}_{\beta\gamma} E^\beta{}_i E^\gamma{}_j \tag{5.7.2}$$

as a consequence of the group invariance, where $C^\alpha{}_{\beta\gamma}$ are the structure constants of the isometry group. These can be decomposed as $C^\alpha{}_{\beta\gamma} = m^{\alpha\delta}\epsilon_{\delta\beta\gamma} + \delta^\alpha{}_{[\beta} a_{\gamma]}$, where $m^{\alpha\delta}$ is symmetric and $a_\gamma = C^\alpha{}_{\alpha\gamma}$. Bianchi class-$A$ models have $a_\gamma = 0$; Bianchi class-B models have $a_\gamma \neq 0$.

It is well known that, for Bianchi class-B models (but not class A), the equations of motion reduced from the full Einstein equations differ from the equations of motion derived from the reduced action. Further, one finds that, for class B, the quantum supersymmetry constraint acquires an extra term proportional to $C^\alpha{}_\alpha{}^\beta n_{AA'} \psi^A{}_\beta$. This should be zero, and so implies

$$C^\alpha{}_{\alpha\beta} = 0, \tag{5.7.3}$$

i.e. that the model must be of class A.

We now follow [Csordás & Graham 1995]. The spatial geometry is described by the tetrad components $e^A{}_\alpha$ and the gravitino field is given by the components $(\psi^A{}_\alpha, \tilde{\psi}^{A'}{}_\alpha)$, with $h_{\alpha\beta}(t) = e^a{}_\alpha e_{a\beta}$ $(\alpha, \beta, \ldots = 1, 2, 3)$. Here $e^a{}_\alpha$ and $(\psi^A{}_\alpha, \tilde{\psi}^{A'}{}_\alpha)$ are all functions of t. One takes the Lagrangian and computes the conjugate momenta $p_a{}^\alpha$ and $(\pi_A{}^\alpha, \tilde{\pi}_{A'}{}^\alpha)$. As usual, there are second-class constraints which must be eliminated by the Dirac procedure. The result is simplified by the introduction of a modified $p_{+a}{}^\alpha$ and modified momentum $\pi_A{}^\alpha$ as variables together with the original $e^a{}_\alpha$ and $\psi^A{}_\alpha$. The only non-zero Dirac brackets are

$$\begin{aligned} [e^a{}_\alpha, p_{+b}{}^\beta]^* &= \delta^a{}_b \delta_\alpha{}^\beta, \\ [\pi_A{}^\alpha, \psi^B{}_\beta]^* &= \epsilon_A{}^B \delta^\alpha{}_\beta. \end{aligned} \tag{5.7.4}$$

The supersymmetry generators $S_A, \tilde{S}_{A'}$ and Lorentz rotation generators $J_{AB}, \tilde{J}_{A'B'}$ have the form

$$S_A = -C^{BA'}{}_{\alpha\beta}\left(V m^{\gamma\delta} e^a{}_\delta + \frac{i}{2} p_+{}^{a\gamma}\right)\sigma_{aAA'}\pi_B{}^\beta, \tag{5.7.5}$$

$$\tilde{S}_{A'} = \left(V m^{\alpha\beta} e^a{}_\beta - \frac{i}{2} p_+{}^{a\alpha}\right)\sigma_{AA'a}\psi^A{}_\alpha, \tag{5.7.6}$$

$$J_{AB} = e_{(A}{}^{A'\alpha} p_{+B)A'\alpha} + \psi_{(A}{}^\alpha \pi_{B)\alpha}, \tag{5.7.7}$$

$$\tilde{J}_{A'B'} = e^A{}_{(A'}{}^\alpha p_{+B')A\alpha}. \tag{5.7.8}$$

Here V is the volume of the three-space with triad $E^\alpha{}_i$ (assumed compactified), and

$$C_{\alpha\beta}{}^{AA'} = \frac{-1}{2Vh^{1/2}}(i h_{\alpha\beta} n^a - \epsilon_{\alpha\beta\gamma} e^{a\gamma})\bar{\sigma}^{AA'}{}_a. \tag{5.7.9}$$

In the quantum theory, one takes

$$p_{+a}{}^{\alpha} = -i\hbar(\partial/\partial e^{a}{}_{\alpha}), \ \pi_{A}{}^{\alpha} = -i\hbar(\partial/\partial\psi^{A}{}_{\alpha}). \tag{5.7.10}$$

Note that there is an operator-ordering ambiguity between $p_{+}{}^{a\gamma}$ and $C^{BA'}{}_{\alpha\beta}$ in S_A, which is chosen to read as in Eq. (5.7.5). The algebra of first-class constraints closes. The commutator of any operator with $J_{AB}, \bar{J}_{A'B'}$ is standard, giving the Lorentz transformation of the operator. Further,

$$[S_A, S_B]_+ = 0, \ [\bar{S}_{A'}, \bar{S}_{B'}]_+ = 0, \tag{5.7.11}$$

$$[S_A, \bar{S}_{A'}]_+ = -\tfrac{1}{2}\hbar\,\mathscr{H}_{AA'}, \tag{5.7.12}$$

$$[\mathscr{H}_{AA'}, S_B]_- = -i\hbar\epsilon_{AB}\bar{D}_{A'}{}^{B'C'}\bar{J}_{B'C'}, \tag{5.7.13}$$

$$\begin{aligned}[\mathscr{H}_{AA'}, \bar{S}_{B'}]_- &= i\hbar\epsilon_{A'B'}J_{BC}D_{A}{}^{BC} \\ &= i\hbar\epsilon_{A'B'}\left[D_{A}{}^{BC}J_{BC} + i\hbar\bar{E}_{A}{}^{C'D'}\bar{J}_{C'D'} \right. \\ &\quad \left. + \frac{i\hbar n^{a}}{Vh^{1/2}}\sigma_{AC'a}\bar{S}^{C'}\right].\end{aligned} \tag{5.7.14}$$

Here the anticommutator (5.7.12) is taken as the definition of $\mathscr{H}_{AA'}$. It can be checked that $\mathscr{H}_{AA'}$ differs from the Hamiltonian constraints $\tilde{\mathscr{H}}_{AA'}$ which arise in the Hamiltonian reduction, as $\tilde{\mathscr{H}}_{AA'} = \sigma_{AA'a}(e^{a}{}_{\alpha}\mathscr{H}^{\alpha} + n^{a}\mathscr{H}_{\perp})$, only by terms proportional to Lorentz generators (see also [Teitelboim 1977a, Henneaux 1983]. Here $D_{A}{}^{BC}$ and $\bar{E}_{A}{}^{C'D'}$ are odd Grassmann operators; their explicit form is not needed.

Because of the chiral invariance of the theory, one can restrict attention to physical states which contain a particular number of fermions, which must be even by Lorentz invariance. Since there are six variables $\psi^{A}{}_{\alpha}$, one has states with 0, 2, 4, and 6 fermions. At the no-fermion level, one can immediately solve $\bar{S}_{A'}\Psi = 0$ to find the unique state

$$\Psi_0 = \text{const.}\ \exp(-I/\hbar), \tag{5.7.15}$$

where

$$I = -Vm^{\alpha\beta}h_{\alpha\beta}. \tag{5.7.16}$$

This was first derived by [Hughes 1986]. As in section 3.3, there is a duality between the ψ^0 and ψ^6 levels, given by a fermionic Fourier transform. One finds analogously that

$$\Psi_6 = \text{const.}\ \exp(I/\hbar)\prod_{\alpha}(\psi_{\alpha})^2. \tag{5.7.17}$$

These expressions for Ψ_0, Ψ_6 show how much more restrictive supergravity can be than ordinary quantum gravity, where one would have had a general solution of the Wheeler-DeWitt equation.

At the two-fermion level, one can consider the wave function

$$\Psi_2 = \bar{S}_{A'}\bar{S}^{A'} f(h_{\alpha\beta}), \tag{5.7.18}$$

with $\bar{S}^{A'} f \neq 0$. Here f is a Lorentz scalar since it only depends on the $h_{\alpha\beta}$. Hence Ψ_2 automatically obeys the Lorentz constraints, and also obeys $\bar{S}_{B'}\Psi_2 = 0$ because of the anticommutator relation (5.7.11). Using Eq. (5.7.12), the remaining constraint $S_A \Psi_2 = 0$ reads

$$[\mathscr{H}_{AA'}, \bar{S}^{A'}]_- f + 2\bar{S}^{A'} \mathscr{H}_{AA'} f = 0. \tag{5.7.19}$$

By Eq. (5.7.14), the first term involves terms proportional to $J_{BC}, \bar{J}_{C'D'}$ and $\bar{S}^{C'}$. Since f is a Lorentz scalar, only the $\bar{S}^{C'}$ term contributes. In the resulting form of Eq. (5.7.19), $\bar{S}^{C'}$ can be factored to the left, using $[\bar{S}^{C'}, \sigma_{AC'a} n^a / h^{1/2}]_- = 0$. Hence Eq. (5.7.19) can be solved if f obeys the Wheeler–DeWitt equation

$$\left(\mathscr{H}^{(0)}_{AA'} - \frac{\hbar^2}{V h^{1/2}} n^a \sigma_{AA'a} \right) f(h_{\alpha\beta}) = 0. \tag{5.7.20}$$

Here $\mathscr{H}^{(0)}_{AA'}$ consists of only the bosonic terms in $\mathscr{H}_{AA'}$, namely the terms which remain if $\pi_A{}^\alpha$ is brought to the right and then set to zero.

Similarly, in the four-fermion sector, one considers

$$\Psi_4 = S^A S_A \, g(h_{\alpha\beta}) \prod_{\alpha=1}^{3} (\psi_\alpha)^2. \tag{5.7.21}$$

The relevant Wheeler-DeWitt equation is

$$\mathscr{H}^{(1)}_{AA'} g(h_{\alpha\beta}) = 0, \tag{5.7.22}$$

where $\mathscr{H}^{(1)}_{AA'}$ consists of those terms in $\mathscr{H}_{AA'}$ which remain if the $\pi_A{}^\alpha$ are brought to the left and then set to zero. This construction gives a large class of solutions of the quantum constraints, at the two- and four-fermion level, although it is not clear whether or not it gives the general solution of the quantum constraints.

5.8 Bianchi-IX model

The Bianchi model closest to the $k = +1$ Friedmann model is the Bianchi-IX model, with S^3 spatial topology. Here the structure constants are $C^\alpha{}_{\beta\gamma} = \epsilon^\alpha{}_{\beta\gamma}$, whence $m^{\alpha\beta} = \delta^{\alpha\beta}$. The coordinates of the general Bianchi-IX model are the six components $h_{\alpha\beta}$ and six quantities $\psi^A{}_\alpha$. As pointed out by [Csordás & Graham 1995], one can restrict attention to *diagonal Bianchi-IX* models, in which $h_{\alpha\beta} = \mathrm{diag}(A^2, B^2, C^2)$: for example, in the four-fermion sector, the three equations

$$e^a{}_\alpha \bar{\sigma}^{AA'}{}_a \mathscr{H}^{(1)}_{AA'}(g) = 0$$

are automatically satisfied if $h_{\alpha\beta}$ is diagonal. The metric is thus

$$h_{ij} = A^2 E^1{}_i E^1{}_j + B^2 E^2{}_i E^2{}_j + C^2 E^3{}_i E^3{}_j, \tag{5.8.1}$$

where $E^1{}_i, E^2{}_i, E^3{}_i$ are a basis of left-invariant one-forms on the unit three-sphere. The general wave function, subject to the Lorentz constraints, must be made out of Lorentz invariants in the $\psi^A{}_\alpha$, leading to

$$\begin{aligned}\Psi &= \phi_0(A,B,C) + C_{\alpha\beta}(A,B,C)\psi^{A\alpha}\psi_A{}^\beta \\ &+ V^{\alpha\beta\gamma}(A,B,C) n_{AA'} e_B{}^{A'}{}_\alpha \psi^A{}_\beta \psi^B{}_\gamma \\ &+ O(\psi^4) + \phi_6(A,B,C)\prod_{\alpha=1}^{3}\psi^{A\alpha}\psi_{A\alpha},\end{aligned} \tag{5.8.2}$$

where $C_{\alpha\beta} = C_{(\alpha\beta)}$ is symmetric and $V^{\alpha\beta\gamma} = V^{\alpha[\beta\gamma]}$ is antisymmetric on its last two indices. The $C_{\alpha\beta}$ and $V^{\alpha\beta\gamma}$ provide 6 and 9 degrees of freedom respectively.

In the subsequent calculations, the following expression, formed from the three-dimensional connection, will be needed repeatedly:

$$\begin{aligned}{}^3\omega_{ABi}\, n^A{}_{B'} e^{BB'j} &= \frac{i}{4}\left(\frac{C}{AB} + \frac{B}{CA} - \frac{A}{BC}\right) E^1{}_i E^{1j} \\ &+ \frac{i}{4}\left(\frac{A}{BC} + \frac{C}{AB} - \frac{B}{CA}\right) E^2{}_i E^{2j} \\ &+ \frac{i}{4}\left(\frac{B}{CA} + \frac{A}{BC} - \frac{C}{AB}\right) E^3{}_i E^{3j}.\end{aligned} \tag{5.8.3}$$

Using this, the solutions for ϕ_0 and ϕ_6 in Eq. (5.8.2) are

$$\phi_0 = \text{const. } \exp\left[\frac{-1}{2\hbar}(A^2 + B^2 + C^2)\right], \tag{5.8.4}$$

$$\phi_6 = \text{const. } \exp\left[\frac{1}{2\hbar}(A^2 + B^2 + C^2)\right]. \tag{5.8.5}$$

It is interesting to note that ϕ_0 was also obtained as a solution of the Wheeler-DeWitt equation for Bianchi IX in pure gravity, with a suitable factor ordering ([Moncrief & Ryan 1991], following [Kodama 1990, Ashtekar & Pullin 1990]). The solution ϕ_0 corresponds to a wormhole quantum state (chapter 6) [Hawking & Page 1990] – see below.

At the more interesting two-fermion level, it turns out that the physical degrees of freedom are carried by $C_{11}, C_{22}, C_{33}, V^{123}, V^{231}$, and V^{312}. The other nine components of $C_{\alpha\beta}$ and $V^{\alpha\beta\gamma}$ are fixed uniquely by the quantum supersymmetry constraints, up to an overall constant factor. Meanwhile, the physical components $C_{11}, \ldots, V^{123}, \ldots$ are possibly given in terms of a general solution of the Wheeler–DeWitt equation (5.7.20), if one assumes Csordás and Graham's Ansatz (5.7.18).

The detailed calculation is too tedious to describe here, but one finds [Cheng & D'Eath 1995] from the $S_A\Psi = 0$ and $\bar{S}_{A'}\Psi = 0$ constraints that

$$\begin{aligned}
C_{12}, C_{13}, C_{23} &\propto \exp\left[\frac{-1}{2\hbar}(A^2+B^2+C^2)\right],\\
V^{112}, V^{113} &\propto \frac{1}{A^3BC}\exp\left[\frac{-1}{2\hbar}(A^2+B^2+C^2)\right],\\
V^{212}, V^{223} &\propto \frac{1}{AB^3C}\exp\left[\frac{-1}{2\hbar}(A^2+B^2+C^2)\right],\\
V^{313}, V^{323} &\propto \frac{1}{ABC^3}\exp\left[\frac{-1}{2\hbar}(A^2+B^2+C^2)\right].
\end{aligned} \tag{5.8.6}$$

The supersymmetry constraints relating the remaining coefficients $C_{11}, \ldots, V^{123}, \ldots$ are coupled partial differential equations

$$\begin{aligned}
&A^2\left[\frac{\hbar}{2}\left(C\frac{\partial}{\partial C} - A\frac{\partial}{\partial A} - B\frac{\partial}{\partial B}\right) + (A^2+B^2-C^2) - \frac{3}{2}\hbar\right]V^{231}\\
&+ A^2\left[\frac{\hbar}{2}\left(C\frac{\partial}{\partial C} + A\frac{\partial}{\partial A} - B\frac{\partial}{\partial B}\right) - (A^2+C^2-B^2) + \frac{3}{2}\hbar\right]V^{321}\\
&+ i\hbar\left(A\frac{\partial}{\partial A} - C\frac{\partial}{\partial C} - B\frac{\partial}{\partial B}\right)C_{11} + i(C^2+B^2-A^2)C_{11} = 0,
\end{aligned} \tag{5.8.7}$$

and cyclic permutations,

$$\begin{aligned}
&\hbar C\frac{\partial C_{11}}{\partial C} + C^2C_{11}\\
&\quad - \frac{i}{2}A^2\left(\hbar A\frac{\partial}{\partial A} + \hbar + A^2\right)V^{213} = 0,
\end{aligned} \tag{5.8.8}$$

$$\begin{aligned}
&\hbar B\frac{\partial C_{11}}{\partial B} + B^2C_{11}\\
&\quad + \frac{i}{2}A^2\left(\hbar A\frac{\partial}{\partial A} + \hbar + A^2\right)V^{312} = 0,
\end{aligned} \tag{5.8.9}$$

$$\begin{aligned}
&\left(\hbar B\frac{\partial}{\partial B} + B^2 + 3\hbar\right)V^{213}\\
&\quad + \left(\hbar C\frac{\partial}{\partial C} + C^2 + 3\hbar\right)V^{312} = 0,
\end{aligned} \tag{5.8.10}$$

and cyclic permutations.

Csordás and Graham's Ansatz (5.7.18) solves all these coupled constraints by taking $f(A,B,C)$ to be a solution of Eq. (5.7.20):

$$\begin{aligned}\hbar^2\Big(&-\frac{A^2}{4}\frac{\partial^2 f}{\partial A^2}-\frac{B^2}{4}\frac{\partial^2 f}{\partial B^2}-\frac{C^2}{4}\frac{\partial^2 f}{\partial C^2}\\&+\frac{AB}{2}\frac{\partial^2 f}{\partial A\partial B}+\frac{BC}{2}\frac{\partial^2 f}{\partial B\partial C}+\frac{CA}{2}\frac{\partial^2 f}{\partial C\partial A}\Big)\\&+\hbar^2\Big(-\frac{A}{4}\frac{\partial f}{\partial A}-\frac{B}{4}\frac{\partial f}{\partial B}-\frac{C}{4}\frac{\partial f}{\partial C}-f\Big)\\&+\frac{9}{4}\hbar ABCf-\frac{3}{4}(ABC)^2 f=0,\end{aligned} \tag{5.8.11}$$

and then defining [Eq. (5.7.18)]

$$\begin{aligned}C_{11}=\hbar^2&\Big(\frac{A^4}{8}\frac{\partial^2 f}{\partial A^2}+\frac{A^3}{2}\frac{\partial f}{\partial A}\Big)\\&+\hbar\Big(\frac{A^4BC}{4}\frac{\partial f}{\partial A}+\frac{5}{8}A^3BCf\Big)\\&+\frac{1}{8}A^4B^2C^2 f,\end{aligned} \tag{5.8.12}$$

and cyclic permutations, and

$$\begin{aligned}V^{123}&=i\hbar^2 BC\frac{\partial^2 f}{\partial B\partial C}\\&+\frac{i}{2}\hbar ABC\Big(B\frac{\partial f}{\partial B}+C\frac{\partial f}{\partial C}+f\Big)\\&+\frac{i}{2}(ABC)^2 f,\end{aligned} \tag{5.8.13}$$

and cyclic permutations. As remarked above, it is not at all clear whether this Ansatz gives the general solution of the quantum constraints (5.8.7-10) and their cyclic permutations.

One solution, given by Csordás and Graham, is

$$f=\text{prefactor}\ \ x\ \exp\Big[\frac{-1}{2\hbar}(A^2+B^2+C^2)\Big], \tag{5.8.14}$$

which again gives a wormhole state. To see this, consider the Hamilton–Jacobi equations, giving the classical flow corresponding to the action $I=\frac{1}{2}(A^2+B^2+C^2)$ appearing through $\exp[-I/\hbar]$ in Eq. (5.8.14). Fortunately, it has been shown [Gibbons & Pope 1979] that these Hamilton–Jacobi equations are first integrals for a particular class of self-dual four-metrics. The four-geometries are asymptotically Euclidean with metrics [Gibbons

& Pope 1979, Belinskii *et al.* 1978]

$$ds^2 = F^{-\frac{1}{2}}d\rho^2 + \frac{1}{4}F^{\frac{1}{2}}\rho^2\left[\left(1 - \frac{a_1{}^4}{\rho^4}\right)^{-1} E^1{}_i E^1{}_j\right.$$
$$+ \left(1 - \frac{a_2{}^4}{\rho^4}\right)^{-1} E^2{}_i E^2{}_j$$
$$\left. + \left(1 - \frac{a_3{}^4}{\rho^4}\right)^{-1} E^3{}_i E^3{}_j\right] dx^i dx^j, \qquad (5.8.15)$$

where

$$F = \left(1 - \frac{a_1{}^4}{\rho^4}\right)\left(1 - \frac{a_2{}^4}{\rho^4}\right)\left(1 - \frac{a_3{}^4}{\rho^4}\right), \qquad (5.8.16)$$

and a_1, a_2, a_3 are constants. Given a Bianchi-IX three-geometry (5.7.1), one can find a four-geometry (5.8.15–16), by suitable choice of a_1, a_2, a_3 which has the three-geometry at its interior boundary. Further, the action of the exterior four-geometry is I in (5.8.14). Hence the quantum state $\Psi_0 = \text{const. } \exp(-I/\hbar)$ is the (ground) quantum wormhole state [Hawking & Page 1990] in the Bianchi-IX case (see also chapter 6), corresponding to a path integral in which one specifies the three-geometry (5.7.1) on an inner S^3 boundary, and requires the four-geometry to be asymptotically Euclidean.

There is a second family of self-dual Bianchi-IX metrics, the Atiyah–Hitchin solutions [Atiyah & Hitchin 1985], which are regular in the interior, although not asymptotically Euclidean. These arise from the Hamilton–Jacobi equations for a different action [Gibbons & Pope 1979]

$$I_1 = \tfrac{1}{2}\left(A^2 + B^2 + C^2\right) - (BC + CA + AB). \qquad (5.8.17)$$

This action corresponds to the Hartle–Hawking state, and one finds that it obeys the Hamilton–Jacobi equation arising from (5.8.11). Thus one expects there to be a Hartle–Hawking state with semi-classical approximation

$$f = \text{prefactor } x \exp(-I_1/\hbar). \qquad (5.8.18)$$

One can verfy that

$$I_2 = \tfrac{1}{2}(A^2 + B^2 + C^2) + AB + AC - BC, \qquad (5.8.19)$$

$$I_3 = \tfrac{1}{2}(A^2 + B^2 + C^2) + AB - AC + BC, \qquad (5.8.20)$$

$$I_4 = \tfrac{1}{2}(A^2 + B^2 + C^2) - AB + AC + BC, \qquad (5.8.21)$$

also obey the Hamilton–Jacobi equation. Hence, analogously one expects that they will give semi-classical approximations to physical quantum states.

One can trade the wormhole state for the Hartle–Hawking state in the Bianchi-IX case [Graham & Luckock 1994], when a modified definition of homogeneous spinor fields is used. Here the spinor components have the opposite sign at antipodal points of the spatial three-manifold, rather than the same sign as used above. In such a case, the Hartle–Hawking state is found in the bosonic sector. The above dependence of physical states on the fermionic boundary conditions might have been expected on the basis of section 2.9. Positive-frequency fermionic boundary data will lead to fermionic classical solutions which are regular in the interior, and hence to the Hartle–Hawking state. Negative-frequency data will lead to fermionic classical solutions which decay at infinity, and hence to the wormhole ground state [Hawking & Page 1990].

Further references on supersymmetric Bianchi models may be found in [Csordás & Graham 1995].

5.9 Supersymmetric $k = +1$ Friedmann model with Λ-term

The $k = +1$ Friedmann model without a Λ-term was treated in sections 5.2–5.4. There are two linearly independent physical quantum states. One is bosonic and corresponds to the wormhole state; the other is at quadratic order in fermions and corresponds to the Hartle–Hawking state. In the Friedmann model with Λ-term, the coupling between the different fermionic levels 'mixes up' this pattern.

In the Friedmann model, the wave function has the form (section 5.4)

$$\Psi = \Psi_0(A) + (\beta_C \beta^C)\Psi_2(A). \tag{5.9.1}$$

As part of the Ansatz of section 5.2, one requires $\psi^A{}_i = e^{AA'}{}_i \tilde{\psi}_{A'}$ and $\tilde{\psi}^{A'}{}_i = e^{AA'}{}_i \psi_A$; this is in order that the form of the one-dimensional Ansatz should be preserved under one-dimensional local supersymmetry, suitably modified by local coordinate and Lorentz transformations. Thus the gravitino field is truncated to spin 1/2. Note that $\beta^A = \frac{3}{4} n^{AA'} \tilde{\psi}_{A'}$.

The $\bar{S}_{A'}\Psi = 0$ and $S_A \Psi = 0$ constraints at level ψ^1 give [Cheng *et al.* 1994]

$$\hbar\kappa^2 \frac{d\Psi_0}{dA} + 48\pi^2 A \Psi_0 + 18\pi^2 \hbar g A^2 \Psi_2 = 0, \tag{5.9.2}$$

$$\hbar^2\kappa^2 \frac{d\Psi_2}{dA} - 48\pi^2 \hbar A \Psi_2 - 256\pi^2 g A^2 \Psi_0 = 0. \tag{5.9.3}$$

These give second-order equations, for example

$$A\frac{d^2\Psi_0}{dA^2}-2\frac{d\Psi_0}{dA}+\left[-\frac{48\pi^2}{\hbar\kappa^2}A-\frac{(48)^2\pi^4}{\hbar^2\kappa^4}A^3+\frac{9\times 512\pi^4 g^2}{\hbar^2\kappa^4}A^5\right]\Psi_0=0. \quad (5.9.4)$$

This has a regular singular point at $A = 0$, with indices $\lambda = 0$ and 3. There are two independent solutions, of the form

$$\begin{aligned}\Psi_0 &= a_0 + a_2A^2 + a_4A^4 + \ldots,\\ \Psi_0 &= A^3(b_0 + b_2A^2 + b_4A^4 + \ldots),\end{aligned} \quad (5.9.5)$$

convergent for all A. They obey complicated recurrence relations, where (e.g.) a_6 is related to a_4, a_2 and a_0.

One can look for asymptotic solutions of the type $\Psi_0 \sim (B_0 + \hbar B_1 + \hbar^2 B_2 + \ldots)\exp(-I/\hbar)$, and finds

$$I = \pm\frac{\pi^2}{g}(1 - 2g^2A^2)^{3/2}, \quad (5.9.6)$$

for $2g^2A^2 < 1$. The minus sign in I corresponds to taking the action of the classical Riemannian solution filling in smoothly inside the three-sphere, namely a portion of the four-sphere S^4 of constant positive curvature. This gives the Hartle–Hawking state [Hartle & Hawking 1983]. For $A^2 > (1/2g^2)$, the Riemannian solution joins onto the Lorentzian solution [Hartle 1986]

$$\Psi \approx \cos\left\{\hbar^{-1}\left[\frac{\pi^2(2g^2A^2-1)^{3/2}}{g}-\frac{\pi}{4}\right]\right\}, \quad (5.9.7)$$

which describes de Sitter spacetime.

These results suggest that there will be analogous quantum states in the Bianchi-IX case with a Λ-term (and not as stated erroneously in [Cheng *et al.* 1994], based on a too-restrictive choice of fermionic states.)

6

Supersymmetric quantum wormhole states

6.1 Wormholes

Wormholes are small tubes or handles which may join otherwise remote regions of spacetime [Hawking 1988]. The possibility of wormholes in any quantum theory of gravity means that effective local interactions are generated at low energies, which may differ greatly from the interactions of the bare theory [Hawking 1988]. In particular, scalar particles are in general expected to develop a mass of the order of the Planck mass. This seems unphysical, and is cured by supersymmetry. It will be seen in sections 6.3, 6.4 that in the locally supersymmetric model of sections 5.5, 5.6, with massless complex scalar and spin-1/2 fermionic partner coupled to supergravity, the mass remains zero in the presence of wormholes [Alty *et al.* 1992]. This is a very strong argument in favour of local supersymmetry in nature.

The introduction given here follows [Hawking 1988]. One works with Euclidean-signature (Riemannian) geometries, and attempts to calculate the effect of closed universes, which branch off from a larger universe, on the behaviour of particles in asymptotically flat space at energies low compared with the Planck energy. Consider, for example, a conformally invariant scalar field ϕ. In order to find the effect of the closed universe or wormhole on the matter field ϕ in the asymptotically flat spaces, one needs to calculate the Green's functions

$$\langle \phi(y_1)\,\phi(y_2)\ldots\phi(y_r)\,\phi(z_1)\,\phi(z_2)\ldots\phi(z_s)\rangle\,,$$

where $y_1,\ldots,y_r$ and $z_1,\ldots,z_s$ are points in the two asymptotic regions (which may be the same region). The Green's functions are found by a path integration over all matter fields ϕ and all metrics $g_{\mu\nu}$ that have one or two asymptotically flat regions and a handle or wormhole connecting them. Let S be a three-sphere, which is a cross-section of the closed universe or wormhole. The path integral can then be factorized into a part

$$\langle 0|\phi(y_1)\ldots\phi(y_r)|\psi\rangle,$$

which depends on the fields on one side of S, and a part

$$\langle\psi|\phi(z_1)\ldots\phi(z_s)|0\rangle,$$

which depends on the fields on the other side of S. Strictly, the path integral can be factorized in this way only when the regions at the two ends of the wormhole are separate asymptotic regions. However, even when they are the same region, one can neglect the interaction between the ends and factorize the path integral if the ends are widely separated.

The state $|0\rangle$ represents the usual particle scattering vacuum state defined by a path integral over asymptotically Euclidean metrics and matter fields that vanish at infinity (see section 2.5). $|\psi\rangle$ represents the quantum state of the closed universe or wormhole on the surface S. $|\psi\rangle$ obeys the quantum constraints of the theory. The solutions of the Wheeler–DeWitt equation that correspond (as with $|\psi\rangle$) to wormholes form a Hilbert space H_w with the inner product

$$\langle\psi_1|\psi_2\rangle = \int \mathscr{D}h_{ij}\mathscr{D}\phi_0 \Psi_1{}^*\Psi_2. \tag{6.1.1}$$

Let $|\psi_i\rangle$ be a basis for H_w. Then the Green's function can be written in the factorized form

$$\langle\phi(y_1)\ldots\phi(y_r)\,\phi(z_1)\ldots\phi(z_s)\rangle$$

$$= \sum_i \langle 0|\phi(y_1)\ldots\phi(y_r)|\psi_i\rangle\langle\psi_i|\phi(z_1)\ldots\phi(z_s)|0\rangle. \tag{6.1.2}$$

For small perturbations of a three-sphere of radius a, the wormhole quantum states $|\psi_i\rangle$ can be written in terms of

$$\Psi = \Psi_0(a, a_i, b_i, c_i, d_i)\,\Pi_n\psi_n(f_n). \tag{6.1.3}$$

Here a_i, b_i, c_i, d_i specify gravitational perturbation modes [Halliwell & Hawking 1985] (section 2.8), and

$$\phi_0 = (\sigma a)^{-1}\sum_n f_n Q_n, \tag{6.1.4}$$

where $\sigma^2 = 2/3\pi\,(m_p)^2$ with m_p the Planck mass, and the Q_n are the standard scalar harmonics on the three-sphere. The part of the Wheeler–DeWitt operator that acts on ψ_n is

$$-\frac{d^2}{df_n{}^2} + \left(n^2+1\right) f_n{}^2.$$

Thus one takes these wave functions to be harmonic-oscillator states

$$\psi_{nm} = \left[\frac{\beta^2}{\pi 2^{2m}(m!)^2}\right]^{1/4} e^{-\beta^2 f_n{}^2/2} H_m(\beta f_n), \tag{6.1.5}$$

where $\beta^4 = \left(n^2+1\right)$ and H_m are Hermite polynomials. The wave functions ψ_{nm} can be interpreted as corresponding to the closed universe containing m scalar particles in the nth harmonic mode. For simplicity, consider only the case in which the inhomogeneous gravitational modes a_i, b_i, c_i, d_i are not excited.

Had one instead studied spin-1/2 particles in a gravitational field, one would have obtained fermionic harmonics which start with $n = \frac{1}{2}$ [D'Eath & Halliwell 1987] (section 2.9), where the Wheeler–DeWitt equation has solutions which are again harmonic-oscillator eigenstates, with frequency independent of a. Similarly, spin-1 particles are described by modes which start with $n = 1$.

The scale factor a appears in the Wheeler–DeWitt equation through the operator

$$\frac{\partial^2}{\partial a^2} - a^2.$$

In a supersymmetric theory, the zero-point energies of the modes will be either subtracted or cancelled by fermions. The total wave function Ψ will satisfy the Wheeler–DeWitt equation if the gravitational part Ψ_0 is a harmonic-oscillator wave function in a with unit frequency, and level equal to the sum E of the energies of the matter-field harmonic oscillators.

Note that the wave function is exponentially damped at large a, whereas a cosmological wave function such as that of a Hartle–Hawking state tends to grow exponentially at large a, as exp(const. $a^2/\hbar$). The difference here is that one is looking at the closed universe from an asymptotically Euclidean region, instead of from a compact Euclidean space, as in the cosmological case. This changes the sign of the trace K surface term [Eq. (2.6.23)] in the action.

Quantum wormhole states can thus be regarded as normalisable solutions of the quantum constraints, defined for compact three-surfaces such as S^3. An alternative path-integral definition of quantum wormhole states [Hawking & Page 1990] will be discussed in section 6.2, and has already been mentioned in sections 5.4, 5.7.

One can now consider the matrix element

$$\begin{aligned}&\langle 0|\phi(y_1)\ldots\phi(y_r)|\psi\rangle \\ &= \int \mathscr{D}h_{ij}\mathscr{D}\phi_0 \Psi(h_{ij},\phi_0) \int \mathscr{D}g_{\mu\nu}\mathscr{D}\phi\ \phi(y_1)\ldots\phi(y_r)\, e^{-I(g,\phi)}.\end{aligned} \tag{6.1.6}$$

One assumes that the gravitational field is asymptotically flat at infinity, and that there is a three-sphere S^3 with induced metric h_{ij} at the inner boundary. The scalar field is taken to be zero at infinity, and to have the value ϕ_0 on S^3. The positions of the points y_i cannot be specified in a gauge-invariant manner. However, suppose that one is only concerned with the effects of wormholes on low-energy particle physics. Then the separation of the points y_i can be taken to be large compared with the Planck length, and they can be taken to lie in flat Euclidean space. Their positions can then be specified up to an overall translation and rotation of Euclidean space.

First consider a wormhole state $|\psi\rangle$ in which only the $n = 0$ homogeneous scalar mode is excited above its ground state. The integral over the wave function Ψ of the wormhole can then be replaced in Eq. (6.1.6) by

$$\int da\, df_0\, \psi_E(a)\psi_{0m}(f_0).$$

The path integral is over asymptotically Euclidean metrics whose inner boundary is a three-sphere S^3 of radius a and scalar fields with the constant value f_0 on S^3. The saddle point for the path integral is flat Euclidean space outside a three-sphere of radius a centred on a point x_0, and the scalar field

$$\phi = \frac{a\sigma f_0}{(x - x_0)^2}. \tag{6.1.7}$$

The energy–momentum tensor of this scalar field is zero. The action of the saddle point is $\left(a^2 + {f_0}^2\right)/2$. The determinant Δ of the small fluctuations about the saddle point will be independent of f_0; its precise form will not be important.

The integral over the coefficient f_0 of the $n = 0$ scalar harmonic will contain a factor of

$$\int df_0\, (f_0)^r\, e^{-(f_0)^2} H_m(f_0).$$

This is zero when m, the number of particles in the mode $n = 0$, is greater than r, the number of points y_i in the correlation function. One expects this, because each particle in the closed universe must be created or annihilated at a point y_i in the asymptotically flat region. If $r > m$, particles may be created at one point y_i and annihilated at another point y_j without going into the closed universe. However, such matrix elements are just products of flat-space propagators with matrix elements with $r = m$. It is therefore sufficient to consider only the case with $r = m$.

The integral over the radius a contains a factor

$$\int da\, a^m e^{-a^2} H_E(a)\Delta(a),$$

where $E = m$ is the level number of the radial harmonic oscillator. For small m, the dominant contribution comes from $a \sim 1$, that is, from wormholes of the Planck size. The value $C(m)$ of the integral will be ~ 1.

The matrix element will then be

$$D(m) \prod_i \frac{\sigma}{(y_i - x_0)^2},$$

where $D(m)$ is another factor ~ 1. One now integrates over the position x_0 of the wormhole, with a measure of the form $(m_p)^4 \, d{x_0}^4$, and over an orthogonal matrix O which specifies the orientation of the wormhole with respect to the points y_i. Here m_p is the Planck mass $(\hbar c/G)^{1/2}$. The $n = 0$ mode is invariant under O, so this second integral will have no effect, but the integral over x_0 will ensure that energy and momentum are conserved in the asymptotically flat region. This is what one would expect, because the Wheeler–DeWitt and momentum constraint equations imply that a closed universe has no energy or momentum.

One can interpret the resulting matrix element as follows: it is the same as if one were in flat space with an effective interaction of the form

$$F(m) \, (m_p)^{4-m} \, \phi^m \left(c_{0m} + c_{0m}^\dagger\right),$$

where $F(m)$ is another coefficient ~ 1 and c_{0m} and $c_{0m}^\dagger$ are the annihilation and creation operators for a closed universe containing m scalar particles in the $n = 0$ homogeneous mode. One can similarly calculate the matrix elements of products of ϕ between the vacuum and a closed-universe state containing m_0 particles in the $n = 0$ mode, m_1 particles in the $n = 1$ mode, and so on. Only the case of two particles in the $n = 1$ modes is significant, giving an effective interaction of the form

$$\nabla\phi . \nabla\phi \left(c_{12} + c_{12}^\dagger\right),$$

with a coefficient ~ 1.

In the case of spin-1/2 particles, to be studied in sections 6.3, 6.4, the matrix elements are equivalent to effective interactions of the form

$$(m_p)^{4-3m/2} \, \psi^m d_m + \text{H. c.}$$

Here ψ^m denotes some Lorentz-invariant combination of m spinor fields ψ (m even) or their adjoints $\bar{\psi}$, and d_m is the annihilation operator for a closed universe containing m spin-1/2 particles in $n = \frac{1}{2}$ modes. One can neglect the effect of closed universes with spin-1/2 particles in higher modes.

Wormholes will not interact significantly with each other, unless their separation is of the order of the Planck length. Thus, the creation and annihilation operators for wormholes are practically independent of the positions in the asymptotically flat region. This means that the effective propagator of a wormhole excited state is $\delta^4(p)$. Using the propagator, one can calculate Feynman diagrams that include wormholes, in the usual manner.

Consider the matrix element for the scalar field, and its complex conjugate, between the vacuum and a closed universe containing a scalar particle and antiparticle in the $n = 0$ mode. One can allow for the possibility of coupling to a Yang–Mills field; then, one has to average over all orientations of the gauge group for the closed universe. The matrix element will be non-zero, because a particle–antiparticle state contains a Yang–Mills singlet. It would give an effective interaction of the form

$$(m_p)^2 \, tr\left(\phi\bar{\phi}\right)\left(c_{011} + c_{011}^{\dagger}\right),$$

where c_{011} is the annihilation operator for a closed universe with one scalar particle and one antiparticle in the $n = 0$ mode. This would have a serious consequence; with two of these vertices, one could make a closed loop consisting of a closed universe (propagator $\delta^4(p)$) and a scalar particle (propagator $1/p^2$). This closed loop would be infrared divergent. One could cut off the divergence by giving the scalar particle a mass, but the effective mass would be the Planck mass. One might be able to remove this mass by renormalization, but the creation of closed universes would mean that a scalar particle would lose quantum coherence within a Planck length.

The difficulty is avoided in the locally supersymmetric model of sections 5.5, 5.6, with supergravity coupled to a massless complex scalar and spin-1/2 fermionic partner. The Lagrangian is chirally invariant (i.e. invariant under the rigid transformation $\psi \to \exp(i\theta)\psi$, $\bar{\psi} \to \exp(-i\theta)\bar{\psi}$ and its analogue for the gravitino), and one finds that this property is preserved in the effective field theory. Thus the spin-1/2 fermion remains massless, and by supersymmetry, the scalar must also remain massless. This is described in sections 6.3, 6.4. In the following section 6.2, we describe the path-integral characterization of quantum wormhole states, which will be needed in section 6.4.

6.2 Quantum wormhole states

The quantum wormhole states studied in the previous section can be described by means of a path integral [Hawking & Page 1990]. They have so far been regarded as normalizable solutions $\Psi(h_{ij}, \phi_0)$ of the quantum constraints. One can also consider theories for which there are classical wormhole solutions [Giddings & Strominger 1989, Hosoya & Ogura 1989, Verbin & Davidson 1989, Rey 1990]. However, classical wormholes only occur for particular models, whereas quantum wormhole states can be defined quite generally. If the wave functions are to correspond to wormholes rather than other kinds of spacetime, they should obey certain boundary conditions when the three-metric h_{ij} degenerates or becomes infinite.

The boundary conditions when h_{ij} degenerates should express the fact that the four-metric is non-singular. It is not clear what these boundary conditions should be in the full superspace of all three-metrics, but in mini-superspace models it seems reasonable to suppose that the wave function should be regular, or (depending on the factor ordering) possibly go as a power of the radius a as a approaches zero. It certainly should not oscillate an infinite number of times.

The boundary conditions on the ground wormhole state when h_{ij} is large should express the property that the four-metric is asymptotically Euclidean. One can interpret this as saying that there are no gravitational excitations in the asymptotic state. The ground state will follow from also imposing the condition that there are no matter excitations in the asymptotic region. As with the no-boundary (Hartle–Hawking) wave function, one can obtain the ground wave function from a path integral

$$\Psi_0\left(h_{ij}, \phi_0\right) = \int \mathscr{D}g_{\mu\nu} \mathscr{D}\phi \exp\left[-I\left(g_{\mu\nu}, \phi\right)\right] . \tag{6.2.1}$$

In the case of the no-boundary state, the path integral is over all compact metrics and matter fields with the given boundary values. But in the case of the vacuum or ground state, the path integral is over all asymptotically Euclidean metrics, and all matter fields that are zero, or gauge equivalent to zero, at infinity. The ground state will decay exponentially fast at large radii.

However, if one studies mini-superspace models (see below), there are other solutions of the Wheeler–DeWitt equation that are also regular at $a = 0$, and are damped at large radius. Some of these solutions can be expressed as superpositions of solutions that have a non-zero flux of a conserved quantity across the three-surface S. Such solutions cannot close off with a compact four-geometry, for then the flux would be zero. The behaviour at large radius indicates that these solutions are asymptotically Euclidean, and the regularity at $a = 0$ indicates that they are non-singular.

The wave functions of these excited-wormhole states can also be represented by path integrals. The metrics in the path integrals can be taken to be asymptotically Euclidean, which means that there are no gravitational excitations asymptotically. (One can also introduce gravitons in the asymptotic metric.) The matter fields have sources at infinity, which can be interpreted as saying that there are matter particles passing through the wormhole. This gives boundary conditions on the Wheeler–DeWitt equation that, at least in mini-superspace examples, allow only a discrete spectrum of solutions.

This may be illustrated in the model studied in [Hawking & Page 1990], in which a Friedmann universe with S^3 spatial sections of radius $a(\tau)$ contains a massless minimally coupled scalar field $\phi(\tau)$. This provides a simpler non-supersymmetric analogue of the theory to be studied in this chapter. The Wheeler–DeWitt equation is

$$\left[\frac{1}{a^2}\frac{\partial}{\partial a}\left(a\frac{\partial}{\partial a}\right) - \frac{1}{a^3}\frac{\partial^2}{\partial \phi^2} - a\right]\Psi(a,\phi) = 0, \tag{6.2.2}$$

where the factor ordering has been chosen such that the derivative terms give the Laplacian in mini-superspace. This has separable solutions of the form

$$\Psi_k(a,\phi) = \gamma_k(a)\, e^{ik\phi}, \tag{6.2.3}$$

where

$$\gamma_k(a) = \beta_1 J_{ik/2}\left(\frac{i}{2}a^2\right) + \beta_2 J_{-ik/2}\left(\frac{i}{2}a^2\right). \tag{6.2.4}$$

Provided that the coefficients are chosen such that $\beta_1 = -e^{k\pi/2}\beta_2$, then $\Psi_k(a,\phi)$ decays like $\exp\left(-\frac{1}{2}a^2\right)$ as $a \to \infty$. However, for small a, the Bessel functions in (6.2.4) behave as $a^{\pm ik}$, so that the solution oscillates an infinite number of times as $a \to 0$.

These separable solutions have a semi-classical interpretation. For $0 < a < k^{1/2}$, the wave function oscillates, corresponding to a classical Lorentzian Friedmann solution which expands from $a = 0$ to a maximum radius $k^{1/2}$ and then collapses to $a = 0$, containing a real scalar field ϕ with conserved flux $q = 2\pi^2 k$. The singularities of the classical solution at $a = 0$ correspond to the infinite number of oscillations of the wave function as $a \to 0$. For $a > k^{1/2}$, the wave function decays exponentially as $\exp\left(-\frac{1}{2}a^2\right)$. In this region, the wave function corresponds to a classical Riemannian solution (the analytical continuation of the preceding Lorentzian solution) in which a wormhole, of minimum radius $k^{1/2}$, connects two asymptotically Euclidean regions. The classical scalar field ϕ is imaginary, with a conserved flux $2\pi^2 ik$.

Because of the infinite oscillation near $a = 0$, the separable solutions (6.2.3, 6.2.4) do not obey the regularity condition at $a = 0$ needed for a quantum wormhole state. Fortunately [Hawking & Page 1990], the desired quantum states can be found using the change of variables

$$x = a\sinh\phi, \qquad y = a\cosh\ \phi, \tag{6.2.5}$$

with the property that $(x \pm y)$ are null coordinates for the Wheeler–DeWitt equation (6.2.2), which becomes

$$\left(\frac{\partial^2}{\partial y^2} - \frac{\partial^2}{\partial x^2} - y^2 + x^2\right)\Psi = 0. \tag{6.2.6}$$

This clearly has solutions which are regular at the origin and damped at infinity, of the form

$$\Psi_n(x, y) = \psi_n(x)\psi_n(y), \tag{6.2.7}$$

where $\psi_n(x)$ is a harmonic oscillator eigenstate. These are the quantum wormhole states for this model. Once such states have been found, one can then proceed along the lines of section 6.1 or [Hawking 1988] to calculate, for example, the contribution of such wormholes to effective particle coupling constants.

The harmonic oscillator states Ψ_n may be expressed in terms of the separable states Ψ_k and *vice versa* [Hawking & Page 1990]. For example, the ground quantum wormhole state ($n = 0$) may be written

$$\Psi_0(a, \phi) = \frac{i}{4\sqrt{\pi}} \int_{-\infty}^{\infty} dk \frac{\left[e^{k\pi/4} J_{ik/2}\left(\frac{ia^2}{2}\right) - e^{-k\pi/4} J_{-ik/2}\left(\frac{ia^2}{2}\right)\right] e^{ik\phi}}{\sinh\left(\frac{k\pi}{2}\right)}. \tag{6.2.8}$$

The integrand in (6.2.8) decays at the rate $\exp\left(-\frac{1}{2}a^2\right)$ as $a \to \infty$. However, the coefficients $\beta_1(k)$ and $\beta_2(k)$ multiplying $J_{\pm ik/2}\left(\frac{ia^2}{2}\right)$ in (6.2.8) are such that $\Psi_0(a, \phi)$ has the correct more rapid decrease at the rate $\exp\left(-I_{\text{class}}\right)$ as $a \to \infty$, where $I_{\text{class}} = \frac{1}{2}a^2 \cosh(2\phi)$ is the Euclidean action of the asymptotically flat solution outside a sphere of radius a with ϕ specified. A similar delicate cancellation is crucial in finding the analogous ground state in the supersymmetric model (section 6.4). Given the ground state (6.2.8), excited states for $n = 1, 2, 3, \ldots$ may be found by repeated application of $\partial/\partial\phi$ together with a Gram–Schmidt construction [Courant & Hilbert 1953] to ensure the states are orthonormal.

The supersymmetric theory to be used here, with $N = 1$ supergravity coupled to a complex scalar $(\phi, \tilde{\phi})$ and spin-1/2 field $(\chi^A, \tilde{\chi}^{A'})$, was described in sections 5.5, 5.6 following [D'Eath & Hughes 1992]. In section 6.3 we describe the quantization of the model. In section 6.4, we find the wormhole ground state, and hence the excited states, with the help of a representation of the general physical quantum state, analogous to (6.2.8) [Alty *et al.* 1992]. We also give an argument showing that chirality is preserved in the effective action, allowing for wormhole effects, so that the spin-1/2 fermion and complex scalar remain massless, thus avoiding the difficulty described in section 6.1.

6.3 Quantization of supergravity–supermatter model

Consider the model studied in sections 5.5, 5.6, with Friedmann $k = +1$ geometry, based on the Lagrangian of [Das *et al.* 1977]. The dynamical variables are the radius a and its canonical momentum π_a, the gravitino field $(\psi^A, \tilde{\psi}^{A'})$, spin-1/2 field $(\chi^A, \tilde{\chi}^{A'})$ and complex scalar $(\phi, \bar{\phi})$ with momenta $\pi_\phi, \pi_{\bar{\phi}}$. Now apply the quantum constraints J^{AB}, S^A and $\bar{S}^{A'}$ to a typical wave function Ψ.

The constraint $J^{AB}\Psi = 0$ implies that Ψ is invariant under the internal rotations generated by J_{AB}. For this constraint, the general solution is a wave function formed from invariants in ψ^A and χ^A [D'Eath & Hughes 1988,1992]:

$$\Psi = A + iB\psi^C\psi_C + iC\psi^C\chi_C + iD\chi^C\chi_C - \tfrac{1}{2}E\psi^C\psi_C\chi^D\chi_D, \tag{6.3.1}$$

where A, B, C, D, E are functions of $\left(a, \phi, \bar{\phi}\right)$ only. Invariants such as $\psi^C\psi_C = \epsilon_{BA}\psi^A\psi^B$ are non-zero since ψ^A and χ^A are odd Grassmann quantities; for the same reason, the series in (6.3.1) terminates. One can think of the complex-valued functions $A, \ldots, E$ as wave functions in different sectors, containing different numbers of spin-1/2 and spin-3/2 fermions (the total number being even).

The factor ordering in the S_A and $\bar{S}_A$ quantum constraints can be restricted by certain reasonable criteria [D'Eath & Hughes 1988,1992]. Note that the operator version of S_A, given classically by (5.5.7), involves only first-order bosonic and fermionic derivatives in the representation used here. If S_A is ordered with these derivatives on the right, as in Eq. (5.5.7), then the quantum constraint $S_A\Psi = 0$ describes the transformation of the wave function Ψ under an infinitesimal supersymmetry transformation of its arguments $(a, \phi, \bar{\phi}, \psi^A, \chi^A)$, generated by S_A [D'Eath 1984, D'Eath & Hughes 1988, 1992]. The transformation is that which would naively be expected from the change in the action (and hence the path integral) produced by a supersymmetry transformation with parameter $\epsilon_A(\tau)$. Similarly, if $\bar{S}_A$ is ordered with $\pi_a, \pi_{\bar{\phi}}, \psi_A$ and χ_A on the right, as in (5.5.8), then the corresponding quantum constraint describes simply the transformation of the wave function under a transformation with parameter $\tilde{\epsilon}_{A'}(\tau)$, when one uses the conjugate representation with arguments $(a, \phi, \bar{\phi}, \bar{\psi}^A, \bar{\chi}^A)$. Recall the definition (section 5.5) $\bar{\psi}_A = 2n_A{}^{B'}\tilde{\psi}_{B'}, \bar{\chi}_A = 2n_A{}^{B'}\tilde{\chi}_{B'}$, and similarly for $\bar{S}_A$. This ordering of S_A and $\bar{S}_A$ was used in section 5.6 [D'Eath *et al.* 1991, D'Eath & Hughes 1992], and leads to a consistent set of quantum constraints (5.6.4). The algebra of quantum constraints closes: i.e. the operator versions of equations such as (5.5.10), involving (anti)commutators of the constraint operators $\mathscr{H}, S_A, \bar{S}_A$ and J_{AB}, give right-hand sides in which the factors of $\mathscr{H}, S, \bar{S}$ or J appear ordered on the right. In this case the constraints $J_{AB}\Psi = 0$, $S_A\Psi = 0$ and $\bar{S}_A\Psi = 0$ automatically imply the remaining constraint $\mathscr{H}\Psi = 0$ (section 5.6) [D'Eath 1984, D'Eath & Hughes 1988, 1992, D'Eath *et al.* 1991, Teitelboim 1977a].

However, this 'naive' factor ordering does not have the property that the operators S_A and $\bar{S}_{A'}$ are Hermitian adjoints in the standard inner product, appropriate to the holomorphic representation being used here for the fermions (section 3.4) [Faddeev & Slavnov 1980, D'Eath & Hughes 1988]. If one allows for the factor-ordering ambiguity in S_A, due to the terms cubic in fermions, and insists that $\bar{S}_{A'}$ be the Hermitian adjoint of S_A, the operators have the form

$$S_{A\,\mathrm{new}} = S_A + i\lambda\psi_A + i\mu\bar{\phi}\chi_A, \tag{6.3.2}$$

$$\bar{S}_{A\,\mathrm{new}} = \bar{S}_A + i\,(10-\lambda)\,\bar{\psi}_A + i\,(18-\mu)\,\phi\bar{\chi}_A. \tag{6.3.3}$$

Here S_A and $\bar{S}_A$ on the right-hand side are the operators with the ordering of (5.5.7),(5.5.8) or (5.6.3),(5.6.4), and λ, μ are real numbers. The consistency of the quantum theory can then be checked, by studying the resulting equations analogous to (6.3.6),(6.3.7) below. These constraints are only consistent provided

$$\lambda = 5, \qquad \mu = 9. \tag{6.3.4}$$

We shall adopt this factor ordering here; in this case the 'quantum corrections' in (6.3.2,6.3.3) modify the naive transformation rules for the wave function under supersymmetry.

When written out in terms of the expansion (6.3.1) of the wave function Ψ, the supersymmetry constraints become a system of eight coupled first-order equations. Here for simplicity the massless case $M = 0$ is considered. The equations are most simply described in terms of

$$\tilde{B} = B\exp\left(-3\phi\bar{\phi}\right),$$
$$\tilde{C} = C\exp\left(-3\phi\bar{\phi}\right), \tag{6.3.5}$$
$$\tilde{D} = D\exp\left(-3\phi\bar{\phi}\right).$$

Then

$$\left(a\frac{\partial}{\partial a} + 6a^2 - 5\right)A = 0, \tag{6.3.6a}$$

$$\left(\frac{\partial}{\partial\phi} - 9\bar{\phi}\right)A = 0, \tag{6.3.6b}$$

$$\left(a\frac{\partial}{\partial a} + 6a^2\right)\tilde{C} - 2\frac{\partial\tilde{B}}{\partial\phi} = 0, \tag{6.3.6c}$$

$$2\left(a\frac{\partial}{\partial a}+6a^2-2\right)\tilde{D}-\frac{\partial\tilde{C}}{\partial\phi}=0, \qquad (6.3.6d)$$

$$2\left(a\frac{\partial}{\partial a}-6a^2-2\right)\tilde{B}-\frac{\partial\tilde{C}}{\partial\bar{\phi}}=0, \qquad (6.3.7a)$$

$$\left(a\frac{\partial}{\partial a}-6a^2\right)\tilde{C}-2\frac{\partial\tilde{D}}{\partial\bar{\phi}}=0, \qquad (6.3.7b)$$

$$\left(\frac{\partial}{\partial\bar{\phi}}-9\phi\right)E=0, \qquad (6.3.7c)$$

$$\left(a\frac{\partial}{\partial a}-6a^2-5\right)E=0. \qquad (6.3.7d)$$

The equations (6.3.6*a*), (6.3.6*b*) and (6.3.7*c*), (6.3.7*d*) for A and E can be solved immediately to give

$$A\left(a,\phi,\bar{\phi}\right)=A_0a^5f(\bar{\phi})\exp\left(9\phi\bar{\phi}-3a^2\right), \qquad (6.3.8)$$

$$E\left(a,\phi,\bar{\phi}\right)=E_0a^5g(\phi)\exp\left(9\phi\bar{\phi}+3a^2\right), \qquad (6.3.9)$$

where $f\left(\bar{\phi}\right)$ is an antiholomorphic function and $g(\phi)$ is a holomorphic function.

The four remaining Dirac-like equations (6.3.6*c*), (6.3.6*d*), (6.3.7*a*), (6.3.7*b*) for the components $\tilde{B},\tilde{C},\tilde{D}$ lead (consistently) to a set of Wheeler–DeWitt equations for $\tilde{B},\tilde{C},\tilde{D}$. For instance

$$a\frac{\partial}{\partial a}\left(a\frac{\partial\tilde{C}}{\partial a}\right)-2a\frac{\partial\tilde{C}}{\partial a}-36a^4\tilde{C}-\frac{\partial^2\tilde{C}}{\partial\phi\partial\bar{\phi}}=0. \qquad (6.3.10)$$

6.4 Quantum wormhole states with supersymmetry

One can examine whether the differential equations (6.3.6), (6.3.7) admit any solutions (A,B,C,D,E) describing wormhole quantum states, as characterized in section 6.2. In particular, we look for the ground state. It is convenient to separate variables when considering the dependence on the scalar field ϕ, writing

$$\phi=\rho e^{i\theta}, \qquad \bar{\phi}=\rho e^{-i\theta}, \qquad (6.4.1)$$

where ρ and θ are real. The structure of the supersymmetry constraints (6.3.6*c*–6.3.7*b*) allows one to consider cases in which B, C and D depend on θ through factors of the form $e^{in\theta}$. Here we expect the ground state to be of the most symmetrical type in which, regarded as functions of (a,ρ,θ), $B\propto e^{i\theta}$, C is independent of θ, and $D\propto e^{-i\theta}$.

The boundary conditions at large a are found from the path-integral description of the ground state. It follows from section 6.2 that for large

a the ground state should have the approximate form $P\exp(-I_c)$, where P is a prefactor and I_c is the Euclidean action of the bosonic classical solution which is asymptotically Euclidean, with $\rho \to 0$ as $a \to \infty$, and which has specified values of (a, ρ) on a three-sphere at its inner boundary. It is easily verified that this action is

$$I_c = 3a^2 \cosh(4\rho). \tag{6.4.2}$$

This is a particular solution of the Hamilton–Jacobi equation

$$\frac{1}{4}\left(\frac{\partial I}{\partial \rho}\right)^2 - a^2 \left(\frac{\partial I}{\partial a}\right)^2 + 36a^4 = 0. \tag{6.4.3}$$

The general solution must, on dimensional grounds, have the form $I = a^2 f(\rho)$, leading to

$$I(a, \rho) = \pm 3a^2 \,, \qquad \pm 3a^2 \cosh(4\rho). \tag{6.4.4}$$

If one only required that the wave function decay as $a \to \infty$, but did not specify the rate of decay, one would typically encounter wave functions with the weaker asymptotic behaviour $\exp\left(-3a^2\right)$. The stronger fall-off requirement $\exp(-I_c)$ will single out the wormhole states. It will also need to be checked that the wormhole states are suitably regular at $a = 0$.

The asymptotic form $\exp\left[-3a^2 \cosh(4\rho)\right]$, together with the general solution (6.3.8),(6.3.9) for A and E, immediately shows that

$$A = E = 0 \tag{6.4.5}$$

for the ground state. One might hope, by analogy with the solution (6.2.7) of the non-supersymmetric model, that the prefactor might be given by a terminating series:

$$\begin{aligned} \tilde{B} &= e^{i\theta} a^p \left[b_0(\rho) + b_1(\rho) a^{-2} + \ldots + b_n(\rho) a^{-2n}\right] \\ &\quad \times \exp\left[-3a^2 \cosh(4\rho)\right], \\ \tilde{C} &= a^p \left[c_0(\rho) + c_1(\rho) a^{-2} + \ldots + c_n(\rho) a^{-2n}\right] \\ &\quad \times \exp\left[-3a^2 \cosh(4\rho)\right], \\ \tilde{D} &= e^{-i\theta} a^p \left[d_0(\rho) + d_1(\rho) a^{-2} + \ldots + d_n(\rho) a^{-2n}\right] \\ &\quad \times \exp\left[-3a^2 \cosh(4\rho)\right], \end{aligned} \tag{6.4.6}$$

for a suitable constant p. By imposing the constraints $(6.3.6c - 6.3.7b)$, one finds

$$\begin{aligned} c_0(\rho) &= \text{const.}\ \ \rho^{-1/2}(\sinh 4\rho)^{1/2(p-1)}, \\ b_0(\rho) &= \tfrac{1}{2}\tanh(2\rho)c_0(\rho), \\ d_0(\rho) &= \tfrac{1}{2}\coth(2\rho)c_0(\rho). \end{aligned} \tag{6.4.7}$$

Further coefficients such as $c_1(\rho)$, $c_2(\rho)$, etc. can be found by solving recurrence relations. One can show that, unfortunately, there is no terminating solution of the form (6.4.6); such an approach would only generate an infinite asymptotic series for the prefactor.

As in the example of section 6.2, one can find solutions to the constraints by separating variables. For Eq. (6.3.10) one obtains the solution

$$\tilde{C}(a,\rho) = aZ_{\nu(k)}\left(3ia^2\right)J_0(k\rho), \tag{6.4.8}$$

depending on the parameter k, where Z denotes a Bessel function, and $\nu(k) = \frac{1}{4}\sqrt{(4-k^2)}$. The corresponding $\tilde{B}$ and $\tilde{D}$ which then provide solutions of the coupled equations $(6.3.6c - 6.3.7b)$ are

$$\tilde{B}\,(a,\rho,\theta) = k^{-1}e^{i\theta}\left(a\frac{\partial}{\partial a} + 6a^2\right)\left[aZ_{\nu(k)}\left(3ia^2\right)\right]J_1(k\rho), \tag{6.4.9}$$

$$\tilde{D}\,(a,\rho,\theta) = k^{-1}e^{-i\theta}\left(a\frac{\partial}{\partial a} - 6a^2\right)\left[aZ_{\nu(k)}\left(3ia^2\right)\right]J_1(k\rho). \tag{6.4.10}$$

These wave functions will decay as $\exp\left(-3a^2\right)$ as $a \to \infty$, provided one chooses

$$Z_{\nu(k)}\left(3ia^2\right) = H^{(1)}_{\nu(k)}\left(3ia^2\right), \tag{6.4.11}$$

up to a multiple. These separated solutions are analogous to those of [Hawking & Page 1990], given in Eqs. (6.2.3,6.2.4). They involve the powers $a^{\pm 2\nu(k)}$ as $a \to 0$, which for $k > 2$ again describe infinite oscillation. For $0 < k < 2$ or for k imaginary, they involve real powers of a as $a \to 0$. In particular, when $k^2 = -16m(m+1)$, for $m = 0, 1, 2, \ldots$, and the positive root $\nu(k) = m + \frac{1}{2}$ is taken, the a-dependence of the solution involves a finite series, since [Abramowitz & Stegun 1965]

$$H^{(1)}_{m+1/2}(z) = \sqrt{\frac{2}{\pi}}\,i^{-m-1}z^{-1/2}e^{iz}\sum_{\ell=0}^{m}\frac{(m+\ell)!}{\ell!\,\Gamma(m-\ell+1)}(-2iz)^{-\ell}. \tag{6.4.12}$$

For large m, these correspond to classical solutions of the field equations which are singular as $a \to 0$. None of the separated solutions corresponds to a quantum wormhole state, since their $\exp\left(-3a^2\right)$ fall-off at large a

is too slow (quite apart from their behaviour as $a \to 0$ being possibly unsuitable).

Wormhole states must be constructed by combining separated solutions for different values of k, by analogy with Eq. (6.2.8). One is led to study the Hankel transforms [Erdélyi *et al.* 1954]

$$\tilde{C}(a,\rho) = a \int_0^\infty dk\; k\; g(k) H^{(1)}_{\nu(k)}\left(3ia^2\right) J_0(k\rho), \tag{6.4.13a}$$

$$\tilde{B}(a,\rho,\theta) = e^{i\theta}\left(a\frac{\partial}{\partial a} + 6a^2\right) a \times \int_0^\infty dk\; g(k) H^{(1)}_{\nu(k)}\left(3ia^2\right) J_1(k\rho), \tag{6.4.13b}$$

$$\tilde{D}(a,\rho,\theta) = e^{-i\theta}\left(a\frac{\partial}{\partial a} - 6a^2\right) a \times \int_0^\infty dk\; g(k) H^{(1)}_{\nu(k)}\left(3ia^2\right) J_1(k\rho), \tag{6.4.13c}$$

for suitable choices of the weighting function $g(k)$. In estimating the asymptotic behaviour as $a \to \infty$, one uses [Abramowitz & Stegun 1965]

$$H^{(1)}_{\nu(k)}\left(3ia^2\right) \sim \left(\frac{2}{3i\pi a^2}\right)^{1/2} \exp\left[-3a^2 - \tfrac{1}{2}i\pi\nu(k) - \tfrac{1}{4}i\pi\right] \tag{6.4.14}$$

as $a \to \infty$, giving in general

$$\tilde{C}(a,\rho) \sim \left(\frac{2}{3i\pi}\right)^{1/2} \exp\left(-3a^2 - \tfrac{1}{4}i\pi\right) \times \int_0^\infty dk\; k\; g(k) \exp\left[-\tfrac{1}{2}i\pi\nu(k)\right] J_0(k\rho) \tag{6.4.15}$$

as $a \to \infty$, and similarly for $\tilde{B}$ and $\tilde{D}$. The integral in Eq. (6.4.15) gives the Hankel transform of the function $g(k)\exp[-\frac{1}{2}i\pi\nu(k)]$. For a typical function $g(k)$, this will be non-zero, and $\tilde{C}(a,\rho)$ will decay only at the rate $\exp\left(-3a^2\right)$ as $a \to \infty$. However, for the choice

$$g(k) = \exp\left[\tfrac{1}{2}i\pi\nu(k)\right], \tag{6.4.16}$$

one has the property

$$\int_0^\infty dk\; k\; \rho\; J_0(k\rho) = \delta(\rho), \tag{6.4.17}$$

which follows from the Hankel–transform inversion formula [Erdélyi *et al.* 1954]

$$\int_0^\infty dk\; k\; \rho\; J_0(k\rho')\, J_0(k\rho) = \delta(\rho - \rho'), \tag{6.4.18}$$

with $\rho' = 0$. Hence, for $\rho > 0$, $\tilde{C}(a,\rho)$ decays faster than $\exp\left(-3a^2\right)$ as $a \to \infty$. From the Hamilton–Jacobi Eq. (6.4.3), the only possibility is that of the desired $\exp\left[-3a^2 \cosh(4\rho)\right]$ fall-off. For $\rho = 0$ with $a \to \infty$, a more careful treatment is needed. An analogous situation occurs in the non-supersymmetric example, where the estimate (6.4.14) used in the integral of Eq. (6.2.8) would give naively $\Psi_0(a,\phi) \sim$ const. $\delta(\phi)a^{-1}\exp(-\frac{1}{2}a^2)$ as $a \to \infty$. The true behaviour in that example is $\Psi_0(a,\phi) =$ const. $\exp[-\frac{1}{2}a^2\cosh(2\phi)]$. Thus this rough estimate succeeds for $\phi \neq 0$ in predicting that Ψ_0 should have the faster fall-off of $\exp[-\frac{1}{2}a^2\cosh(2\phi)]$, but fails badly to show that Ψ_0 still has this fall-off for $\phi = 0$. One expects that a similar situation may hold in the present example, so that $\tilde{C}(a,\rho)$ does have the correct $\exp\left[-3a^2\cosh(4\rho)\right]$ fall-off as $a \to \infty$, for all values of ρ. A similar analysis leads to the same fall-off for $\tilde{B}(a,\rho,\theta)$ and $\tilde{D}(a,\rho,\theta)$ in Eqs. (6.4.13*b*, *c*), with $g(k)$ given by Eq. (6.4.16). Thus we have found a possible wormhole quantum state, which from its symmetry, and by analogy with the non-supersymmetric ground state (6.2.8), we expect also to be the ground state.

It remains to be verified that this quantum state is suitably regular as $a \to 0$. Consider first $\tilde{C}(a,\rho)$. By a change of variables, $\tilde{C}$ may be written as

$$\tilde{C}(a,\rho) = iNa\int_{-i}^{\infty} dK\, K\, e^{-K\pi/4} H^{(1)}_{iK/2}\left(3ia^2\right) J_0\left(2\rho\sqrt{1+K^2}\right), \qquad (6.4.19)$$

where N is a normalisation constant. This is split into the sum of the integrals $\int_{-i}^{0}(\quad)$ and $\int_0^{\infty}(\quad)$. The latter integral may be rewritten as

$$iNa\int_{-\infty}^{\infty} dK \frac{K e^{K\pi/4}}{\sinh\left(\frac{1}{2}K\pi\right)} J_{iK/2}\left(3ia^2\right) J_0\left(2\rho\sqrt{1+K^2}\right), \qquad (6.4.20)$$

on using [Abramowitz & Stegun 1965]

$$e^{-K\pi/4} H^{(1)}_{iK/2}\left(3ia^2\right) = \frac{\left[e^{K\pi/4} J_{iK/2}\left(3ia^2\right) - e^{-K\pi/4} J_{-iK/2}\left(3ia^2\right)\right]}{\sinh\left(\frac{1}{2}K\pi\right)}. \qquad (6.4.21)$$

As described in the Appendix of [Alty *et al.* 1992], for $ae^{\rho} < \sqrt{\frac{2}{3}}$, the integral (6.4.20) can be rewritten as a sum over the poles at $K = -2i, -4i, \ldots$. This shows that, for $ae^{\rho} < \sqrt{\frac{2}{3}}$,

$$\begin{aligned}\tilde{C}(a,\rho) = {} & iNa\int_{-i}^{0} dK\, K\, e^{-K\pi/4} H^{(1)}_{iK/2}\left(3ia^2\right) J_0\left(2\rho\sqrt{1+K^2}\right) \\ & - 8Na\sum_{m=1}^{\infty} i^{m+1} m J_m\left(3ia^2\right) I_0\left(2\rho\sqrt{4m^2-1}\right). \end{aligned} \qquad (6.4.22)$$

The infinite sum is analytic in a and ρ. The integral is analytic except possibly at $a = 0$. To estimate its behaviour near $a = 0$, one uses [Abramowitz & Stegun 1965]

$$H_\lambda^{(1)}(z) \sim \frac{-i}{\pi}\Gamma(\lambda)\left(\tfrac{1}{2}z\right)^{-\lambda} \tag{6.4.23}$$

as $z \to 0$, valid (in particular) in the relevant range $0 < \lambda < \frac{1}{2}$. This shows that $\tilde{C}$ is $O(1)$ as $a \to 0$, with $\partial\tilde{C}/\partial a$ of $O\left(a^{-1}\right)$, $\partial^2\tilde{C}/\partial a^2$ of $O\left(a^{-2}\right)$, etc.

Similarly, for $ae^\rho < \sqrt{\frac{2}{3}}$, $\tilde{B}(a,\rho,\theta)$ can be written as

$$\begin{aligned}
&\tilde{B}(a,\rho,\theta) \\
&= \tfrac{1}{2}iNe^{i\theta}\left(a\frac{\partial}{\partial a} + 6a^2\right)a \\
&\quad\times \int_{-i}^{0} dK \frac{Ke^{-K\pi/4}}{\sqrt{1+K^2}} H_{iK/2}^{(1)}\left(3ia^2\right) J_1\left(2\rho\sqrt{1+K^2}\right) \\
&\quad - 4Ne^{i\theta}\left(a\frac{\partial}{\partial a} + 6a^2\right)a \\
&\quad\times \sum_{m=1}^{\infty} \frac{i^{m+1}m}{\sqrt{4m^2-1}} J_m\left(3ia^2\right) I_1\left(2\rho\sqrt{4m^2-1}\right).
\end{aligned} \tag{6.4.24}$$

$\tilde{D}(a,\rho,\theta)$ is given by the corresponding expression with $e^{i\theta}$ replaced by $e^{-i\theta}$, and $(a\frac{\partial}{\partial a} + 6a^2)$ replaced by $(a\frac{\partial}{\partial a} - 6a^2)$. Just as with $\tilde{C}$, it can be seen that $\tilde{B}$ and $\tilde{D}$ are $O(1)$ as $a \to 0$, with $\partial\tilde{B}/\partial a$ and $\partial\tilde{D}/\partial a$ of $O\left(a^{-1}\right)$, etc.

Excited wormhole quantum states can be found by repeated application of the operators $\partial/\partial\phi$ and $\partial/\partial\bar{\phi}$ to $(\tilde{B},\tilde{C},\tilde{D})$ for the ground state, and then writing $(B,C,D) = \exp(3\phi\bar{\phi})\ (\tilde{B},\tilde{C},\tilde{D})$. This will give a discrete family of states, labelled by two integers. They can be made orthonormal in the standard inner product by the Gram–Schmidt process [Courant & Hilbert 1953].

We have seen how the locally supersymmetric theory of sections 5.5, 5.6, 6.3, with supergravity coupled to a massless complex scalar and its spin-1/2 partner, admits quantum wormhole states $|m,n\rangle$ when attention is restricted to the homogeneous modes. The wave function has the rapid decay $\exp(-I_{\rm class})$, where $I_{\rm class}$ is large and positive for large 3-geometries. The countable set $|m,n\rangle$ is expected to form a basis for the set of all homogeneous quantum wormhole states. The states have the form

$$\Psi = iB\psi^C\psi_C + iC\psi^C\chi_C + iD\chi^C\chi_C. \tag{6.4.25}$$

In particular, for the ground state, B, C, D are given by Eqs. (6.3.5), (6.4.13),(6.4.16). The possibility of quantum states such as (6.4.25), which do not mix different powers of fermions, only arises in the massless theory, with Lagrangian invariant under rigid chiral symmetry $\psi^A, \chi^A \rightarrow e^{i\theta}\psi^A, e^{i\theta}\chi^A; \tilde{\psi}^{A'}, \tilde{\chi}^{A'} \rightarrow e^{-i\theta}\tilde{\psi}^{A'}, e^{-i\theta}\tilde{\chi}^{A'}$. In the massive version of this model [D'Eath & Hughes 1992], there is coupling between B, C, D and the coefficients A, E appearing in the contribution

$$A - \frac{1}{2}E\psi^C\psi_C\chi^D\chi_D$$

which must be added to the wave function (6.4.25.).

Inhomogeneities in the gravitational and matter fields would have to be included as perturbations [Lyons 1989]. One might expect that quantum wormhole states still exist in the full theory, and have a form similar to that above.

The effective mass of the scalar-spin-1/2 matter multiplet, when wormhole contributions are included along the lines of section 6.1 [Hawking 1988, Lyons 1989], must remain zero in the massless model. This is easily seen in the fermionic sector. A spin-1/2 mass would only arise if matrix elements of the form $\langle 0|\chi^A\chi_A|m,n\rangle$ were non-zero. But the quantum states are of the form (6.4.25), and the inner product, which for homogeneous fields is [D'Eath & Hughes 1988, Faddeev & Slavnov 1980]

$$\langle\Psi|\Phi\rangle = \int \bar{\Psi}\Phi \exp\left[2n_{AA'}\left(\psi^A\bar{\psi}^{A'} - \chi^A\bar{\chi}^{A'}\right)\right] da\, d\phi\, d\bar{\phi}\, d\psi^A d\bar{\psi}^{A'} d\chi^A d\bar{\chi}^{A'}, \tag{6.4.26}$$

using Berezin integration for fermions [Eqs. (3.3.8–9)] [Berezin 1966], shows that such matrix elements are zero. Since the effective interactions should respect local supersymmetry, the mass of the scalar should also remain zero. It is not obvious at present how to check more directly in the framework presented here that the scalar mass remains zero; a manifestly supersymmetric formulation would help.

Planckian masses are thus absent in this locally supersymmetric theory. This is a significant result, showing how the difficulties of a typical non-supersymmetric theory, as in section 6.1, are avoided here; this is a powerful argument for local supersymmetry. It would be interesting to see this mechanism at work in more detail by studying the massive model in the limit of small mass. It would also be of interest to know the locally supersymmetric corrections induced in the low-energy effective Lagrangian by wormholes, by generalizing the above calculation.

Ashtekar variables

7.1 General relativity

In general relativity, the Ashtekar canonical variables [Ashtekar 1986,1988, 1991] are given by a helpful change of canonical coordinates to a version $\tilde{\sigma}^{ABi}$ of the spatial triad, and the spatial components A_{ABi} of the four-dimensional connection. As we shall see, this change of variables considerably simplifies the algebraic form of the classical constraints; these become polynomial in the basic variables. Analogous simplifications take place in supergravity (section 7.2), following [Jacobson 1988].

In supergravity, in the case $\Lambda > 0$ of a positive cosmological constant, Ashtekar variables allow one to find a semi-classical approximation to a new quantum state [Sano & Shiraishi 1993], of the form $\Psi = \exp(-i S_{CS})$, where S_{CS} is the Chern–Simons action. It is conceivable that $\exp(-i S_{CS})$ may give an exact quantum state. This state and its interpretation are discussed in section 7.3 in the simplest case of a mini-superspace describing Robertson–Walker spacetime models.

7.1.1 *Necessary results*

Following [Jacobson 1988], we begin by describing some of the basic variables which appear in the Ashtekar form of general relativity. Define an unprimed version of the tetrad $e^{AA'}{}_i$:

$$\sigma^{AB}{}_i = \sqrt{2} i e^A{}_{A'i} n^{BA'}. \tag{7.1.1.1}$$

Because $e_{AA'i} n^{AA'} = 0$ [Eq. (3.2.16)], $\sigma^{AB}{}_i$ is symmetric:

$$\sigma^{AB}{}_i = \sigma^{(AB)}{}_i. \tag{7.1.1.2}$$

The inverse to Eq. (7.1.1.1) is

$$e^{AA'}{}_i = \sqrt{2} i \sigma^A{}_{Bi} n^{BA'}. \tag{7.1.1.3}$$

There is a way of testing unprimed spinors for 'reality' with respect to a metric, here taken to be $\sqrt{2}n^{AA'}$. For any unprimed spinor such as M^{AB}, one defines the Hermitian conjugate

$$(M^{AB})^{\dagger} = 2n^{A}{}_{A'}n^{B}{}_{B'}\bar{M}^{A'B'}, \tag{7.1.1.4}$$

and one says that M^{AB} is Hermitian if $(M^{AB})^{\dagger} = M^{AB}$. In particular, $\sigma^{AB}{}_{i}$ will be Hermitian (i.e. $(\sigma^{AB}{}_{i})^{\dagger} = \sigma^{AB}{}_{i}$) provided $e^{AA'}{}_{i}$ is Hermitian ($\bar{e}^{AA'}{}_{i} = e^{AA'}{}_{i}$). The spatial metric h_{ij} is given by

$$h_{ij} = \sigma^{AB}{}_{i}\sigma_{ABj}. \tag{7.1.1.5}$$

The inverse $\sigma_{AB}{}^{i}$ of $\sigma^{AB}{}_{i}$ obeys

$$\sigma_{ABj}\sigma^{MNj} = \delta_{(A}{}^{M}\delta_{B)}{}^{N}, \tag{7.1.1.6}$$

$$\sigma_{ABi}\sigma^{ABj} = \delta_{i}{}^{j}. \tag{7.1.1.7}$$

Here $\sigma_{AB}{}^{i}$ is defined as

$$\sigma_{AB}{}^{i} = h^{ij}\sigma_{ABj}. \tag{7.1.1.8}$$

It will be useful later to note that

$$\epsilon^{ijk} = -\sqrt{2}h^{1/2}\mathrm{tr}(\sigma^{i}\sigma^{j}\sigma^{k}), \tag{7.1.1.9}$$

where $h = \det h_{ij}$, and the convention $(\sigma^{i}\sigma^{j})_{A}{}^{C} = \sigma_{A}{}^{Bi}\sigma_{B}{}^{Cj}$ is adopted. Similarly

$$\epsilon^{ijk}\sigma_{i} = \sqrt{2}h^{1/2}\sigma^{[j}\sigma^{k]}, \tag{7.1.1.10}$$

$$\epsilon^{ijk}\sigma_{i}\sigma_{j} = \sqrt{2}h^{1/2}\sigma^{k}. \tag{7.1.1.11}$$

Further, define

$$\tilde{\sigma}^{ABi} = h^{1/2}\sigma^{ABi}. \tag{7.1.1.12}$$

This variable $\tilde{\sigma}^{ABi}$ turns out to be canonically conjugate to the connection variable A_{ABi}, and is used in the Ashtekar canonical treatment of gravity.

It will at certain points in this chapter be convenient to use the language of differential forms [Misner *et al.* 1973, Wald 1984]. A differential p-form ω corresponds to

$$\omega_{\mu_1\cdots\mu_p} = \omega_{[\mu_1\cdots\mu_p]}. \tag{7.1.1.13}$$

If ω is a p-form and μ is a q-form, then their wedge product $\omega\wedge\mu$ is given by

$$(\omega\wedge\mu)_{\mu_1\cdots\mu_p\nu_1\cdots\nu_q} = \frac{(p+q)!}{p!q!}\omega_{[\mu_1\cdots\mu_p}\mu_{\nu_1\cdots\nu_q]}. \tag{7.1.1.14}$$

Given a p-form ω, one defines the exterior derivative $d\omega, a(p+1)$-form, by

$$d\omega = (p+1)\partial_{[\nu}\omega_{\mu_1\cdots\mu_p]}. \tag{7.1.1.15}$$

In the case that ω is a 0-form, i.e. a function, $d\omega$ is the gradient of ω.

Given the connection form A_{AB}, one can define the covariant exterior derivative $\mathscr{D}$ on an unprimed spinor-valued form, such as ϕ_A, by

$$\mathscr{D}\phi_A = d\phi_A + A_A{}^B \wedge \phi_B. \tag{7.1.1.16}$$

This is the antisymmetrized version of Eq. (3.2.3). The antisymmetrization ensures that $\mathscr{D}\phi_A$ is again a spinor-valued form. The curvature two-form $F_A{}^B$ is given by

$$\mathscr{D}^2\phi_A = F_A{}^B \wedge \phi_B. \tag{7.1.1.17}$$

Explicitly,

$$F_A{}^B = dA_A{}^B + A_A{}^C \wedge A_C{}^B \tag{7.1.1.18}$$

corresponds to the self-dual part of the Riemann tensor (3.2.10).

In the Lagrangian versions of general relativity and supergravity, the action can be written as a four-form. The orientation (four-form) ϵ on a four-manifold with metric $g_{\mu\nu}$ is given by [Wald 1984]

$$\epsilon = \frac{1}{4!} g^{1/2} \epsilon_{\lambda\mu\nu\rho} dx^\lambda \wedge dx^\mu \wedge dx^\nu \wedge dx^\rho. \tag{7.1.1.19}$$

Equivalently one can write

$$\epsilon = g^{1/2} dx^1 \wedge dx^2 \wedge dx^3 \wedge dx^4, \tag{7.1.1.20}$$

and a general four-form can be written as

$$\alpha = a(x^1, x^2, x^3, x^4) dx^1 \wedge dx^2 \wedge dx^3 \wedge dx^4. \tag{7.1.1.21}$$

One defines the integral of α as

$$\int_U \alpha = \int_{\psi(U)} a\, dx^1 dx^2 dx^3 dx^4, \tag{7.1.1.22}$$

where ψ denotes a single coordinate system for U, which is an open set in the manifold.

7.1.2 Lagrangian form

To simplify the form of the Lorentzian action, let us use units such that $\kappa^2 = \frac{1}{2}$. The Einstein–Hilbert Lagrangian density of general relativity is [Jacobson 1988]

$$\begin{aligned} \mathscr{L} &= \sqrt{-g}\ {}^4R \\ &= \sqrt{-g}\, g^{\lambda\nu} g^{\mu\rho}\ {}^4R_{\lambda\mu\nu\rho} \\ &= \sqrt{-g}\, g^{\lambda\nu} g^{\mu\rho} ({}^4R_{\lambda\mu\nu\rho} + \frac{i}{2} \epsilon_{\lambda\mu}{}^{\alpha\beta}\ {}^4R_{\alpha\beta\nu\rho}) \\ &= \sqrt{-g}\, g^{\lambda\nu} g^{\mu\rho}\ {}^{+4}R_{\lambda\mu\nu\rho}. \end{aligned} \tag{7.1.2.1}$$

The third line follows because of the identity ${}^4R_{[\lambda\mu\nu]\rho} = 0$. In the fourth line, ${}^{+4}R_{\lambda\mu\nu\rho}$ denotes the self-dual part of the curvature. In spinor terms, this gives

$$\mathscr{L} = \sqrt{-g}\, e^{\lambda}{}_{A}{}^{A'}\, e^{\mu}{}_{BA'}\, {}^4R_{\lambda\mu}{}^{AB}. \tag{7.1.2.2}$$

Equivalently, in the language of forms (subsection 7.1.1), it can be shown that one has [Jacobson 1988]

$$\mathscr{L} = i e^{AA'}{}_{\wedge} e_{BA'} \wedge F_{A}{}^{B}, \tag{7.1.2.3}$$

where one uses the relation (3.3.20) for $\epsilon_{\lambda\mu\nu\rho}$.

One takes a Palatini approach, regarding $\mathscr{L}$ as a function of two independent variables, $e^{AA'}{}_{\mu}$ and the complex unprimed connection $A_{AB\mu}$. The Lagrangian (7.1.2.2) or (7.1.2.3) is at first sight complex, but manages to be real without including the $\bar{A}_{A'B'\mu}$ part of the connection, which appears in the first two lines of Eq. (7.1.2.1), because of the identity ${}^4R_{[\lambda\mu\nu]\rho} = 0$. Varying Eq. (7.1.2.3) with respect to $A_{AB\mu}$, one obtains

$$\mathscr{D}(e^{AA'}{}_{\wedge} e^{B}{}_{A'}) = 0. \tag{7.1.2.4}$$

Here, let us extend the definition of $\mathscr{D}$ to act also on primed spinor indices, using the conjugate connection $\bar{A}_{A'B'\mu}$. Provided that $e^{AA'}{}_{\mu}$ is real, one can show that Eq. (7.1.2.4) is equivalent to

$$\mathscr{D} e^{AA'}{}_{i} = 0. \tag{7.1.2.5}$$

This is the torsion equation of Eq. (3.2.4), with zero torsion $S^{AA'}{}_{\mu\nu}$. By analogy with section 3.2, it gives $A_{AB\mu}$ as the torsion-free metric connection, corresponding to the tetrad $e^{AA'}{}_{\mu}$.

The variation of the action with respect to $e^{AA'}{}_{\mu}$ can be shown to give the vacuum Einstein equations [Ashtekar 1988]

$${}^4R_{\mu\nu} = 0. \tag{7.1.2.6}$$

7.1.3 *Hamiltonian formulation*

In index form, the Lagrangian density of Eq. (7.1.2.3) gives

$$\mathscr{L} = \tfrac{1}{2} i \epsilon^{\mu\nu\rho\sigma} e^{AA'}{}_{\mu} e_{BA'\nu} F_{A}{}^{B}{}_{\rho\sigma}. \tag{7.1.3.1}$$

As usual, this Lagrangian will be decomposed with respect to a family of hypersurfaces $x^0 = \text{const}$. One obtains

$$\mathscr{L} = \sqrt{2}\, i (\tilde{\sigma}^{ABi} \dot{A}_{ABi} - \mathscr{H}), \tag{7.1.3.2}$$

where

$$\mathscr{H} = e_{AA'0}\, \mathscr{H}^{AA'} + A_{AB0}\, J^{AB} + \text{total divergence}. \tag{7.1.3.3}$$

The canonical momentum is (see subsection 7.1.1)

$$\tilde{\sigma}^{ABk} = \frac{1}{\sqrt{2}}\epsilon^{ijk}e^{AA'}{}_i e^B{}_{A'j}. \tag{7.1.3.4}$$

The constraints are

$$\mathscr{H}^{AA'} = \frac{1}{\sqrt{2}}\epsilon^{ijk}e^{BA'}_i F_B{}^A{}_{jk}, \tag{7.1.3.5}$$

$$J^{AB} = \mathscr{D}_k\tilde{\sigma}^{ABk}. \tag{7.1.3.6}$$

From Eq. (7.1.3.2), one sees that A_{ABk} and $\tilde{\sigma}^{ABk}$ are conjugate variables. Their Poisson brackets are

$$[\tilde{\sigma}^{ABk}(x), A_{MNj}(y)] = \delta_j{}^k\delta_{(M}{}^A\delta_{N)}{}^B\delta^3(x,y). \tag{7.1.3.7}$$

The $\tilde{\sigma}^{ABk}$ are constructed from the $e^{AA'}{}_i$ as in subsection 7.1.1 or as in Eq. (7.1.3.4). If one is to use the $\tilde{\sigma}^{ABk}$ as half of the canonical variables, it is necessary to show that the variables $e^{AA'}{}_i$ of the usual ADM representation can be recovered from the $\tilde{\sigma}^{ABk}$. Following Eqs. (7.1.1.3),(7.1.1.12), one has

$$e^{AA'}{}_i = ih^{-1/2}n^{BA'}\tilde{\sigma}_B{}^A{}_i. \tag{7.1.3.8}$$

Now $n^{BA'}$ is determined by σ_{ABi}, through the reality condition (7.1.1.4) with respect to the metric $\sqrt{2}n^{AA'}$:

$$2n^A{}_{A'}n^B{}_{B'}\bar{\sigma}^{A'B'k} = \sigma^{ABk}. \tag{7.1.3.9}$$

Two different $n^{AA'}$ choices would give

$$2n_1{}^A{}_{A'}n_1{}^B{}_{B'}\bar{\sigma}^{A'B'k} = \sigma^{ABk} = 2n_2{}^A{}_{A'}n_2{}^B{}_{B'}\bar{\sigma}^{A'B'k}, \tag{7.1.3.10}$$

hence

$$n_1{}^A{}_{A'}n_2{}^B{}_{B'} = n_2{}^A{}_{A'}n_1{}^B{}_{B'}, \tag{7.1.3.11}$$

and hence

$$n_1{}^A{}_{A'} = \pm n_2{}^A{}_{A'}. \tag{7.1.3.12}$$

If the normal is required to be future-pointing, one obtains

$$n_1{}^A{}_{A'} = n_2{}^A{}_{A'}. \tag{7.1.3.13}$$

Further, h is determined by $\tilde{\sigma}_B{}^A{}_i$ through

$$\sigma^{AB}{}_i\sigma_{ABj} = h_{ij}, \tag{7.1.3.14}$$

$$\tilde{\sigma}^{AB}{}_i\tilde{\sigma}_{ABj} = h_{ij}. \tag{7.1.3.15}$$

Hence, from Eq. (7.1.3.8), $e^{AA'}{}_i$ is uniquely determined by $\tilde{\sigma}^{ABi}$.

The constraints (7.1.3.5),(7.1.3.6) are in polynomial form. For comparison with the ADM constraints [Arnowitt *et al.* 1962] (i.e. the constraints of section 2.3), write

$$\mathscr{H}^{AA'} = -n^{AA'}\mathscr{H}_\perp + e^{AA'i}\mathscr{H}_i. \tag{7.1.3.16}$$

Then

$$\mathscr{H}_\perp = \mathrm{tr}(\tilde{\sigma}^i\tilde{\sigma}^j F_{ij}), \tag{7.1.3.17}$$

$$\mathscr{H}_i = \mathrm{tr}(\tilde{\sigma}^j F_{ij}), \tag{7.1.3.18}$$

with the index summation convention following Eq. (7.1.1.9). Together with the constraint $J^{AB} = \mathscr{D}_k\tilde{\sigma}^{ABk}$, all these constraints are polynomials in the basic variables $\tilde{\sigma}^{ABk}, A_{ABk}$ and their spatial derivatives. Equivalently, one can write

$$e_{AA'0}\mathscr{H}^{AA'} = e_{AB0}\mathscr{H}^{AB}, \tag{7.1.3.19}$$

where

$$e_{AB0} = ih^{-1/2}e_A{}^{A'}{}_0 n_{BA'}, \tag{7.1.3.20}$$
$$\mathscr{H}^{AB} = (\tilde{\sigma}^j\tilde{\sigma}^k F_{jk})^{BA}, \tag{7.1.3.21}$$

from which Eqs. (7.1.3.17),(7.1.3.18) can be derived.

One can verify [Ashtekar 1986,1988] that J^{AB} generates local rotations of unprimed spinor variables, $\mathscr{H}_\perp$ generates normal deformations, and $\mathscr{H}_i$ spatial diffeomorphisms (modified by local rotations).

7.1.4 Reality conditions

One needs to be certain that the canonical theory described above corresponds to a real theory with Hermitian tetrad $e^{AA'}{}_\mu$, and one would like to understand the reality (Hermiticity) properties of the connection $A_{AB\mu}$. First, note that $\tilde{\sigma}^{ABi}$ is Hermitian with respect to the metric $\sqrt{2}n^{AA'}$ [Eq. (7.1.1.4) and following]. This implies that the spatial tetrad $e^{AA'}{}_i$ is Hermitian, using the definition (7.1.1.1),(7.1.1.12).

For the Lagrange multipliers e_{AB0} of (7.1.3.20), one can see that $e_{AB0} = (e_{AB0})^\dagger$ because of the reality of the $e_{AA'0}$. One satisfies this condition by writing

$$e_{AB0} = iN\epsilon_{AB} + n^i\sigma_{ABi}, \tag{7.1.4.1}$$

where N and N^i are real scalar and vector densities of weight -1. Conversely, if $e_{AB0} = (e_{AB0})^\dagger$, then $e_{AA'0}$ is Hermitian.

One can also ask about reality conditions corresponding to the connection A_{ABi}. Since the torsion is zero, the spatial components of the four-dimensional connection $A_{AB\mu}$ or $\omega_{AB\mu}$ are of the form $A_{ABi} = {}^{3s}\omega_{ABi} +$

a second fundamental form piece. Here ${}^{3s}\omega_{ABi}$ is the torsion-free metric connection, as used in section 3.4, and the second fundamental form piece appears through the $p_{AA'}{}^{i}$ term in Eq. (3.4.5), for example. The vanishing of the torsion of the four-dimensional connection $A_{AB\mu}$ and $\bar{A}_{A'B'\mu}$ can be shown to imply [Jacobson 1988] that the above equation for A_{ABi} reads

$$A_{ABi} = {}^{3s}\omega_{ABi} + i\,\Pi_{ABi}, \tag{7.1.4.2}$$

where both ${}^{3s}\omega_{ABi}$ and Π_{ABi} are Hermitian. When the constraint $J^{AB} = 0$ is satisfied, $\Pi_{ji} = \sigma^{AB}{}_{j}\Pi_{ABi}$ is the second fundamental form K_{ij}. Hence the conjugate of A_{ABi} is $(A_{ABi})^{\dagger} = {}^{3s}\omega_{ABi} - i\Pi_{ABi}$. The Hermitian part of A_{ABi} is thus the torsion-free metric connection ${}^{3s}\omega_{ABi}$. In the quantum theory, the reality conditions should be incorporated by an appropriate choice of inner product [Ashtekar 1988,1991].

If, instead of the Lorentzian theory, one studies the Euclidean version, the relevant conditions are that both $\sigma_{AB}{}^{i}$ and A_{ABi} should be Hermitian.

7.1.5 Algebra of constraints

Given a generator $\mathcal{N}_{A}{}^{B}(x)$ of $SU(2)$ rotations, with $\mathcal{N}_{AB} = \mathcal{N}_{(AB)}$, define [Ashtekar 1988]

$$C_{\mathcal{N}}(\tilde{\sigma}, A) = \int_{\Sigma} d^{3}x \operatorname{tr}(\mathcal{N}\mathcal{D}_{i}\tilde{\sigma}^{i}), \tag{7.1.5.1}$$

where Σ is a spacelike hypersurface. One can verify, as described at the end of subsection 7.1.3, that $J_{AB} = \mathcal{D}_{i}\tilde{\sigma}^{ABi}$ or equivalently $C_{\mathcal{N}}(\tilde{\sigma}, A)$ generates local rotations on unprimed spinor indices.

Turning to the vector constraint (7.1.3.18), one has to modify $\mathcal{H}_{i}$ by a field-dependent local Lorentz transformation, to obtain the generator of spatial diffeomorphisms. Given a shift vector field $N^{i}(x)$ or $\mathbf{N}(x)$, define

$$C_{\mathbf{N}}(\tilde{\sigma}, A) = \int_{\Sigma} d^{3}x\, N^{i} \operatorname{tr}(\tilde{\sigma}^{j}F_{ij} - A_{i}\mathcal{D}_{j}\tilde{\sigma}^{j}). \tag{7.1.5.2}$$

This can be shown [Ashtekar 1988] to generate spatial diffeomorphisms.

For the scalar constraint (7.1.3.17), take the lapse $N(x)$ to be a scalar density of weight -1, and define

$$C_{N}(\tilde{\sigma}, A) = \int_{\Sigma} d^{3}x N \operatorname{tr}(\tilde{\sigma}^{i}\tilde{\sigma}^{j}F_{ij}). \tag{7.1.5.3}$$

This generates the correct time evolution of $\tilde{\sigma}$ and A, provided that the constraints are already satisfied.

The Poisson bracket algebra of the above generators is

$$
\begin{aligned}
[C_{\mathcal{N}}, C_{\mathcal{M}}] &= -C_{[\mathcal{N},\mathcal{M}]},\\
[C_{\mathbf{N}}, C_{\mathcal{M}}] &= -C_{\mathscr{L}_{\mathbf{N}}\mathcal{M}},\\
[C_{\mathbf{N}}, C_{\mathbf{M}}] &= -C_{[\mathbf{N},\mathbf{M}]},\\
[C_{\mathcal{N}}, C_{M}] &= 0\\
[C_{\mathbf{N}}, C_{M}] &= -C_{\mathscr{L}_{\mathbf{N}}M},\\
[C_{N}, C_{M}] &= C_{\mathbf{K}} - C_{A_m K^m}.
\end{aligned}
\tag{7.1.5.4}
$$

Here $\mathscr{L}_{\mathbf{N}}$ denotes the Lie derivative with respect to the vector field $\mathbf{N}$ [Misner *et al.* 1973], and $K^i = 2(N\partial_j M - M\partial_j N)\mathrm{tr}(\tilde{\sigma}^i\tilde{\sigma}^j)$. Hence the constraints form a first-class system, with structure functions. The brackets, up to contributions on the right-hand side linear in J^{AB} [Henneaux 1983], follow the form described by [Teitelboim 1977a].

7.1.6 Quantization

One can attempt to impose the first-class quantum constraints on physical quantum states. For example, one can use the $\Psi(A_{ABi})$ representation of a wave function, writing $\tilde{\sigma}^{ABi}$ as the operator $\delta/\delta A_{ABi}$. Then the J^{AB} constraints imply the invariance of $\Psi(A_{ABi})$ under local Lorentz transformations. With a certain choice of factor ordering, one has

$$\mathscr{H}_{\perp}\Psi(A) = F_{ABij}\frac{\delta^2}{\delta A^{AC}{}_i\,\delta A^{B}{}_{Cj}}\Psi(A) = 0, \tag{7.1.6.1}$$

$$\mathscr{H}_{i}\Psi(A) = F_{ABij}\frac{\delta}{\delta A^{AB}{}_j}\Psi(A) = 0. \tag{7.1.6.2}$$

By comparison with the constraints (2.3.57),(2.3.58) in the ADM representation, one still has the second-order functional differential equation $\mathscr{H}_{\perp}\Psi = 0$, with its attendant difficulties. One can solve the $\mathscr{H}_{\perp}$ and $\mathscr{H}_i$ constraints in terms of states defined on loops [Jacobson & Smolin 1988]; however, it is unclear what relation this has to the usual quantum states containing gravitons, as prescribed by wave-like perturbations in the initial and final data $e^{AA'}{}_i$.

7.2 Supergravity

To derive the canonical form of supergravity in the Ashtekar foundations, we follow [Jacobson 1988].

7.2.1 Lagrangian form

As above, the convention $\kappa^2 = \frac{1}{2}$ is used, and, for consistency with [Jacobson 1988], the gravitino field is rescaled by a factor of $1/\sqrt{2}$. The Lagrangian density is then taken to be

$$\mathscr{L} = ie^{AA'} \wedge e_{BA'} \wedge F_A{}^B - e^{AA'} \wedge \bar{\psi}_{A'} \wedge \mathscr{D}\psi_A. \tag{7.2.1.1}$$

By analogy with subsection 7.1.2 for general relativity, one must check that the complex Lagrangian (7.2.1.1) gives the correct equations of motion. In fact, $\mathrm{Im}\mathscr{L}$ becomes a total derivative when $A_{AB\mu}$ obeys its field equation. This can be seen by varying $A_{AB\mu}$ to obtain its field equation

$$\mathscr{D}(e^{AA'} \wedge e^B{}_{A'}) + ie^{A'(A} \wedge \psi^{B)} \wedge \bar{\psi}_{A'} = 0. \tag{7.2.1.2}$$

This only involves the unprimed connection $A_{AB\mu}$. One can use $\bar{A}_{A'B'\mu}$ to extend the action of $\mathscr{D}$ to primed indices. Provided $e^{AA'}{}_\mu$ is real, it can be shown from Eq. (7.2.1.2) that

$$\mathscr{D}e_{AA'} = \tfrac{1}{2}i\psi_A \wedge \bar{\psi}_{A'}. \tag{7.2.1.3}$$

This is the torsion equation of Eqs. (3.2.4),(3.2.5), and does depend on the primed connection $\bar{A}_{A'B'\mu}$. As in section 3.2, one can solve Eq. (7.2.1.3) to obtain $A_{AB\mu}$ in terms of $e_{AA'\mu}$ and $\psi_{A\mu}$. One can see that the imaginary part of the Lagrangian is a total derivative, using Eq. (7.2.1.3):

$$\begin{aligned} \mathscr{L} - \bar{\mathscr{L}} &= ie^{AA'} \wedge \mathscr{D}^2 e_{AA'} + e^{AA'} \wedge \mathscr{D}(\bar{\psi}_{A'} \wedge \psi_A) \\ &=^* d(-\tfrac{1}{2}e^{AA'} \wedge \bar{\psi}_{A'} \wedge \psi_A), \end{aligned} \tag{7.2.1.4}$$

where $=^*$ means equality subject to (7.2.1.3).

Hence one can equivalently use the action $\mathrm{Re}\,\mathscr{L}$, which agrees (modulo the conventions above) with section 3.2, and which has the local symmetries of section 3.2. This corresponds to using a second-order formalism for supergravity. The first-order form, including the symmetry variations $\delta A_{AB\mu}$, can also be given following [Deser & Zumino 1976]. For a real supersymmetry variation, one uses conjugate parameters $\epsilon_A, \bar{\epsilon}_{A'}$. For subsequent work, where one takes A_{ABi} as a complex coordinate in phase space, one can take ϵ_A and $\tilde{\epsilon}_{A'}$ to be independent, and similarly one takes independent $\psi^A{}_\mu, \tilde{\psi}^{A'}{}_\mu$ and complex $e^{AA'}{}_\mu$. Then $\tilde{A}_{A'B'\mu}$ is independent of $A_{AB\mu}$; this situation is described in section 3.5 and in chapter 4.

7.2.2 Hamiltonian formulation

When written in index form, by analogy with section 7.1.3, the Lagrangian density of $\mathscr{L}$ is

$$\mathscr{L} = \epsilon^{\mu\nu\rho\sigma}(\tfrac{1}{2}ie^{AA'}{}_\mu e_{BA'\nu}F_A{}^B{}_{\rho\sigma} - e^{AA'}{}_\mu\bar{\psi}_{A'\nu}\mathscr{D}_\rho\psi_{A\sigma}). \tag{7.2.2.1}$$

This may be decomposed in 3 + 1 form, as

$$\mathscr{L} = \sqrt{2}\, i(\tilde{\sigma}^{ABi}\dot{A}_{ABi} + \tilde{\pi}^{Ai}\dot{\psi}_{Ai} - \mathscr{H}), \tag{7.2.2.2}$$

where $\mathscr{H}$ has the form

$$\begin{aligned}\mathscr{H} = e_{AA'0}\mathscr{H}^{AA'} + \psi_{A0}S^{A} + \hat{S}^{A'}\bar{\psi}_{A'0} + A_{AB0}J^{AB} \\ + \text{total divergence.}\end{aligned} \tag{7.2.2.3}$$

The canonical coordinates may be taken to be A_{ABi} and ψ_{Ai}, in which case the canonical momenta are $\tilde{\sigma}^{ABi}$ and

$$\tilde{\pi}^{Ai} = \frac{i}{\sqrt{2}}\epsilon^{ijk}e^{AA'}{}_{j}\bar{\psi}_{A'k}. \tag{7.2.2.4}$$

The constraints are

$$\mathscr{H}^{AA'} = \frac{1}{\sqrt{2}}\epsilon^{ijk}(e^{BA'}{}_{i}F_{B}{}^{A}{}_{jk} - i\bar{\psi}^{A'}{}_{i}\mathscr{D}_{j}\psi^{A}{}_{k}), \tag{7.2.2.5}$$

$$S^{A} = \mathscr{D}_{i}\tilde{\pi}^{Ai} \tag{7.2.2.6}$$

$$\hat{S}^{A'} = \frac{i}{\sqrt{2}}\epsilon^{ijk}e^{AA'}{}_{i}\mathscr{D}_{j}\psi_{Ak} \tag{7.2.2.7}$$

$$J^{AB} = \mathscr{D}_{i}\tilde{\sigma}^{ABi} - \tilde{\pi}^{(Ai}\psi^{B)}{}_{i}. \tag{7.2.2.8}$$

Here $\hat{S}^{A'}$ is not equal to $\bar{S}^{A'}$, but differs from it by

$$\hat{S}^{A'} - \bar{S}^{A'} = \frac{i}{\sqrt{2}}\epsilon^{ijk}(\mathscr{D}_{k}e^{AA'}{}_{i})\psi_{Aj}. \tag{7.2.2.9}$$

Because we only use the fermionic variables $(\tilde{\pi}^{Ai}, \psi_{Ai})$, while the Hermitian conjugate variables are to be regarded as functions of $(\tilde{\pi}^{Ai}, \psi_{Ai})$ (see below), there are no second-class constraints and hence there is no need for Poisson brackets to be replaced by Dirac brackets, as in section 3.2. The fundamental non-zero Poisson brackets are

$$[\tilde{\sigma}^{ABi}(x), A_{MNj}(y)] = \delta_{j}{}^{i}\delta_{(M}{}^{A}\delta_{N)}{}^{B}\delta^{3}(x,y), \tag{7.2.2.10}$$

$$[\tilde{\pi}^{Ai}(x), \psi_{Mj}(y)] = -\delta_{j}{}^{i}\delta_{M}{}^{A}\delta^{3}(x,y). \tag{7.2.2.11}$$

The constraints (7.2.2.5–8) involve the variables $e^{AA'}{}_{i}$ and $\bar{\psi}_{A'i}$, and so it is necessary to show that these are determined in terms of the basic variables – in fact in terms of $\tilde{\sigma}^{ABi}$ and $\tilde{\pi}^{Ai}$. Now it was already shown in subsection 7.1.3 how $e^{AA'}{}_{i}$ is determined in terms of $\tilde{\sigma}^{ABi}$. To show further that $\bar{\psi}_{A'i}$ is determined in terms of $\tilde{\pi}^{Ai}$ and $\tilde{\sigma}^{ABi}$, one must invert Eq. (7.2.2.4). One multiplies both sides by $\sigma_{k}\sigma_{m}$ and uses

$$\epsilon^{ijk}\sigma_{k}\sigma_{m}\sigma_{i} = \frac{1}{\sqrt{2}}h^{1/2}\delta_{m}{}^{j}. \tag{7.2.2.12}$$

This leads to

$$\bar{\psi}^{A'}{}_{i} = -2h^{-3/2}n^{BA'}(\tilde{\sigma}_{m}\tilde{\sigma}_{i}\tilde{\pi}^{m})_{B}. \tag{7.2.2.13}$$

As in subsection 7.1.3, one can rewrite the Lagrange multiplier $e_{AA'0}$ as

$$e_{AB0} = ih^{-1/2} e_A{}^{A'}{}_0 n_{BA'} \tag{7.2.2.14}$$

Then the term $e_{AA'0}\mathscr{H}^{AA'}$ in $\mathscr{H}$ [Eq. (7.2.2.3)] can be rewritten as $e_{AB0}\mathscr{H}^{AB}$, with

$$\mathscr{H}^{AB} = (\tilde{\sigma}^j \tilde{\sigma}^k F_{jk})^{BA} + \sqrt{2}h^{-1}\epsilon^{jkl}(\tilde{\sigma}_m \tilde{\sigma}_l \tilde{\pi}^m)^B \mathscr{D}_j \psi_k^A, \tag{7.2.2.15}$$

with the help of Eq. (7.1.1.10) in the bosonic term. A corresponding re-definition can be made in the term $\hat{S}^{A'}\bar{\psi}_{A'0}$, which may be written as $S^{\dagger A}\psi^\dagger{}_{A0}$, with

$$\psi^\dagger{}_{A0} = h^{-1/2} n_A{}^{A'} \bar{\psi}_{A'0}, \tag{7.2.2.16}$$

$$S^{\dagger A} = (\tilde{\sigma}^j \tilde{\sigma}^k \mathscr{D}_{[j}\psi_{k]})^A. \tag{7.2.2.17}$$

In the case of both multipliers $e_{AA'0}$ and $\bar{\psi}_{A'0}$, the expression $h^{-1/2}n^{AA'}$ has been absorbed, leading to a simplification in the corresponding constraints.

The constraints $S^A, S^{\dagger A}$ and J^{AB} of Eqs. (7.2.2.6),(7.2.2.8),(7.2.2.17) are polynomial functions of the basic canonical variables of Eqs. (7.2.2.10–11). The other constraint $\mathscr{H}^{AB}$ is polynomial in the bosonic variables, but the fermionic term involves h^{-1}. Note that $\tilde{\sigma}_m$ is a polynomial in $\tilde{\sigma}^i$, as may be seen from Eq. (7.1.1.11). In fact, there is a field-dependent linear combination of $\mathscr{H}^{AB}$ and $S^{\dagger A}$ which is also polynomial in the fermionic variables. This will be described in more detail in subsection 7.1.4. This must occur since the Poisson bracket $[S^A, S^{\dagger B}]$ is polynomial and must close on $\mathscr{H}^{AB}$ together with a (field-dependent) linear combination of other constraints [Teitelboim 1977a]. The calculation of subsection 7.1.4 below shows that the combination

$$\ddot{\mathscr{H}}^{AB} \equiv \mathscr{H}^{AB} + C^{AB}{}_M S^{\dagger M}, \tag{7.2.2.18}$$

with

$$C^{AB}{}_M \equiv 2h^{-1}\tilde{\sigma}_M{}^A{}_i \tilde{\pi}^{Bi}, \tag{7.2.2.19}$$

is polynomial, having the form

$$\begin{aligned}\ddot{\mathscr{H}}^{AB} = {} & (\tilde{\sigma}^j \tilde{\sigma}^k F_{jk})^{BA} + 2(\tilde{\pi}^j \tilde{\sigma}^k \mathscr{D}_{[j}\psi_{k]})\epsilon^{AB} \\ & + 2(\tilde{\pi}^j \mathscr{D}_{[j}\psi_{k]})\tilde{\sigma}^{ABk}.\end{aligned} \tag{7.2.2.20}$$

To summarize, all the constraints $\ddot{\mathscr{H}}^{AB}, S^A, S^{\dagger A}$ and J^{AB} of supergravity [Eqs. (7.2.2.6,8,17,20)] can be put in a form which is polynomial in the canonical coordinates of Eqs. (7.2.2.10–11).

7.2.3 Reality conditions

As in section 7.1.4, one would like to understand the reality properties of the variables $\tilde{\sigma}^{ABi}$ and A_{ABi}. As before, $\tilde{\sigma}^{ABi}$ corresponds to a real $e^{AA'}{}_i$

because $\tilde{\sigma}^{ABi}$ is Hermitian. Similarly, the Hermiticity of e_{AB0} leads to the reality of the Lagrange multipliers $e_{AA'0}$.

To understand the reality conditions on $A_{AB\mu}$, start from its field equation

$$\mathscr{D}_{[\mu}e^{AA'}{}_{\nu]} = S^{AA'}{}_{\mu\nu}, \tag{7.2.3.1}$$

[Eq. (3.2.4)], where the torsion is given with the present conventions by

$$S^{AA'}{}_{\mu\nu} = \tfrac{1}{2}i\psi^{A}{}_{[\mu}\bar{\psi}^{A'}{}_{\nu]}. \tag{7.2.3.2}$$

Note that Eq. (7.2.3.1) involves $\bar{A}_{A'B'\mu}$ as well as $A_{AB\mu}$. Now the $0i$ components of Eq. (7.2.3.1) are evolution equations, and can be determined from Hamilton's equations. But the spatial components are constraints on the initial data:

$$T^{AA'}{}_{ij} \equiv \mathscr{D}_{[i}e^{AA'}{}_{j]} - S^{AA'}{}_{ij} = 0. \tag{7.2.3.3}$$

There are twelve real constraints here, of which six do not involve $\bar{A}_{A'B'i}$; these six are the local rotation constraint

$$\epsilon^{ijk}e^{(A}{}_{A'k}T^{B)A'}{}_{ij} = -\frac{1}{\sqrt{2}}J^{AB} = 0. \tag{7.2.3.4}$$

The reality conditions reside in the remaining six constraints. They are given by

$$0 = -in^{(A}{}_{A'}T^{B)A'}{}_{ij} = \mathscr{D}_{[i}e^{AB}{}_{j]} + in^{(A}{}_{A'}S^{B)A'}{}_{ij} + i(\mathscr{D}_{[i}n^{(A}{}_{A'})e^{B)A'}{}_{j]}. \tag{7.2.3.5}$$

The constraints (7.2.3.4–5) are together equivalent to (7.2.3.3). One can write $T^{AA'}{}_{ij} = \epsilon_{ijm}(T^{m}n^{AA'} + T^{mn}e^{AA'}{}_{n})$, where T^m and T^{mn} are real. The nine real equations (7.2.3.5) imply $T^{mn} = 0$, and the three complex equations (7.2.3.4) imply $T^m = 0$ and $T^{[mn]} = 0$ (again!). If the local rotation constraint (7.2.3.4) is satisfied, then the remaining six equations $T^{(mn)} = 0$ are the reality conditions.

Now Eq. (7.2.3.5) gives the torsion of A_{ABi} for the triad $\sigma^{AB}{}_{i}$:

$$\mathscr{D}_{[i}\sigma^{AB}{}_{j]} = S^{AB}{}_{ij} + iK^{AB}{}_{ij}, \tag{7.2.3.6}$$

$$S^{AB}{}_{ij} = -in^{(A}{}_{A'}S^{B)A'}{}_{ij}, \tag{7.2.3.7}$$

$$K^{AB}{}_{ij} = -(\mathscr{D}_{[i}n^{(A}{}_{A'})e^{B)A'}{}_{j]}. \tag{7.2.3.8}$$

One can check that both $S^{AB}{}_{ij}$ and $K^{AB}{}_{ij}$ are Hermitian. Hence the reality condition (7.2.3.5) is equivalent to the condition that the Hermitian part of the torsion of A_{ABi} is $S^{AB}{}_{ij}$. Thus

$$A_{ABi} = \omega_{ABi} + i\Pi_{ABi}, \tag{7.2.3.9}$$

where Π_{ABi} is Hermitian and $\omega_{ABi} = \omega_{ABi}(\tilde{\sigma}, \tilde{\pi}, \psi)$ is the Hermitian three-dimensional spin connection of section 3.2 with torsion $S^{AB}{}_{ij}$. Note that this reality condition is non-polynomial in the canonical coordinates, at the same time as the constraints are polynomial (subsection 7.2.2).

7.2.4 Algebra of constraints

By the general argument of [Teitelboim 1977a], since the constraints generate local symmetries of the action, the Poisson brackets of the constraints give terms linear in the constraints, with structure functions as coefficients. Thus the constraints are first-class. The algebra can be determined by Teitelboim's method, up to terms in the local rotation generator J^{AB} [Henneaux 1983], from the symmetries of the action.

The complete set of Poisson brackets is not known. Here, following [Jacobson 1988], the brackets among the constraints S^A and $S^{\dagger A}$ are treated. They close on themselves together with $\mathscr{H}^{AB}$ (or $\ddot{\mathscr{H}}^{AB}$ of subsection 7.2.2). To show that all brackets close on constraints, one must also check $[S^A, \ddot{\mathscr{H}}^{BC}]$, $[S^{\dagger A}, \ddot{\mathscr{H}}^{BC}]$ and $[\ddot{\mathscr{H}}^{AB}, \ddot{\mathscr{H}}^{CD}]$. This calculation has not yet been performed, but it is in principle straightforward since the constraints are simple polynomial functions of the canonical variables. As in chapter 3, the brackets of constraints with J^{AB} are standard, since J^{AB} simply rotates the constraint.

One finds [Jacobson 1988]

$$[S^A(x), S^B(y)] = 0, \tag{7.2.4.1}$$

$$[S^{\dagger A}(x), S^{\dagger B}(y)] = C_1{}^{AB}{}_M S^{\dagger M}(x)\delta^3(x,y), \tag{7.2.4.2}$$

$$C_1{}^{AB}{}_M = 2\tilde{\sigma}_M{}^{(Ak}\psi^{B)}{}_k - \tilde{\sigma}^{ABk}\psi_{Mk}, \tag{7.2.4.3}$$

$$[S^A(x), S^{\dagger B}(y)] = \tfrac{1}{2}\ddot{\mathscr{H}}^{AB}(x)\delta^3(x,y), \tag{7.2.4.4}$$

$$\ddot{\mathscr{H}}^{AB} = (\tilde{\sigma}^j\tilde{\sigma}^k F_{jk})^{BA} + 2(\tilde{\pi}^j\tilde{\sigma}^k\mathscr{D}_{[j}\psi_{k]})\epsilon^{AB} + 2(\tilde{\pi}^j\mathscr{D}_{[j}\psi_{k]})\tilde{\sigma}^{ABk}. \tag{7.2.4.5}$$

As in Eqs. (7.2.2.18–19), the fermionic part of $\ddot{\mathscr{H}}^{AB}$ is different from the fermionic part of $\mathscr{H}^{AB}$. These results show that the supersymmetry constraints S^A and $S^{\dagger A}$ in the Ashtekar formalism close on themselves together with $\mathscr{H}^{AB}$ or $\ddot{\mathscr{H}}^{AB}$. The structure is somewhat different from that in Eq. (3.4.15) for supergravity in the 'ADM formulation', partly because of the definition (7.2.2.17) of $S^{\dagger A}$.

7.2.5 Quantization

Following the treatment above, and the example of quantization of general relativity in subsection 7.1.6, let us consider the quantization in which A_{ABi} and ψ_{Aj} are taken as coordinates. Thus a wave function takes the form $\Psi[A_{ABi}, \psi_{Aj}]$. The reality conditions of subsection 7.2.3 on $\tilde{\sigma}^{ABk}$ and A_{ABk} should become Hermiticity conditions with respect to the quantum inner product. They may be quite difficult to realize, because they are non-polynomial. However, it may be possible in part to separate the problem of finding solutions (wave functions annihilated by the constraint operators) from the problem of finding an inner product which expresses the reality conditions.

The constraint

$$J^{AB}\Psi = 0 \tag{7.2.5.1}$$

implies that Ψ is invariant under local rotations applied to its arguments. The constraint $S^A = 0$ becomes

$$S^A\Psi = \mathscr{D}_i \frac{\delta}{\delta\psi_{Ai}}\Psi = 0. \tag{7.2.5.2}$$

This shows that Ψ is invariant under a supersymmetry transformation

$$\delta\psi_{Ai} = \mathscr{D}_i\epsilon_A. \tag{7.2.5.3}$$

Finally, with a particular choice of operator ordering, the $S^{\dagger A}$ constraint reads

$$S^{\dagger A}\Psi = \mathscr{D}_{[i}\psi_{cj]} \frac{\delta^2\Psi}{\delta A_{ABi}\delta A_C{}^B{}_j} = 0. \tag{7.2.5.4}$$

The remaining constraint $\ddot{\mathscr{H}}^{AB}\Psi = 0$ will automatically be satisfied provided the structure of Eq. (7.2.4.4) still holds in the quantum theory.

It is interesting to compare these constraints with those of the formulation of supergravity given in chapter 3. In the $\Psi(e^{AA'}{}_i, \psi^A{}_i)$ representation used there, the constraint $\bar{S}^{A'}\Psi = 0$ was of first order in the bosonic derivative $\delta\Psi/\delta e^{AA'}{}_i$, giving a simple description of the transformation property of the wave function under an infinitesimal primed supersymmetry transformation with parameter $\tilde{\epsilon}^{A'}(x)$. The conjugate constraint $S^A\Psi$ in that representation involves a mixed second-derivative term $\delta^2\Psi/\delta e\delta\psi$, which only has one bosonic derivative, and might help determine whether the theory is finite (chapters 4 and 8). In the present section, it is the 'unprimed' constraint $S^A\Psi = 0$ which is simple, corresponding to an infinitesimal supersymmetry ϵ_A, whereas the constraint $S^{\dagger A}\Psi = 0$ involves two bosonic derivatives at the same spatial point, which may make it harder to understand any potential finiteness properties.

7.2.6 Cosmological constant

As in section 5.9, one can study a generalization of $N = 1$ supergravity which includes a cosmological constant Λ [Townsend 1977, Jacobson 1988]. The Lagrangian is

$$\mathscr{L}_m = \mathscr{L} + \mathscr{L}', \tag{7.2.6.1}$$

where $\mathscr{L}$ is the Lagrangian (7.2.1.1) of supergravity, and

$$\begin{aligned}\mathscr{L}' = & -2im^2 e^{AA'}{}_{\wedge} e^{B}{}_{A'} \wedge e_B{}^{B'} \wedge e_{AB'} \\ & + ime^{AA'}{}_{\wedge} e^{B}{}_{A'} \wedge \psi_A \wedge \psi_B + ime^{AA'}{}_{\wedge} e_A{}^{B'} \wedge \bar{\psi}_{A'} \wedge \bar{\psi}_{B'},\end{aligned} \tag{7.2.6.2}$$

with $\Lambda = 6m^2$. Here $\mathscr{L}_m$ is invariant under a modified unprimed supersymmetry transformation

$$\delta\psi_A = 2\mathscr{D}\epsilon_A,\ \delta\bar{\psi}_{A'} = -4ime^{A}{}_{A'}\epsilon_A,\ \delta e_{AA'} = -i\bar{\psi}_{A'}\epsilon_A. \tag{7.2.6.3}$$

Note that $\mathscr{L}'$ is real and independent of A_{ABi}; hence the total Lagrangian $\mathscr{L} + \mathscr{L}'$ is real subject to the field equation (7.2.1.3) for A_{ABi}. Hence $\mathscr{L}'$ is also invariant under the conjugate of Eq. (7.2.6.3), provided again that A_{ABi} obeys (7.2.1.3).

In comparison with the pure supergravity case (subsection 7.2.2), there are extra terms in the constraints $S^A, S^{\dagger A}$, and $\mathscr{H}^{AB}$, but the general structure is otherwise similar. The supersymmetry constraints with Λ term are

$$S^{A}{}_{m} = S^A + 2\sqrt{2}im(\tilde{\sigma}^k\psi_k)^A, \tag{7.2.6.4}$$

$$S^{\dagger A}{}_{m} = S^{\dagger A} - 2\sqrt{2}im(\tilde{\sigma}_k\tilde{\pi}^k)^A. \tag{7.2.6.5}$$

Here $S^{A}{}_{m}$ is clearly polynomial. And $S^{\dagger A}{}_{m}$ is also polynomial, since $\tilde{\sigma}_k = (1/\sqrt{2})\epsilon_{kmn}\tilde{\sigma}^m\tilde{\sigma}^n$. As in subsections 7.2.2, 7.2.4, one can also work with the polynomial constraint $\ddot{\mathscr{H}}^{AB}_m$, defined by the bracket

$$[S^A_m(x), S^{\dagger B}_m(y)] = \tfrac{1}{2}\ddot{\mathscr{H}}^{AB}(x)\delta^3(x,y),$$

instead of the constraint $\mathscr{H}^{AB}_m$.

7.3 Chern–Simons state

$N = 1$ supergravity with a non-zero cosmological constant, as in the preceding subsection 7.2.6, was studied by [Sano & Shiraishi 1993]. Using Jacobson's representation with a wave function $\Psi(A_{ABi}, \psi_{Aj})$, they found a candidate for a simple physical state, given by $\Psi(A,\psi) = \exp(iS_{CS}) = \exp(-I_{CS})$, where the Euclidean Chern–Simons action is given by

$$I_{CS} = \frac{3}{2g^2}\int (A_{AB} \wedge dA^{AB} + \frac{2}{3}A_{AC} \wedge A_B{}^C \wedge A^{AB} - \lambda g\psi^A \wedge \mathscr{D}\psi_A). \tag{7.3.1}$$

Here, using Sano and Shiraishi's notation, $g = \sqrt{6}m$, where m appears in the action of Eqs. (7.2.6.1–2). The cosmological constant is $\Lambda = g^2 = 6m^2$, and $\lambda = \sqrt{2/3}$ in the present units is a measure of the gravitational constant. Note that the Chern–Simons action appears naturally in the (A,ψ) representation, but would never have been seen using the $(e^{AA'}{}_i, \psi^A{}_i)$ representation of chapter 3. It is not clear whether $\exp(-I_{CS})$ is an exact solution of the constraints, because the $S^{\dagger A}$ constraint of Eqs.

(7.2.5.4), (7.2.6.5) is of second order in bosonic derivatives. Note [Sano & Shiraishi 1993] that $\exp(-I_{CS})$ does obey the S^A quantum constraint of Eqs. (7.2.5.2),(7.2.6.4), since I_{CS} is invariant under the infinitesimal local supersymmetry transformation

$$\delta A_{AB} = -\lambda g \epsilon_{(A}\psi_{B)}, \tag{7.3.2}$$

$$\delta \psi_A = \mathscr{D}\epsilon_A. \tag{7.3.3}$$

However [Sano & Shiraishi 1993], $\exp(-I_{CS})$ is a solution of the constraints in the semi-classical approximation.

To understand this better, consider the state when restricted to the mini-superspace in which the geometry has Robertson–Walker symmetry (written here in the Lorentzian case):

$$ds^2 = -N^2 dt^2 + \frac{1}{8} e^{2\alpha} d\Omega^2. \tag{7.3.4}$$

Here $d\Omega^2$ is the metric of a reference three-sphere, given [Eq. (7.1.1.5)] by $d\Omega^2 = \sigma_{ABi}\sigma^{AB}{}_j dx^i dx^j$. The corresponding spatial triad obeys the Maurer–Cartan relation (5.2.2); this gives

$$d\sigma^{AB} = \sigma^A{}_C \wedge \sigma^{CB}. \tag{7.3.5}$$

The four canonical variables $(A_{ABi}, \tilde{\sigma}^{ABi}; \psi_{Ai}, \tilde{\pi}_A{}^i)$ appearing in section 7.2 can be decomposed as:

$$A_{ABi} = i\omega\sigma_{ABi}, \tag{7.3.6}$$

$$\tilde{\sigma}^{ABi} = -\frac{1}{3V}\sigma h^{1/2}\sigma^{ABi}, \tag{7.3.7}$$

$$\psi_{Ai} = \frac{i}{\sqrt{24V}}\sigma_{ABi}\theta^B + \frac{i}{\sqrt{8V}}\Theta_{ABC}\sigma^{BC}{}_i, \tag{7.3.8}$$

$$\tilde{\pi}^i_A = \sqrt{\frac{32}{3V}} h^{1/2}\sigma_{AB}{}^i\eta^B + \sqrt{\frac{8}{V}} h^{1/2} X_{ABC}\sigma^{BCi}, \tag{7.3.9}$$

where $\omega, \sigma, \theta_A, \Theta_{ABC}, \eta_A$ and X_{ABC} depend only on time. The volume of the S^3 is normalized as $V = \int d^3x|\sigma| = \pi^2/4$. The variable σ is given in terms of α [Eq. (7.3.4)] by

$$\sigma = 12Ve^{2\alpha}. \tag{7.3.10}$$

As in section 5.2, if one tries to derive the above model by making an Ansatz as in Eqs. (7.3.6)–(7.3.9) and requires that local supersymmetry be inherited from the four-dimensional theory, one finds that the spin-3/2 part of Eqs. (7.3.8),(7.3.9) is absent:

$$\Theta_{ABC} = X_{ABC} = 0. \tag{7.3.11}$$

Alternatively, one can leave Θ_{ABC} and X_{ABC} in the calculation, and find that the quantum constraint $S^A\Psi = 0$ is not satisfied unless (7.3.11) holds.

We shall set Θ_{ABC} and X_{ABC} to zero at this point. Finally, in describing the Ansatz, assume that all the Lagrange multipliers

$$A_{AB0}\,, \psi_{A0},\ M_A(\text{for} S^{\dagger A}),\ N \text{and}\ N^i$$

depend only on time.

The Robertson–Walker Chern–Simons quantum state is then

$$\Psi[\omega, \theta_A] = \exp(iS), \tag{7.3.12}$$

where

$$S = \frac{12}{g'^2}\left[\frac{\omega^3}{3} - \frac{i}{2}\omega^2 + \frac{1}{4}\lambda' g'(\omega - i)\theta_A\theta^A\right], \tag{7.3.13}$$

where g', λ' are rescaled from g, λ. Once the spatial coordinates are integrated over, the Lagrangian has the form

$$L = \dot{\omega}\sigma + \dot{\theta}_A\eta^A - H. \tag{7.3.14}$$

The Hamiltonian H will be given implicitly in a particular gauge in Eq. (7.3.28) below. The basic Poisson brackets are then

$$[\omega, \sigma] = 1, \tag{7.3.15}$$

$$[\theta_A, \eta^B] = -\delta_A{}^B. \tag{7.3.16}$$

In the quantum theory, one represents the 'momentum' variables by

$$\sigma = -i\frac{\partial}{\partial\omega}, \tag{7.3.17}$$

$$\eta^A = -i\frac{\partial}{\partial\theta_A}. \tag{7.3.18}$$

It can be verified that the state $\Psi[\omega, \theta_A]$ of Eqs. (7.3.12),(7.3.13) obeys all the quantum constraints $J^{AB}\Psi = 0$, $S^A\Psi = 0$, $S^{\dagger A}\Psi = 0$, $\mathscr{H}_\perp\Psi = 0$, $\mathscr{H}_i\Psi = 0$, where

$$J^{AB} = \frac{i}{6V}h^{1/2}(\theta^A\eta^B + \theta^B\eta^A), \tag{7.3.19}$$

$$S^A = i\sqrt{\frac{6}{V}}h^{1/2}(2\omega\eta^A - \lambda' g'\theta^A\sigma), \tag{7.3.20}$$

$$S^{\dagger A} = \frac{1}{12V^2\sqrt{6V}}h\sigma^2\left[2(i - \omega)\theta^A + \frac{g'}{3\lambda'}\eta^A\right], \tag{7.3.21}$$

$$\mathscr{H}_\perp = \frac{2}{3V^2}h\sigma\left\{\sigma(i\omega - \omega^2 + \frac{g'^2}{12}\sigma - \frac{1}{4}\lambda' g'\theta_A\theta^A)\right.$$

$$\left. + \eta_A\left[2(i - \omega)\theta^A + \frac{g'}{3\lambda'}\eta^A\right]\right\}, \tag{7.3.22}$$

$$\mathscr{H}_i = -\frac{2}{3V}h^{1/2}\eta_A\left[2(i - \omega)\theta_B + \frac{g'}{3\lambda'}\eta_B\right]\sigma^{AB}{}_i. \tag{7.3.23}$$

One can consider the classical Robertson–Walker universes described by the Chern–Simons action S of Eq. (7.3.13). The Hamiltonian H obeys classically

$$H(\omega, \sigma, \theta_A, \eta^A) = 0. \tag{7.3.24}$$

The remaining Hamilton–Jacobi equations are

$$\sigma = \frac{\partial S}{\partial \omega} = \frac{12}{g'^2}(\omega^2 - i\omega + \frac{1}{4}\lambda' g' \theta_A \theta^A), \tag{7.3.25}$$

$$\eta^A = \frac{\partial S}{\partial \theta_A} = \frac{6\lambda'}{g'}(\omega - i)\theta^A. \tag{7.3.26}$$

In studying the classical evolution, pick for simplicity the gauge

$$A_{AB0} = 0, \psi_{A0} = 0, M_A = 0, N = 1, N^i = 0. \tag{7.3.27}$$

Then the Hamiltonian takes the form

$$H = 4\sqrt{12V}\sigma^{1/2}\left(i\omega - \omega^2 + \frac{g'^2}{12}\sigma - \frac{1}{4}\lambda' g' \theta_A \theta^A\right)$$
$$+ 4\sqrt{12V}\sigma^{-1/2}\eta_A\left[2(i - \omega)\theta^A + \frac{g'}{3\lambda'}\eta^A\right]. \tag{7.3.28}$$

The classical evolution is governed by Hamilton's equations:

$$\frac{d\omega}{dt} = [\omega, H] = \frac{g'^2}{3}\sqrt{12V}\sigma^{1/2}, \tag{7.3.29}$$

$$\frac{d\theta_A}{dt} = [\theta_A, H] = 8\sqrt{12V}\sigma^{-1/2}(\omega - i)\theta_A, \tag{7.3.30}$$

$$\frac{d\sigma}{dt} = [\sigma, H] = 4\sqrt{12V}[\sigma^{1/2}(2\omega - i) + 2\sigma^{-1/2}\eta_A\theta^A], \tag{7.3.31}$$

$$\frac{d\eta^A}{dt} = [\eta^A, H] = 4\sqrt{12V}\sigma^{-1/2}[\tfrac{1}{2}\lambda' g' \sigma \theta^A + 2(\omega - i)\eta^A], \tag{7.3.32}$$

where Eqs. (7.3.25),(7.3.26) have been used. As described in section 4.1, following [Bao *et al.* 1985], because of the Grassmann nature of θ_A and η^A, one begins by solving the bosonic part of Eqs. (7.3.29),(7.3.31), corresponding to Einstein's equations with a cosmological constant Λ, in the Robertson–Walker case. One next works at first order in fermions, solving Eqs. (7.3.30),(7.3.32) at linear order in fermions, giving the gravitino evolution in the above gravitational background. Then one feeds $(\theta^A)_{\text{lin}}$, $(\eta^A)_{\text{lin}}$ back in to Eqs. (7.3.29),(7.3.31) to obtain the second-order fermionic corrections to ω and σ. Following these remarks, one expects that the bosonic solution for ω and σ will give a de Sitter spacetime, and that the fermions then propagate linearly in this background.

The coupled equations for the bosonic parts of ω and σ, obtained from Eqs. (7.3.29),(7.3.31), by deleting the last term in Eq. (7.3.31), simplify when one uses the variables ω and $\sigma^{1/2}$, and can be solved. There are different cases depending on the signature of the spacetime [Sano & Shiraishi 1993]. In particular, there is a Riemannian solution at bosonic order

$$ds^2 = \frac{1}{16g'^2 V}(d\xi^2 + \tfrac{1}{2}\sin^2 \xi d\Omega^2), \qquad (7.3.33)$$

giving positive-definite de Sitter space – i.e. a four-sphere. The linearized fermion fields are

$$\theta_A = \sin^4(\xi/2)\theta_{0A}, \qquad (7.3.34)$$

$$\eta^A = -\frac{6i\lambda'}{g'}\cos^2(\xi/2)\sin^4(\xi/2)\theta_0^A. \qquad (7.3.35)$$

There is also a Lorentzian solution at bosonic order

$$ds^2 = \frac{1}{16g'^2 V}(-d\xi^2 + \tfrac{1}{2}\cosh^2 \xi d\Omega^2), \qquad (7.3.36)$$

giving Lorentzian de Sitter spacetime. The linearized fermion fields are

$$\theta_A = \cosh^2\xi \ \exp(-2i\arctan \sinh\xi)\theta_{0A}, \qquad (7.3.37)$$

$$\eta^A = \frac{3\lambda'}{g'}(\sinh\xi - i)\cosh^2\xi\exp(-2i\arctan \sinh\xi)\theta_0^A. \qquad (7.3.38)$$

There are further solutions [Sano & Shiraishi 1993] with different signatures, but they all correspond at the bosonic level to continuations of de Sitter spacetime.

Thus, if the boundary data for the Chern–Simons state are set to data for a Robertson–Walker spacetime, one obtains the classical solutions as forms of de Sitter spacetime (e.g. Lorentzian or Euclidean). It would be of considerable interest to know whether the general Chern–Simons state $\exp(iS_{CS}) = \exp(-I_{CS})$, with S_{CS} given by Eq. (7.3.1), is a solution of the full quantum constraint equations. This may be a difficult question to decide, because of the second-order nature of the quantum constraint (7.2.42). One would also like to know the form of $\exp(-I_{CS})$ for boundary data which are perturbed away from Robertson–Walker data. What, if any, is the relation between this state and the Hartle–Hawking or wormhole states?

The wave function $\exp(iS_{CS})$ in gravity (without fermions), was studied by [Kodama 1990], using Ashtekar variables. In this way Kodama found (e.g.) exact solutions of the quantum constraints with Λ-term for Robertson–Walker and Bianchi-IX symmetry, of the form $\exp(-I_{CS})$. This state has also been discussed for gravity, without fermions, by [Louko 1995], who shows that $\exp(iS_{CS})$ gives a semi-classical estimate of the Hartle–Hawking wave function in the Bianchi IX case.

Other treatments involving supergravity and Ashtekar variables are by [Capovilla & Guven 1994], [Capovilla & Obregón 1994], [Matschull 1994], [Obregón, Pullin & Ryan 1993], and [Sano 1992], who has generalized Jacobson's treatment of section 7.2 to $N = 2$ supergravity.

8
Further developments

8.1 Local boundary conditions and possible finiteness

This section follows from chapter 4 in treating the local boundary conditions for $N = 1$ supergravity in which $\tilde{\psi}^{A'}{}_{iI}$ is specified on an initial three-surface, and $\psi^A{}_{iF}$ is specified on a final three-surface. The amplitude to go from initial to final data in Euclidean time τ at spatial infinity is given by a path integral (4.1.1) with auxiliary fields included in the boundary data, and denoted by

$$K(e^{AA'}{}_{iF}, \psi^A{}_{iF}, b_{AA'F}; e^{AA'}{}_{iI}, \tilde{\psi}^{A'}{}_{iI}, b_{AA'I}; \tau, \ldots).$$

Here an argument will be presented which indicates that the amplitude K for local boundary conditions may be finite, at least for fairly weak gravitational fields.

One begins, following chapter 4, by noting that general boundary data, regarded now as the data for a classical boundary-value problem, will not give a solution of the classical supersymmetry constraint equations $\tilde{S}_{A'} = 0, S_A = 0$. One approach is simply to choose the fermionic data $\psi^A{}_{iF} = {}^s\psi^A{}_{iF}$ and $\tilde{\psi}^{A'}{}_{iI} = {}^s\tilde{\psi}^{A'}{}_{iI}$ to give classical solutions of the supersymmetry constraints. Alternatively, following section 4.3, this question can be resolved by making use of the four independent complex components of the auxiliary field $b_{AA'}$ to solve the four constraints $\tilde{S}_{A'} = 0$, $S_A = 0$. One can, following section 4.3, also use the Rarita–Schwinger equation with auxiliary fields $b_{AA'}$ or $\bar{b}_{AA'}$, to find the evolution of $\psi^A{}_\mu$ and $\tilde{\psi}^{A'}{}_\mu$ (up to gauge) and to give $b_{AA'}$ in the infilling spacetime. Thus the data $b_{AA'F}, b_{AA'I}$ are not freely specifiable, but depend on the other boundary data.

In this part of the argument, one considers data in which $\tilde{\psi}^{A'}{}_{iI} = 0$ but $\psi^A{}_{iF} \neq 0$. Then [Eq. (4.3.7)] one studies the wave function

$$\Psi = \exp(-I_B/\hbar). \tag{8.1.1}$$

Here $I_B = I_B(e^{AA'}{}_{iF}; e^{AA'}{}_{iI}; \tau, \ldots)$ is the classical gravitational action to go from the initial three-geometry $e^{AA'}{}_{iI}(x)$ to the final three-geometry $e^{AA'}{}_{iF}(x)$ in Euclidean time τ at spatial infinity (and possibly with a spatial translation X^i between the surfaces). For each choice of final data $\psi^A{}_{iF}$, one chooses the auxiliary field $b_{AA'}$ as described above, such that the classical constraint $\tilde{S}_{A'} = 0$ holds for the classical boundary-value problem. Then, as seen in section 4.3, $\Psi = \exp(-I_B/\hbar)$ obeys all the quantum constraints at the final surface, and is a candidate for a bosonic quantum amplitude. One checks this by verifying that $\Psi = \exp(-I_B/\hbar)$ obeys the heat equation of section 4.4 with the correct asymptotic behaviour as $\tau \to 0_+$.

An error in an early argument [D'Eath 1994] was pointed out by [Carroll *et al.* 1994]; this error, however, was consequent upon a too-restrictive assumption in [D'Eath 1984] mentioned above, namely that any choice of fermionic boundary data yields a classical solution. As described above and in section 4.3, the use of auxiliary fields in the present argument means that one can choose $\psi^A{}_{iF}$ and $\tilde{\psi}^{A'}{}_{iI}$ freely. One thus takes a different path from [D'Eath 1994], to which the remark of [Carroll *et al.* 1994] does not apply.

Turning to the heat equation, note that $\Psi = \exp(-I_B/\hbar)$ obeys the equation (4.4.1):

$$\hbar \frac{\partial K}{\partial \tau} + \hat{H}_E K = 0, \tag{8.1.2}$$

satisfied by the full propagator K, where $\hat{H}_E$ is the quantum version of the classical Euclidean Hamiltonian (4.4.2). Replacing K by $\exp(-I_B/\hbar)$ in Eq. (8.1.2), one has for the first term $-(\partial I_B/\partial \tau)\exp(-I_B/\hbar)$. But [Eq. (4.1.5)]

$$\frac{\partial I_B}{\partial \tau} = M_B, \tag{8.1.3}$$

where $M_B = M_B(e^{AA'}{}_{iF}; e^{AA'}{}_{iI}; \tau, \ldots)$ is the classical bosonic mass. Hence the first term in Eq. (8.1.2) gives $-M_B \exp(-I_B/\hbar)$. The second term in Eq. (8.1.2) has zero contribution from the volume terms in Eq. (4.4.2) since $\Psi = \exp(-I_B/\hbar)$ obeys the quantum constraints with $\mathscr{H}_{AA'}$ given by the anticommutator of S_A and $\bar{S}_{A'}$, and a contribution $M_B \exp(-I_B/\hbar)$ from the surface terms in Eq. (4.4.2). Hence $\Psi = \exp(-I_B/\hbar)$ obeys the heat equation for $\tau > 0$:

$$\hbar \frac{\partial \Psi}{\partial \tau} + \hat{H}_E \Psi = 0. \tag{8.1.4}$$

One now has to check that $\Psi = \exp(-I_B/\hbar)$ obeys the correct initial conditions as $\tau \to 0_+$, as described in section 4.4. But the asymptotic form

(4.4.29) of the full propagator as $\tau \to 0_+$ is

$$K \sim \exp[-\tau^{-1} v(h_{ijF}; h_{ijI}; \ldots)/\hbar], \tag{8.1.5}$$

where the action $I_B(h_{ijF}; h_{ijI}; \tau, \ldots)$ has the asymptotic form (4.4.22)

$$I_B \sim \tau^{-1} v(h_{ijF}; h_{ijI}; \ldots) \tag{8.1.6}$$

as $\tau \to 0_+$. Hence $\Psi = \exp(-I_B/\hbar)$ and the full propagator K are asymptotic to each other as $\tau \to 0_+$. By uniqueness for the heat equation, the full propagator K is then given by

$$K = \exp(-I_B/\hbar), \tag{8.1.7}$$

for boundary data $(e^{AA'}{}_{iF}, \psi^A{}_{iF}, b_{AA'F}; e^{AA'}{}_{iI}, 0, b_{AA'I}; \tau, \ldots)$. This is a semi-classical expression without any quantum corrections.

Alternatively, following Eqs. (4.4.27,28), one can study the equation $\bar{S}_{A'}K = 0$, applied to the semi-classical expansion (4.1.13)

$$K \sim (A + \hbar A_1 + \hbar^2 A_2 + \ldots)\exp(-I_{\text{class}}/\hbar). \tag{8.1.8}$$

In the case of boundary data

$$(e^{AA'}{}_{iF}, \psi^A{}_{iF}, b_{AA'F}; e^{AA'}{}_{iI}, 0, b_{AA'I}; \tau \ldots),$$

one has [Eqs. (4.1.17)–(4.1.19)]

$$\begin{aligned} \psi^A{}_i(x)\frac{\delta A}{\delta e^{AA'}{}_i(x)} &= 0, \\ \psi^A{}_i(x)\frac{\delta A_1}{\delta e^{AA'}{}_i(x)} &= 0, \\ &\ldots, \end{aligned} \tag{8.1.9}$$

evaluated at the final surface. Since $\psi^A{}_i(x)$ can be chosen arbitrarily at each point x, one has

$$\frac{\delta A}{\delta e^{AA'}{}_i(x)} = 0, \quad \frac{\delta A_1}{\delta e^{AA'}{}_i(x)} = 0, \ldots. \tag{8.1.10}$$

Hence A and all A_n do not depend on $e^{AA'}{}_{iF}(x)$. By symmetry, A and all A_n do not depend on $e^{AA'}{}_{iI}(x)$. Further, since $\tilde{\psi}^{A'}{}_{iI} = 0$, they do not depend on $\psi^A{}_{iF}$, and they do not depend on the auxiliary field data $b_{AA'I}, b_{AA'F}$, since $b_{AA'}$ has zero conjugate momentum. Hence $A, A_1, A_2, \ldots$ are functions of τ only, and can be evaluated by choosing the boundary data to be flat. This gives, in the case $\tilde{\psi}^{A'}{}_{iI} = 0$,

$$A = 1, \; A_1 = A_2 = \cdots = 0. \tag{8.1.11}$$

The equation $A = 1$ is immediate, since (section 4.2) A is given by the super-determinant of the perturbation operators for gravity and the gravitino, taken about flat Euclidean spacetime. Now consider the higher-loop terms $A_1, A_2, \ldots$, seen above to be functions of τ only. These are given by the extent to which the approximation $A\exp(-I_B/\hbar)$ fails to satisfy the evolution equation (3.5.11). This amounts to evolving the amplitude K by considering the commutator between

$$\left[\psi^A{}_0(\infty)Q_A + \int d^3x\, \psi^A{}_0 S_A\right] \tag{8.1.12}$$

and

$$\left[\tilde{Q}_{A'}\tilde{\psi}^{A'}{}_0(\infty) + \int d^3x\, \tilde{S}_{A'}\tilde{\psi}^{A'}{}_0\right] \tag{8.1.13}$$

applied to K for suitable choices of $\psi^A{}_0(\infty)$ and $\tilde{\psi}_0^{A'}(\infty)$. Here Q_A and $\tilde{Q}_{A'}$ are the supercharge operators. This leads in general to the multiple-scattering expansion [D'Eath 1981,1984], based on [Balian & Bloch 1971,1974] in the case of non-relativistic quantum mechanics. Write

$$K_0(e,\psi,b;e_I,0,b_I;T,\ldots) = A\exp(-I_B/\hbar), \tag{8.1.14}$$

where $A = 1$ in our case. Then the amount by which K_0 fails to satisfy the heat equation (8.1.2),

$$\left[-\hat{H}_E - \hbar\frac{\partial}{\partial\tau}\right]K_0 = \hbar\Lambda(e,\psi,b;e_I,0,b_I;T,\ldots) \tag{8.1.15}$$

of course has $\Lambda = 0$ here.

In the multiple-scattering expansion, one uses the quantity Λ as a source for higher-order terms in an asymptotic series

$$K \sim K_0 + K_1 + K_2 + \cdots. \tag{8.1.16}$$

One can further expand out $K_1, K_2, \ldots$ around a classical path, as

$$\begin{aligned} K_1 &\sim (\hbar A_{11} + \hbar^2 A_{21} + \hbar^3 A_{31} + \cdots)\exp(-I_B/\hbar), \\ K_2 &\sim (\qquad\quad\; \hbar^2 A_{22} + \hbar^3 A_{32} + \cdots)\exp(-I_B/\hbar), \\ &\cdots \end{aligned} \tag{8.1.17}$$

with $A_{11} = A_1, A_{21} + A_{22} = A_2, \ldots$. Since $\Lambda = 0$, one has $K_1 = 0$ [D'Eath 1984]. Similarly, by iteration in the multiple-scattering expansion, one has $K_2 = K_3 = \cdots = 0$. Hence $A_1 = A_2 = A_3 = \cdots = 0$, as in Eq. (8.1.11). This shows by an alternative route that the full bosonic amplitude is given by

$$K = \exp(-I_B/\hbar). \tag{8.1.18}$$

Now consider the general quantum amplitude including non-zero fermionic data $\psi^A{}_{iF}$ and $\tilde{\psi}^{A'}{}_{iI}$, where one takes only data $\psi^A{}_{iF} = {}^s\psi^A{}_{iF}$ and $\tilde{\psi}^{A'}{}_{iI} = {}^s\tilde{\psi}^{A'}{}_{iI}$ which obey the classical constraints. Starting at three loops, there will be possible local bosonic counterterms such as Eq. (4.2.36), which when combined with fermionic and surface partners [D'Eath 1986a,b] may give supersymmetric invariants. There may also conceivably be pure surface invariant counterterms, although this is unlikely, given the results of section 4.2. Each such independent counterterm appears with a coefficient. Since [Eq. (8.1.18)] the bosonic amplitude is exactly semi-classical, all the coefficients must be zero, indicating that the whole fermionic amplitude, with the present local boundary conditions, may be finite. The only way to avoid this would be for there to exist supersymmetrically invariant local counterterms which have no purely bosonic part, but start with a term quadratic or higher in fermions. This would be difficult to achieve, since the variation in the bosonic direction of the leading term in fermions would have to be zero. Nevertheless, this possibility should be borne in mind.

One should also keep in mind the possibility, mentioned in section 4.1, that the Riemannian Einstein equations cease to be well-posed for bounding three-geometries $e^{AA'}{}_{iI}, e^{AA'}{}_{iF}$ which are deformed too far away from flatness. In that case the bosonic expression $K = \exp(-I/\hbar)$ and consequent fermionic finiteness would have to be modified. Subject to these caveats, however, the above results indicate that amplitudes with local boundary data $\psi^{A}{}_{iF}$ and $\tilde{\psi}^{A'}{}_{iI}$ might conceivably be finite.

8.2 Spectral boundary conditions and possible finiteness

The approach of section 8.1 used local boundary data $\psi^{A}{}_{i}$ and $\tilde{\psi}^{A'}{}_{i}$. But in scattering theory [Itzykson & Zuber 1980], one uses spectral boundary conditions, with positive-frequency fermionic data for the 'in' state and negative-frequency data for the 'out' state. Here we shall see how, with these boundary conditions, one can give a treatment of supergravity analogous to that in section 8.1.

Consider the Rarita–Schwinger equation for the evolution of $\psi^{A}{}_{i}$, at linear order in fermions (i.e. excluding the torsional contribution to the connection):

$$\begin{aligned} &\epsilon^{ijk}(Nn_{AA'} + N^{l} e^{AA'}{}_{l})\,{}^{4s}D_{j}\psi^{A}{}_{k} \\ &+\epsilon^{ijk} e_{AA'j}({}^{4s}D_{k}\psi^{A}{}_{0} - {}^{4}D_{0}\psi^{A}{}_{k}) = 0. \end{aligned} \tag{8.2.1}$$

Here ${}^{4s}D_{j}$ denotes the torsion-free four-dimensional covariant derivative. It is assumed that the local four-geometry is known, e.g. as the solution of a bosonic classical boundary-value problem. Consider Eq. (8.2.1) on a spacelike three-surface. Assume for simplicity that the Lagrange multipliers of the Hamiltonian approach obey, in a neighbourhood of the surface,

$$N^i = 0,\ \psi^A{}_0 = 0,\ \omega^{AB}{}_0 = 0,\ \tilde{\omega}^{A'B'}{}_0 = 0. \tag{8.2.2}$$

Then

$$\begin{aligned} N\epsilon^{ijk} n_{AA'}{}^{4s}D_j \psi^A{}_k &= \epsilon^{ijk} e_{AA'j}{}^{4s}D_0 \psi^A{}_k \\ &= \epsilon^{ijk} e_{AA'j} \dot{\psi}^A_k, \end{aligned} \tag{8.2.3}$$

where $\dot{\psi}^A_k = \partial_0 \psi^A{}_k$. In the Lorentzian régime, positive-frequency data have $\psi^A{}_k \propto \exp(-i\lambda t)$ proper-time dependence, with $\lambda > 0$. This corresponds to

$$\epsilon^{ijk} n_{AA'}{}^{4s}D_j \psi^{A+}{}_k = -i\lambda \epsilon^{ijk} e_{AA'j} \psi^{A+}{}_k \tag{8.2.4}$$

with $\lambda > 0$. Negative-frequency data $\psi^{A-}{}_k$ have $\lambda < 0$. In the Euclidean régime, with unit normal

$${}_e n_{AA'} = -i n_{AA'}, \tag{8.2.5}$$

one has

$$\epsilon^{ijk}\,{}_e n_{AA'}{}^{4s}D_j \psi^{A+}{}_k = -\lambda \epsilon^{ijk} e_{AA'j} \psi^{A+}{}_k, \tag{8.2.6}$$

for $\lambda > 0$, and

$$\epsilon^{ijk}\,{}_e n_{AA'}{}^{4s}D_j \psi^{A-}{}_k = -\mu \epsilon^{ijk} e_{AA'j} \psi^{A}{}_k, \tag{8.2.7}$$

for $\mu < 0$. Since $t = -i\tau$ in the asymptotic region, the eigenvalue λ corresponds to $e^{-\lambda\tau}$ Euclidean-time dependence. Analogously, for the primed fields, one has

$$\epsilon^{ijk}\,{}_e n_{AA'}{}^{4s}D j \tilde{\psi}^{A'-}{}_k = \lambda \epsilon^{ijk} e_{AA'j} \tilde{\psi}^{A'-}{}_k, \tag{8.2.8}$$

for $\lambda > 0$, and

$$\epsilon^{ijk}\,{}_e n_{AA'}{}^{4s}D_j \tilde{\psi}^{A'+}{}_k = \mu \epsilon^{ijk} e_{AA'j} \tilde{\psi}^{A'+}{}_k, \tag{8.2.9}$$

for $\mu < 0$. One can, alternatively, study the operator ${}^{3s}D_j$ instead of ${}^{4s}D_j$ in a similar way.

Note that the four-dimensional connection ${}^{4s}\omega^{AB}{}_j$ is related to the three-dimensional connection ${}^{3s}\omega^{AB}{}_j$ by

$${}^{4s}\omega^{AB}{}_j - {}^{3s}\omega^{AB}{}_j = n^A{}_{B'} e^{BB'i}\, {}^sK_{ij}. \tag{8.2.10}$$

Here ${}^sK_{ij}$ is (section 3.2) the torsion-free Lorentzian second fundamental form, which is pure imaginary for a Riemannian geometry. Hence

$$\begin{aligned} \epsilon^{ijk}\,{}_e n_{AA'}{}^{4s}D_j \psi^A{}_k = {} & \epsilon^{ijk}\,{}_e n_{AA'}{}^{3s}D_j \psi^A{}_k \\ & + \tfrac{1}{2} i \epsilon^{ijk} e_{BA'j}\, {}^sK_{kl} \psi^{Bl}. \end{aligned} \tag{8.2.11}$$

Further, for a real three-metric h_{ij} one has Hermitian $e^{AA'}{}_i$, with $\tilde{\omega}^{A'B'}{}_i = \bar{\omega}^{A'B'}{}_i$. Hence the operators involved in Eqs. (8.2.4) and following are Hermitian in the Euclidean régime, giving real eigenvalues λ and orthogonality, in the sense

$$\int d^3x e^{ijk} e_{AA'k} \psi^A{}_{i\lambda_1} \tilde{\psi}^{A'}{}_{j\lambda_2} = 0, \ \lambda_1 \neq \lambda_2, \tag{8.2.12}$$

where λ_1 is defined by Eq. (8.2.6) and λ_2 by Eq. (8.2.8). Further, the primed harmonics can be derived from the unprimed harmonics as

$$\begin{aligned} \tilde{\psi}^{A'-}{}_i &= (\overline{\psi^{A+}{}_i}), \\ \tilde{\psi}^{A'+}{}_i &= (\overline{\psi^{A-}{}_i}). \end{aligned} \tag{8.2.13}$$

Assuming that there are no zero modes (see below), one can expand a general $\psi^A{}_i(x)$ and $\tilde{\psi}^{A'}{}_i(x)$ as

$$\psi^A{}_i = \psi^{A+}{}_i + \psi^{A-}{}_i, \tag{8.2.14}$$

$$\tilde{\psi}^{A'}{}_i = \tilde{\psi}^{A'-}{}_i + \tilde{\psi}^{A'+}{}_i, \tag{8.2.15}$$

where $\psi^{A+}{}_i$ is a sum over positive-frequency harmonics (8.2.6), $\psi^{A-}{}_i$ is a sum over negative-frequency harmonics (8.2.7), etc. This ignores the contribution of any zero mode, with $\lambda = 0$, which would anyway be of measure zero in the asymptotically flat case (see below). (Compare section 2.9.3 in the spin-1/2 case. We shall use the symbols $m, \tilde{m}$ and $r, \tilde{r}$ schematically to denote typical harmonic coefficients by analogy with Eqs. (2.9.3.10–11), where the m and r correspond to positive frequencies.) Then one can (e.g.) take the m and r coefficients as coordinates, in which case $\tilde{m}$ and $\tilde{r}$ are momentum-like. More precisely, consider the Hamiltonian decomposition of the supergravity action (cf. section 3.2). Using the orthogonality (8.2.12), this gives

$$\begin{aligned} S = \sum (\tilde{m}\dot{m} + m\dot{\tilde{m}} + \tilde{r}\dot{r} + r\dot{\tilde{r}}) \text{terms} \\ + p\,\dot{e}\,\text{term} + \text{constraint terms}. \end{aligned} \tag{8.2.16}$$

Hence, when the momenta conjugate to $m, \tilde{m}, r$ and $\tilde{r}$ are written out and then eliminated, following the Dirac procedure, one has as the only non-zero Dirac brackets among the fermionic variables:

$$[m, \tilde{m}]^* = -i, \tag{8.2.17}$$

$$[r, \tilde{r}]^* = -i. \tag{8.2.18}$$

In the case that the four-geometry is exactly flat, consider the decomposition of Eq. (3.3.22), with $\psi_{Ai} = -e^{BB'}{}_i \psi_{ABB'}$:

$$\psi_{ABB'} = \rho_{ABB'} + \frac{2}{3}(n_A{}^{A'} \tilde{\beta}_{A'} n_{BB'} + n_B{}^{A'} \tilde{\beta}_{A'} n_{AB'}) + \epsilon_{AB} \tilde{\beta}_{B'}, \tag{8.2.19}$$

where

$$\rho_{ABB'} = n^C{}_{B'}\rho_{ABC} \tag{8.2.20}$$

with ρ_{ABC} totally symmetric:

$$\rho_{ABC} = \rho_{(ABC)}. \tag{8.2.21}$$

The 'physical gravitinos' are described by the spin-3/2 modes $\rho_{(ABC)}$ and $\tilde{\rho}_{(A'B'C')}$, where for a harmonic, ρ_{ABC} and $\tilde{\rho}_{A'B'C'}$ obey

$$e^{AA'i}\,{}^{4s}D_i\rho_{ABC} = i\nu n^{AA'}\rho_{ABC} \tag{8.2.22}$$

$$e^{AA'i}\,{}^{4s}D_i\tilde{\rho}_{A'B'C'} = -i\nu n^{AA'}\tilde{\rho}_{A'B'C'}. \tag{8.2.23}$$

Defining

$$\psi^A{}_i = e^{DD'}{}_i n^E{}_{D'}\rho^A{}_{DE}, \tag{8.2.24}$$

and similarly for $\tilde{\psi}^{A'}{}_i$, one can check that $\psi^A{}_i$ obeys

$$\epsilon^{ijk}n_{AA'}{}^{4s}D_j\psi^A{}_k = \tfrac{3}{2}i\nu\epsilon^{ijk}e_{AA'j}\psi^A{}_k, \tag{8.2.25}$$

with a similar equation for $\tilde{\psi}^{A'}{}_k$. Thus, in the Euclidean régime, $\psi^A{}_i$ so constructed from ρ_{ABC} obeys Eq. (8.2.6) with

$$\lambda = -\tfrac{3}{2}\nu. \tag{8.2.26}$$

Further, one can verify that $\psi^A{}_i$ in Eq. (8.2.25) obeys the $\tilde{S}_{A'}$ constraint at linear order in fermions:

$$\epsilon^{ijk}e_{AA'i}{}^{4s}D_i\psi^A{}_k = 0 \tag{8.2.27}$$

and correspondingly for $\tilde{\psi}^{A'}{}_i$, verifying the interpretation of $\rho^{\pm}{}_{(ABC)}$, $\tilde{\rho}^{\pm}_{A'B'C'}$ as 'physical gravitino data' in the flat case, and equivalent in the flat case to specifying the transverse-traceless $\psi^{\mathrm{TT}A}{}_i$ in Eq. (4.2.38). These spin-3/2 modes have two degrees of freedom per space point.

For a flat four-geometry, the gauge modes

$$\psi^A{}_i = \alpha^A{}_{,i}, \tag{8.2.28}$$

$$\tilde{\psi}^{A'}{}_i = \tilde{\alpha}^{A'}{}_{,i}, \tag{8.2.29}$$

are zero modes of Eqs. (8.2.6),(8.2.8), with $\lambda = 0$. These carry two degrees of freedom per space point. In the decomposition (8.2.19), they involve linear combinations of ρ_{ABC} and $\tilde{\beta}_{A'}$. The remaining two degrees of freedom per space point in $\psi^A{}_i$ (which has six degrees of freedom altogether) are carried by different linear combinations of ρ_{ABC} and $\tilde{\beta}_{A'}$, with non-zero eigenvalues λ.

It is important to note that the infinite number of zero modes associated with local supersymmetry transformations vanish as soon as one has the slightest amount of curvature. To see this, consider an infinitesimal linear-order supersymmetry transformation

$$\delta\psi^{A}{}_{i} = 2\kappa^{-1\,4s}D_i\epsilon^A. \tag{8.2.30}$$

This gives for the left-hand side of Eq. (8.2.4)

$$\text{const. } \epsilon^{ijk} n_{AA'}{}^{4s}R^{A}{}_{Bjk}\epsilon^{B}, \tag{8.2.31}$$

which will only give zero at exceptional spatial points, if at all. One expects that these modes join the remaining degrees of freedom in the splitting $(\psi^{A+}{}_i, \psi^{A-}{}_i)$ and $(\tilde{\psi}^{A'+}{}_i, \tilde{\psi}^{A'-}{}_i)$, so that all six degrees of freedom per space point contribute to the positive- and negative-frequency decomposition.

Quantum-mechanically, one may then consider the amplitude

$$K(e^{AA'}{}_{iF}, m_F, r_F, b_{AA'F}\,; e^{AA'}{}_{iI}, \tilde{m}_I, \tilde{r}_I, b_{AA'I}\,; \tau, \ldots) \tag{8.2.32}$$

where one takes positive-frequency data on the final surface, associated with $e^{-\lambda\tau}$ behaviour with $\lambda > 0$, and negative-frequency data on the initial surface. Here one assumes that the infilling solution $g_{\mu\nu}$ of the classical Einstein equation has been found with bosonic boundary data $e^{AA'}{}_{iF}, e^{AA'}{}_{iI}$, and is used in defining the notion of positive and negative frequency through Eqs. (8.2.6–9). The bracket relations of the equations (8.2.17–18) show that the operators $\bar{m}$ and $\bar{r}$ are given by

$$\bar{m} = \hbar\partial/\partial m, \tag{8.2.33}$$

$$\bar{r} = \hbar\partial/\partial r. \tag{8.2.34}$$

For application to a scattering problem, one takes τ to be large, with weak curvature in the initial and final gravitational data, and one takes the initial and final gravitino field $(\psi^{A-}{}_i, \tilde{\psi}^{A'-}{}_i)_I$ and $(\psi^{A+}{}_i, \tilde{\psi}^{A'+}{}_i)_F$ to be approximately (or asymptotically) a sum of 'physical gravitino' spin-3/2 modes, which obey the supersymmetry constraints.

One then proceeds by analogy with sections 4.3, 4.4 and 8.1. Consider the case in which $\psi^{A-}{}_{iI} = 0$ and $\tilde{\psi}^{A'-}{}_{iI} = 0$, i.e. $\tilde{m}_I = \tilde{r}_I = 0$. Take the wave function

$$\Psi = \exp(-I_B/\hbar), \tag{8.2.35}$$

where I_B is the bosonic gravitational action of the Riemannian solution of the Einstein field equations, with only gravitational boundary data $(e^{AA'}{}_{iF}\,; e^{AA'}{}_{iI}\,; \tau \ldots)$. One examines the dependence of Ψ on the final data. Here

$$\begin{aligned}\bar{S}_{A'}\Psi = [&\epsilon^{ijk} e_{AA'i}{}^{3s}D_j\psi^{A+}{}_k + \tfrac{1}{2}i\kappa^2 p_{AA'}{}^{i}\psi^{A+}{}_i + b\psi^{+}\,\text{terms}] \\ &\times \exp(-I_B/\hbar),\end{aligned} \tag{8.2.36}$$

where $b_{AA'}$ is the auxiliary field of section 4.3, and the $b\psi$ contribution can be found from that section. Similarly,

$$S_A\Psi = [\epsilon^{ijk} e_{AA'i}{}^{3s}D_j\tilde{\psi}^{A'}{}_{+k} - \tfrac{1}{2}i\kappa^2 p_{AA'}{}^i\tilde{\psi}^{A'}{}_{+i} + \bar{b}\tilde{\psi}^+\text{terms}] \times \exp(-I_B/\hbar). \tag{8.2.37}$$

Apart from exceptional cases (as commented upon in section 4.3), one can solve the four constraint equations $\bar{S}^{A'}\Psi = 0$ and $S_A\Psi = 0$ at the final surface, by choice of the four quantities $b_{AA'}$. Hence $\Psi = \exp(-I_B/\hbar)$ obeys all the quantum constraints at the final surface:

$$S_A\Psi = 0, \quad \bar{S}_{A'}\Psi = 0,$$
$$\mathscr{H}_{AA'}\Psi = 0, \quad J_{AB}\Psi = 0, \quad \bar{J}_{A'B'}\Psi = 0.$$

Here, as usual, $\mathscr{H}_{AA'}$ is given by the anticommutator of S_A and $\bar{S}_{A'}$; also, the J constraints are automatic since I is a scalar.

One then needs to verify that $\Psi = \exp(-I_B/\hbar)$ is not just a solution of the quantum constraints, but also is precisely the quantum amplitude for the boundary data $(e^{AA'}{}_{iF}; e^{AA'}{}_{iI}; \tau, \ldots)$, where the final fermionic data $\psi^{A}{}_{+i}, \tilde{\psi}^{A'}{}_{+i}$ have additionally been set to zero. This proceeds as in sections 4.4 and 8.1. When verifying that the one-loop amplitude $A\exp(-I_B/\hbar)$ has $A \to 1$ as $\tau \to 0_+$, one must use the first of the two arguments in section 4.4, showing that the eigenvalues for linearized gravitational and gravitino perturbations about the Kasner-like classical solution tend to the flat-space eigenvalues as $\tau \to 0_+$. Thus A, being given by the ratio of two determinantal expressions, tends to 1 as $\tau \to 0_+$. This shows that the amplitude $K(e^{AA'}{}_{iF}; e^{AA'}{}_{iI}; \tau, \ldots)$ for purely gravitational (weak-field) boundary data should again be given by

$$K = \exp(-I_B/\hbar). \tag{8.2.38}$$

As in section 8.1, one can then consider the general quantum amplitude including spectral fermionic data $\psi^{A}{}_{-iI}, \tilde{\psi}^{A'}{}_{-iI}, \psi^{A}{}_{+iF}, \tilde{\psi}^{A'}{}_{+iF}$, given by 'physical gravitino' spin-3/2 modes in the asymptotic region with large Euclidean time-separation τ, together with bosonic data where $e^{AA'}{}_{iI}$ and $e^{AA'}{}_{iF}$ contain only weak transverse-traceless perturbations of flat three-space. One can then form incoming and outgoing gravitino and graviton states by summing over the propagators K for suitably weighted boundary data, following [Kuchař 1970]. Unless there are counterterms in the evaluation of K which start at quadratic or higher order in fermions, having no bosonic part, this work again suggests that all such fermionic amplitudes, involving spectral gravitino boundary data, might well turn out to be finite. One should also bear in mind for this result that the classical four-geometry should not be too far from flatness, as in section 8.1 for local boundary conditions.

8.3 Cosmology

As described in section 2.9, quantization involving fermions on compact spatial sections is most naturally treated using spectral boundary conditions. The local boundary conditions of section 8.1, in which either $\psi^A{}_i$ or $\tilde{\psi}^{A'}{}_i$ is specified on a compact three-surface, will lead to a meaningless Hartle–Hawking state, at least semi-classically, since half of the classical fermionic variables will diverge in the middle of the spacetime.

The Friedmann model – the simplest cosmological model – was studied in section 5.4, where the Hartle–Hawking state for a $k = +1$ model was found to be

$$\Psi_{HH}(a, \psi^A) = D\exp(3a^2/\hbar)\psi_A\psi^A, \tag{8.3.1}$$

where D is a constant and a is the radius of the three-sphere. One can then treat perturbations of this model by a harmonic decomposition.

The spin-1/2 harmonics for the Einstein-Dirac system are described in section 2.9. For the gravitino field on the three-sphere, the harmonics are constructed [Hughes 1990a] from ρ^{np}_A and $\sigma^{np}_{A'}$, with $n = 1, 2, \ldots$ and $p = 1, \ldots, (n+2)(n+1)$, from $\rho^{np}_{(ABC)}$ and $\sigma^{np}_{(A'B'C')}$, with $n = 1, 2, \ldots$ and $p = 1, \ldots, (n+4)(n+1)$, and from $\rho^{np}_{(AB)B'}$ and $\sigma^{np}_{A'B'B}$, with $n = 1, 2, \ldots$, and $p = 1, \ldots, (n+3)(n+2)$. Any spinor $\psi_{ABB'}$ can be expanded on S^3 in terms of the harmonics:

$$\begin{array}{llll} \rho_{ABB'}, & \rho_{ABC}n^C{}_{B'}, & \rho_{(A}n_{B)B'}, & \epsilon_{AB}\,\rho_C\,n^C{}_{B'}, \\ \bar{\sigma}_{ABB'}, & \bar{\sigma}_{ABC}n^C{}_{B'}, & \bar{\sigma}_{(A}n_{B)B'}, & \epsilon_{AB}\,\bar{\sigma}_C\,n^C{}_{B'}. \end{array} \tag{8.3.2}$$

Similarly for any spinor $\tilde{\psi}_{ABB'}$. One also finds that

$$\begin{aligned} \rho^{np}_{ABB'}n^{BB'} &= i\,\rho^{(n+1)p}_A, \\ \sigma^{np}_{A'B'B}n^{BB'} &= i\,\sigma^{(n+1)p}_{A'}. \end{aligned} \tag{8.3.3}$$

The harmonic equations, with the index p dropped for simplicity, are

$$\begin{aligned} e^{AA'i}\,{}^{3s}D_i\rho^n_A &= i(n+\tfrac{3}{2})n^{AA'}\,\rho^n_A, \\ e^{AA'i}\,{}^{3s}D_i\sigma^n_{A'} &= i(n+\tfrac{3}{2})n^{AA'}\,\sigma^n_{A'}. \\ e^{AA'i}\,{}^{3s}D_i\rho^n_{ABC} &= i(n+\tfrac{5}{2})n^{AA'}\,\rho^n_{ABC}, \\ e^{AA'i}\,{}^{3s}D_i\sigma^n_{A'B'C'} &= i(n+\tfrac{5}{2})n^{AA'}\,\sigma^n_{A'B'C'}, \\ e^{AC'i}\,{}^{3s}D_i\rho^n_{ABA'} &= i(n+\tfrac{3}{2})n^{AC'}\,\rho^n_{ABA'} - \epsilon_{A'}{}^{C'}\,\rho^{(n+1)}_B, \\ e^{CA'i}\,{}^{3s}D_i\sigma^n_{A'B'A} &= i(n+\tfrac{3}{2})n^{CA'}\,\sigma^n_{A'B'A} - \epsilon_A{}^C\,\sigma^{(n+1)}_{B'}, \\ e^{CA'i}\,{}^{3s}D_i\rho^n_{ABA'} &= i(n+\tfrac{1}{2})n^{CA'}\,\rho^n_{ABA'} - 2\epsilon_{(A}{}^C\,\rho^{(n+1)}_{B)}, \\ e^{AC'i}\,{}^{3s}D_i\sigma^n_{A'B'A} &= i(n+\tfrac{1}{2})n^{AC'}\,\sigma^n_{A'B'A} - 2\epsilon_{(A'}{}^{C'}\,\sigma^{(n+1)}_{B')}. \end{aligned} \tag{8.3.4}$$

There are also orthogonality and normalization conditions obeyed by these harmonics, analogous to those in section 2.9.

One then takes as 'coordinate' variables the coefficients

$$a^{np}(t), b^{np}(t), t^{np}(t)$$

appearing in

$$\begin{aligned}\psi^A{}_i = h^{-1/4} \sum_{np} \{ia^{np}(t)[\rho^{npABB'} - \frac{4i}{3}\rho^{(n+1)p(A}{}_n{}^{B)B'}] \\ + b^{np}(t)\rho^{npAB}{}_C n^{CB'} \\ + t^{np}(t)[\rho^{npA}{}_n{}^{BB'} - 2\rho^{npB}{}_n{}^{AB'}]\}e_{BB'i} \\ + \bar{\sigma}\,\text{terms},\end{aligned} \tag{8.3.5}$$

together with the coefficients of analogous σ terms in $\tilde{\psi}^{A'}{}_i$. The 'momentum' variables will be the components of the $\bar{\sigma}$ terms in $\psi^A{}_i$ and the components of the $\bar{\rho}$ terms in $\tilde{\psi}^{A'}{}_i$. One can then proceed by analogy with section 2.9, calculating the action in terms of a^{np}, b^{np}, t^{np}, etc. Now in section 2.9 it was found in the case of massless spin-1/2 fermions that, for the Hartle–Hawking state, the contribution at quadratic order in fermions was through a multiplicative factor

$$\Psi_F = 1. \tag{8.3.6}$$

The same will happen here for supergravity, so that the fermionic contribution I_F at quadratic order to the classical action $I = I_B + I_F$ (where I_B is the bosonic action) is

$$I_F = 0. \tag{8.3.7}$$

In this way, one can start to build up an approximation to the full Hartle–Hawking state in supergravity. It will also be interesting to repeat this in the case of supergravity with a Λ-term, where the 'mass' term will lead, as in the spin-1/2 case, to $I_F \neq 0$ [D'Eath & Halliwell 1987].

One might ask whether it is possible to set up boundary data of positive and negative frequency in general in supersymmetric quantum cosmology, so that a simple form for the Hartle–Hawking state might emerge. In the case that there are no zero modes, one can proceed as in section 8.2, splitting $\psi^A{}_i = \psi^{A+}{}_i + \psi^{A-}{}_i$ and $\tilde{\psi}^{A'}{}_i = \tilde{\psi}^{A'-}{}_i + \tilde{\psi}^{A'+}{}_i$, and taking the positive-frequency modes $(\psi^{A+}{}_i, \tilde{\psi}^{A'+}{}_i)$ as fermionic data on a compact three-surface. The full data on the three-surface, which bounds a compact manifold-with-boundary, are then

$$(e^{AA'}{}_i, \psi^{A+}{}_i, \tilde{\psi}^{A'+}{}_i, b_{AA'}).$$

As in section 8.2, $\Psi = \text{const.} \exp(-I_B/\hbar)$ then obeys all the quantum constraints, provided that $b_{AA'}$ is suitably chosen, where I_B is the gravitational action of the classical gravitational field on the compact manifold-with-boundary, with boundary data $e^{AA'}{}_i$. Note that one can only know the Hartle–Hawking state inside a topologically spherical boundary up to a constant factor. The path-integral approach at one loop shows that there is a one-loop divergence in the amplitude, proportional to the Euler number χ (section 4.2), which is 1 for a four-ball with spherical boundary [Gibbons 1979, Hawking 1979, Eguchi *et al.* 1980]. This induces a factor $(2\pi\mu^2)^{\zeta(0)/2}$ multiplying $\exp(-I_B/\hbar)$ – see Eq. (4.2.15), where μ is an unknown regularization parameter with dimensions of mass (which is then divided by the Planck mass to obtain a dimensionless expression). Here the quantity $\zeta(0)$, which arises from zeta-function regularization (section 4.2), is $(106/45)\chi$ in $N = 1$ supergravity [Duff 1982]. This difficulty can be avoided by working with a suitable version of $N = 8$ supergravity [Duff 1982], where $\zeta(0) = 0$. In the $N = 1$ case, this suggests strongly that, in the case without zero modes, $\Psi = \text{const.} \exp(-I_B/\hbar)$ should be the Hartle–Hawking state. Further work would be needed, however, to prove such a result, i.e. to show that, in this case, const. $\exp(-I_B/\hbar)$ is indeed the Hartle–Hawking path integral. For example, one might study $\Psi = \text{const.} \exp(-I_B/\hbar)$ in the limit that the three-surface shrinks down towards a point, and compare this with a semi-classical expansion of the path integral. If it could be shown that $\Psi = \text{const.} \exp(-I_B/\hbar)$ and the path integral were asymptotically equal for small three-geometries, then one might be able to extend this, using the quantum constraints, to all three-geometries. This would be a step corresponding to the study of the quantum amplitude for data $(e^{AA'}{}_{iF}; e^{AA'}{}_{iI}; \tau, \ldots)$ in the asymptotically flat case in section 4.4, in the limit $\tau \to 0_+$, and then studying the evolution of the amplitude. Alternatively, one might study the limit of large three-geometries in a similar way.

In general, however, there may be zero modes of Eqs. (8.2.6) and (8.2.7) for the fields $(\psi^A{}_i, \tilde{\psi}^{A'}{}_i)$ on the bounding three-surface with geometry given by $e^{AA'}{}_i$. These would be discrete zero modes, rather than possible zero modes in a continuum of eigenvalues, as in the scattering case of section 8.2. For example, zero modes occur in the Friedmann case of sections 5.2–4; the spin-1/2 homogeneous mode given by the quantity ψ^A in sections 5.2–4 is a zero mode. One can see this since there is no kinetic term for ψ^A in the Hamiltonian generator $\mathscr{H} = -a^{-1}({\pi_a}^2 + 36a^2)$ of Eq. (5.3.13). This corresponds to the property that the Hartle–Hawking wave function $\Psi_{HH}(a, \psi^A) = D\exp(3a^2/\hbar)\psi_A\psi^A$ of Eq. (8.3.1) is not simply of the form const. $\exp(-I_B/\hbar)$, with $I_B = -3a^2$ being the classical Euclidean action inside the sphere of radius a. Rather, the field ψ^A appears as a factor through the invariant product $\psi_A\psi^A$ in Eq. (8.3.1). One would expect some analogous behaviour for the case of boundary conditions with no symmetries present, where one could take as the fermionic data

$\psi^{A+}{}_i, \tilde{\psi}^{A'+}{}_i$ and $\psi^{A0}{}_i, \tilde{\psi}^{A'0}{}_i$, where $\psi^{A0}{}_i$ and $\tilde{\psi}^{A'0}{}_i$ describe zero modes at the boundary of the classical four-geometry corresponding to boundary data $e^{AA'}{}_i$.

Note added in proof: It has now been shown that the Hartle–Hawking state in supergravity is const.exp$(-I_B/\hbar)$, in the case that there are no zero modes [Cheng & D'Eath 1996]. The argument involves the study of the large three–geometry limit.

8.4 Supergravity with supermatter

One would like to extend the results of section 8.1 and 8.2 to more general supergravity models involving lower-spin fields. One extension involves studying higher-N gauged supergravity models [van Nieuwenhuizen 1981]. These may well be tractable in the approach given here, since the auxiliary fields which generalize $M(x), \tilde{M}(x)$ of section 4.3 should enable one, if necessary, to cancel out the non-chiral part of the supersymmetry constraints, associated with the Λ term in the action. This would be a very interesting possibility.

Another approach, taken here, is to study the model of $N = 1$ supergravity coupled to supermatter [Wess & Bagger 1992], and in particular its supersymmetry constraints, especially in the case with zero potential $P(A^I)$ [Cheng, *et al.* 1995b]. The Lagrangian of the general such model depends on the tetrad $e^{AA'}{}_\mu$, the odd (anticommuting) gravitino field $(\psi^A{}_\mu, \tilde{\psi}^{A'}{}_\mu)$, a vector field $v^{(a)}_\mu$ labelled by an index (a), its odd spin-$\frac{1}{2}$ partner $(\lambda^{(a)}_A, \tilde{\lambda}^{(a)}_{A'})$, a family of scalars (A^I, A^{J^*}),their odd spin-$\frac{1}{2}$ partner $(\chi^I_A, \tilde{\chi}^{J^*}_{A'})$, and auxiliary fields. The indices $I, \ldots, J^*, \ldots$ are Kähler indices, and there is a Kähler metric

$$g_{IJ^*} = K_{IJ^*} \tag{8.4.1}$$

on the space of (A^I, A^{J^*}), where K_{IJ^*} is a shorthand for $\partial^2 K/\partial A^I \partial A^{J^*}$ with K the Kähler potential. Each index (a) corresponds to an independent Killing vector field of the Kähler geometry. Such Killing vectors are holomorphic vector fields:

$$X^{(b)} = X^{I(b)}\left(A^J\right) \frac{\partial}{\partial A^I},$$
$$X^{*(b)} = X^{I^*(b)}\left(A^{J^*}\right) \frac{\partial}{\partial A^{I^*}}. \tag{8.4.2}$$

Killing's equation implies that there exist real scalar functions

$$D^{(a)}\left(A^I, A^{I^*}\right)$$

known as Killing potentials, such that

$$g_{IJ^*} X^{J^*(a)} = i \frac{\partial}{\partial A^I} D^{(a)},$$
$$g_{IJ^*} X^{I(a)} = -i \frac{\partial}{\partial A^{J^*}} D^{(a)}. \quad (8.4.3)$$

The Lagrangian is given in Eq. (25.12) of [Wess & Bagger 1992]. It is too long to write out here; however, we shall consider the supersymmetry generators S_A and $\tilde{S}_{A'}$ in the Hamiltonian density, which in full has the form

$$H = N\mathscr{H}_\perp + N^i\mathscr{H}_i + \psi^A{}_0 S_A + \tilde{S}_{A'}\tilde{\psi}^{A'}{}_0$$
$$+ v_0^{(a)} Q_{(a)} + M_{AB}J^{AB} + \tilde{M}_{A'B'}\tilde{J}^{A'B'}, \quad (8.4.4)$$

expected for a theory with the corresponding gauge invariances. Here N and N^i are the lapse function and shift vector, while $\mathscr{H}_\perp$ and $\mathscr{H}_i$ are the (modified) generators of deformations in the normal and tangential directions. S_A and $\tilde{S}_{A'}$ are the local supersymmetry generators, $Q_{(a)}$ is the generator of gauge invariance, and J^{AB} and $\tilde{J}^{A'B'}$ are the generators of local Lorentz rotations, while M_{AB} and $\tilde{M}_{A'B'}$ are Lagrange multipliers giving the amount of Lorentz rotation applied per unit time. Classically, the constraints $\mathscr{H}_\perp, \mathscr{H}_i$, etc. vanish, and the set of (first-class) constraints forms an algebra. Quantum-mechanically, the constraints become operators which annihilate physical states Ψ:

$$\mathscr{H}_\perp\Psi = 0, \qquad \mathscr{H}_i\Psi = 0, \qquad S_A\Psi = 0, \qquad \bar{S}_{A'}\Psi = 0,$$
$$Q_{(a)}\Psi = 0, \qquad J^{AB}\Psi = 0, \qquad \bar{J}^{A'B'}\Psi = 0. \quad (8.4.5)$$

Starting with the simplest of these, the J^{AB} and $\bar{J}^{A'B'}$ quantum constraints imply that Ψ is constructed from Lorentz invariants. The $Q_{(a)}$ constraint, derived below, is of first order in functional derivatives, and implies that the wave function Ψ is gauge invariant. The S_A and $\bar{S}_{A'}$ constraints will be derived and discussed below. The $\mathscr{H}_\perp$ and $\mathscr{H}_i$ constraints can be defined through the anticommutator of S_A and $\bar{S}_{A'}$, as in the case of $N = 1$ supergravity without matter fields (section 3.4). Thus the remaining constraints imply $\mathscr{H}_\perp\Psi = 0$, $\mathscr{H}_i\Psi = 0$; if one were able to find a solution of the remaining quantum constraints, the $\mathscr{H}_\perp$ and $\mathscr{H}_i$ constraints would follow (with a certain choice of factor-ordering).

In the Hamiltonian decomposition, the variables are split into the spatial components $e^{AA'}{}_i$, $\psi^A{}_i$, $\tilde{\psi}^{A'}{}_i$, $v_i^{(a)}$, $\lambda_A^{(a)}$, $\bar{\lambda}_{A'}^{(a)}$, χ_A^I, $\bar{\chi}_{A'}^{J^*}$, A^I, A^{J^*}, which together with the bosonic momenta are the basic dynamical variables of the theory, the Lagrange multipliers

$$N,\ N^i,\ \psi^A{}_0,\ \tilde{\psi}^{A'}{}_0,\ v_0^{(a)},\ M_{AB},\ \tilde{M}_{A'B'}$$

of Eq. (8.4.4), where N, N^i are formed from the $e^{AA'}{}_0$ and the $e^{AA'}{}_i$ as in section 3.2, and $M_{AB}, \tilde{M}_{A'B'}$ involve the zero components $\omega_{AB0}, \tilde{\omega}_{A'B'0}$ of the connection, and the auxiliary fields. One computes the canonical momenta conjugate to the dynamical variables listed above in the usual way. The constraint generators $\mathscr{H}_\perp, \mathscr{H}_i$, etc. are functions of the basic dynamical variables. For the gravitino and spin-1/2 fields, the canonical momenta give second-class constraints of the types seen in [D'Eath 1984, Nelson & Teitelboim 1978, D'Eath & Halliwell 1987]. These are eliminated when Dirac brackets are introduced instead of the original Poisson brackets. In particular, one obtains non-trivial Dirac brackets for $p_{AA'}{}^i$, the momentum conjugate to $e^{AA'}{}_i$, for $\psi^A{}_i$ and $\tilde{\psi}^{A'}{}_i$, for $\lambda_A^{(a)}$ and $\tilde{\lambda}_{A'}^{(a)}$, for χ_A^I and $\tilde{\chi}_{A'}^{J^*}$, and for π_L, π_{L^*}, the momenta conjugate to A^L, A^{L^*}. These can be made into simple brackets by three steps.

First, the brackets involving $p_{AA'}{}^i$, $\psi^A{}_i$ and $\tilde{\psi}^{A'}{}_i$ can be simplified as in the case of pure $N = 1$ supergravity (section 3.3). One redefines

$$p_{AA'}{}^i \rightarrow \hat{p}_{AA'}{}^i = p_{AA'}{}^i - \frac{1}{\sqrt{2}} \epsilon^{ijk} \psi_{Aj} \tilde{\psi}_{A'k}. \tag{8.4.6}$$

This gives the Dirac brackets

$$\begin{aligned} \left[\hat{p}_{AA'}{}^i, \hat{p}_{BB'}{}^j\right]^* &= \text{independent of } \psi^A{}_i \text{ and } \tilde{\psi}^{A'}{}_i, \\ \left[\hat{p}_{AA'}{}^i, \psi^B{}_j\right]^* &= 0, \\ \left[\hat{p}_{AA'}{}^i, \tilde{\psi}^{B'}{}_j\right]^* &= 0. \end{aligned} \tag{8.4.7}$$

Next, one must deal with a complication caused by the dependence on the scalars A^I, A^{J^*} of the Kähler metric K_{IJ^*} in the second-class constraints. Defining π_{IA} to be the momentum conjugate to χ^{IA}, and $\tilde{\pi}_{I^*A'}$ to be the momentum conjugate to $\tilde{\chi}^{I^*A'}$, one has

$$\begin{aligned} \pi_{IA} + \frac{ie}{\sqrt{2}} K_{IJ^*} n_{AA'} \tilde{\chi}^{J^*A'} &= 0, \\ \tilde{\pi}_{J^*A'} + \frac{ie}{\sqrt{2}} K_{IJ^*} n_{AA'} \chi^{IA} &= 0, \end{aligned} \tag{8.4.8}$$

where $e = h^{1/2}$, with h the determinant of the spatial metric h_{ij}. Here $n^{AA'}$ is the spinor version of the unit future-directed normal vector n^μ, obeying

$$n_{AA'} n^{AA'} = 1, \qquad n_{AA'} e^{AA'}{}_i = 0. \tag{8.4.9}$$

The A^K and A^{K^*} dependence of K_{IJ^*} is responsible for the unwanted Dirac brackets among χ_A^I, $\tilde{\chi}_A^{J^*}$, π_L and π_{L^*}. One cures this by using the square root of the Kähler metric, $K_{IJ^*}^{1/2}$, obeying

$$K_{IJ^*}^{1/2} \delta^{KJ^*} K_{KL^*}^{1/2} = K_{IL^*}. \tag{8.4.10}$$

This may be found by diagonalizing K_{IJ^*} via a unitary transformation, assuming that the eigenvalues are all positive. One needs to assume that there is an 'identity metric' δ^{KJ^*} defined over the Kähler manifold; this will be true if a positive-definite vielbein field can be introduced. One then introduces the modified variables

$$\hat{\chi}_{IA} = e^{1/2} K_{IJ^*}^{1/2} \delta^{KJ^*} \chi_{KA},$$
$$\hat{\tilde{\chi}}_{I^*A'} = e^{1/2} K_{JI^*}^{1/2} \delta^{JK^*} \tilde{\chi}_{K^*A'}, \tag{8.4.11}$$

where the factor of $e^{1/2}$ has been introduced for later use (in the time gauge). Then the second-class constraints of Eq. (8.4.8) read

$$\hat{\pi}_{IA} + \frac{i}{\sqrt{2}} \delta_{IJ^*} n_{AA'} \hat{\tilde{\chi}}^{J^*A'} = 0,$$
$$\hat{\tilde{\pi}}_{I^*A'} + \frac{i}{\sqrt{2}} \delta_{IJ^*} n_{AA'} \hat{\chi}^{JA} = 0. \tag{8.4.12}$$

The resulting Dirac brackets now give

$$[\pi_L, \pi_M]^* = 0, \text{ etc.},$$
$$\left[\pi_L, \hat{\chi}^A{}_I\right]^* = 0, \text{ etc.} \tag{8.4.13}$$

Finally, there are the brackets among $\hat{p}_{AA'}{}^i$, $\lambda_A^{(a)}$, $\tilde{\lambda}_{A'}^{(a)}$, $\hat{\chi}_A^I$ and $\hat{\tilde{\chi}}_{A'}^{J^*}$, which are just as in the case studied by [Nelson & Teitelboim 1978]. These are dealt with by first defining

$$\hat{\lambda}_A^{(a)} = e^{1/2} \lambda_A^{(a)}, \quad \hat{\tilde{\lambda}}_{A'}^{(a)} = e^{1/2} \tilde{\lambda}_{A'}^{(a)}. \tag{8.4.14}$$

Then one goes to the time gauge, in which the tetrad component n^a of the normal vector n^μ is henceforward restricted by

$$n^a = \delta^a{}_0, \tag{8.4.15}$$

or equivalently

$$e^0{}_i = 0. \tag{8.4.16}$$

Thus the original Lorentz rotation freedom becomes replaced by that of spatial rotations. In the time gauge, the geometry is described by the triad $e^\alpha{}_i (\alpha = 1, 2, 3)$, and the conjugate momentum is $\hat{p}_\alpha{}^i$. One has, following [Nelson & Teitelboim 1978],

$$\left[\hat{p}_\alpha{}^i, \hat{p}_\beta{}^j\right]^{**} = 0,$$
$$\left[\hat{p}_\alpha{}^i, \hat{\lambda}_A^{(a)}\right]^{**} = 0, \text{ etc.},$$
$$\left[\hat{p}_\alpha{}^i, \hat{\chi}^I{}_A\right]^{**} = 0, \text{ etc.} \tag{8.4.17}$$

The remaining brackets are standard; the non-zero fermionic brackets are

$$\left[\hat{\lambda}^{(a)}{}_{A}(x),\ \hat{\tilde{\lambda}}^{(b)}{}_{A'}(x')\right]^{**} = \sqrt{2}in_{AA'}\delta^{(a)(b)}\delta\left(x,x'\right), \tag{8.4.18}$$

$$\left[\hat{\chi}^{I}{}_{A}(x),\ \hat{\tilde{\chi}}^{J^*}{}_{A'}\left(x'\right)\right]^{**} = \sqrt{2}in_{AA'}\delta^{IJ^*}\delta\left(x,x'\right), \tag{8.4.19}$$

$$\left[\psi^{A}{}_{i}(x),\ \tilde{\psi}^{A'}{}_{j}\left(x'\right)\right]^{**} = \frac{1}{\sqrt{2}}D^{AA'}{}_{ij}\delta\left(x,x'\right), \tag{8.4.20}$$

where

$$D^{AA'}{}_{ij} = -2ie^{-1}e^{AB'}{}_{j}e_{BB'i}n^{BA'}. \tag{8.4.21}$$

The supersymmetry constraint $\tilde{S}_{A'}$ is then found to be

$$\begin{aligned}
&\tilde{S}_{A'} = \\
&-\sqrt{2}i\hat{\pi}^{ij}e_{AA'i}\psi^{A}{}_{j} + \sqrt{2}\epsilon^{ijk}e_{AA'i}\,{}^{3s}\tilde{D}_{j}\psi^{A}{}_{k} \\
&+\left(2\sqrt{2}\right)^{-1}e\tilde{\psi}_{B'}{}^{[j}\psi_{B}{}^{i]}n^{BB'}\psi^{A}{}_{j}e_{AA'i} \\
&-\left(\sqrt{2}\right)^{-1}e\tilde{\psi}_{B'}{}^{[i}\psi_{B}{}^{j]}e^{BB'}{}_{j}\psi^{A}{}_{i}n_{AA'} \\
&-\frac{i}{\sqrt{2}}\pi^{n(a)}e_{BA'n}\lambda^{(a)B} + \frac{1}{2\sqrt{2}}\epsilon^{ijk}e_{BA'k}\lambda^{(a)B}F^{(a)}_{ij} \\
&+\frac{1}{\sqrt{2}}egD^{(a)}n^{A}{}_{A'}\lambda^{(a)}{}_{A} \\
&+\frac{1}{\sqrt{2}}\Big[\pi_{J^*} - \frac{ie}{2\sqrt{2}}n^{BB'}\lambda^{(a)}{}_{B}\tilde{\lambda}^{(a)}{}_{B'}K_{J^*} + \frac{i}{\sqrt{2}}eg_{LM^*}\Gamma^{M^*}{}_{J^*N^*}n^{BB'}\tilde{\chi}^{N^*}{}_{B'}\chi^{L}{}_{B} \\
&\qquad -\frac{ie}{2\sqrt{2}}K_{J^*}g_{MM^*}n^{BB'}\tilde{\chi}^{M^*}{}_{B'}\chi^{M}{}_{B} - \frac{1}{2\sqrt{2}}\epsilon^{ijk}K_{J^*}e^{BB'}{}_{j}\psi_{kB}\tilde{\psi}_{iB'} \\
&\qquad -\sqrt{2}eg_{IJ^*}\chi^{IB}e_{BB'}{}^{m}n^{CB'}\psi_{mC}\Big]\tilde{\chi}^{J^*}{}_{A'} \\
&\quad -\sqrt{2}eg_{IJ^*}\left(\tilde{\mathscr{D}}_{i}A^{I}\right)\tilde{\chi}^{J^*}_{B'}n^{BB'}e_{BA'}{}^{i} + \frac{i}{2}g_{IJ^*}\epsilon^{ijk}e_{AA'j}\psi^{A}{}_{i}\tilde{\chi}^{J^*B'}e_{BB'k}\chi^{IB} \\
&\quad +\frac{1}{4}e\psi_{Ai}\left(e_{BA'}{}^{i}n^{AC'} - e^{AC'i}n_{BA'}\right)g_{IJ^*}\tilde{\chi}^{J^*}_{C'}\chi^{IB} \\
&\quad +\frac{1}{4}e\psi_{Ai}\left(e_{BA'}{}^{i}n^{AC'} - e^{AC'i}n_{BA'}\right)\tilde{\chi}^{(a)}_{C'}\lambda^{(a)B} \\
&\quad -e\exp(K/2)\left[2Pn^{A}{}_{A'}e_{AB'}{}^{i}\tilde{\psi}^{B'}{}_{i} + i\left(D_{I}P\right)n_{AA'}\chi^{IA}\right], \\
&\qquad\qquad \text{plus auxiliary field contributions,}
\end{aligned} \tag{8.4.22}$$

where $\lambda^{(a)}_{A}$, $\tilde{\lambda}^{(a)}_{A'}$ and χ_{IA}, $\tilde{\chi}_{I^*A'}$ should be redefined as in Eqs. (8.4.11), (8.4.14). The auxiliary field terms are like those in Eqs. (4.3.4),(4.3.5), plus possible contributions from the matter auxiliary fields.

Here $e_{AA'i} = \sigma^{\alpha}_{AA'}e_{\alpha i}$, where $\sigma^{\alpha}_{AA'}(\alpha = 1,2,3)$ are Infeld–van der Waerden symbols (section 2.9.1) and $\hat{\pi}^{ij} = -\frac{1}{2}e^{\alpha(i}\hat{p}_{\alpha}{}^{j)}$. Following [Wess & Bagger 1992]

$$\begin{aligned}{}^{3s}\tilde{\mathscr{D}}_j\psi^A{}_k &= \partial_j\psi^A{}_k + {}^{3s}\omega^A{}_{Bj}\psi^B{}_k \\ &+\frac{1}{4}\left(K_K\tilde{\mathscr{D}}_jA^K - K_{K^*}\tilde{\mathscr{D}}_jA^{K^*}\right)\psi^A{}_k \\ &+\frac{1}{2}gv_j^{(a)}\left(\mathrm{Im}F^{(a)}\right)\psi^A{}_k, \end{aligned} \tag{8.4.23}$$

where ${}^{3s}\omega_{ABj}$, ${}^{3s}\tilde{\omega}_{A'B'j}$ give the torsion-free three-dimensional connection, and

$$\tilde{\mathscr{D}}_iA^K = \partial_iA^K - gv_i^{(a)}X^{K(a)}, \tag{8.4.24}$$

with g the gauge coupling constant and $X^{K(a)}$ the ath Killing vector field, as in Eq. (8.4.2). Further, the analytic functions $F^{(a)}(A^J)$ and $F^{*(a)}(A^{I^*})$ arise [Wess & Bagger 1992] from the transformation of the Kähler potential K under an isometry generated by the Killing vectors $X^{(a)}$ and $X^{*(a)}$:

$$\delta K = \left(\epsilon^{(a)}X^{(a)} + \epsilon^{*(a)}X^{*(a)}\right)K. \tag{8.4.25}$$

One obtains

$$\delta K = \epsilon^{(a)}F^{(a)} - \epsilon^{*(a)}F^{*(a)} - i\left(\epsilon^{(a)} - \epsilon^{*(a)}\right)D^{(a)}, \tag{8.4.26}$$

where $D^{(a)}$ is the Killing potential of Eq. (8.4.3). Also, in Eq. (8.4.22), $\pi^{n(a)}$ is the momentum conjugate to $v_n^{(a)}$, K_{J^*} denotes $\partial K/\partial A^{J^*}$, $\Gamma^{M^*}{}_{J^*N^*}$ denotes the starred Christoffel symbols [Wess & Bagger 1992] of the Kähler geometry, and $P = P(A^I)$ gives the potential of the theory, with

$$D_IP = (\partial P/\partial A^I) + (\partial K/\partial A^I)P. \tag{8.4.27}$$

The gauge generator $Q^{(a)}$ is given classically by

$$\begin{aligned} Q^{(a)} &= -\partial_n\pi^{n(a)} - gf^{abc}\pi^{n(b)}v_n^{(c)} \\ &+ g\left(\pi_IX^{I(a)} + \pi_{I^*}X^{I^*(a)}\right) \\ &+ \sqrt{2}iegK_{MI^*}n^{AA'}X^{J^*(a)}\Gamma^{I^*}_{J^*N^*}\tilde{\chi}^{N^*}_{A'}\chi^M_A \\ &- \sqrt{2}iegn^{AA'}\tilde{\lambda}^{(b)}_{A'}\left[f^{abc}\lambda^{(c)}_A + \frac{1}{2}i\left(\mathrm{Im}F^{(a)}\right)\lambda^{(b)}_A\right] \\ &+ \sqrt{2}iegn^{AA'}K_{IJ^*}\tilde{\chi}^{J^*}_{A'}\left[\frac{\partial X^{I(a)}}{\partial A^J}\chi^J_A + \frac{1}{2}i\left(\mathrm{Im}F^{(a)}\right)\chi^I_A\right] \\ &- \frac{i}{\sqrt{2}}g\left(\mathrm{Im}F^{(a)}\right)\epsilon^{ijk}\tilde{\psi}_{iA'}e^{AA'}{}_j\psi_{Ak}, \end{aligned} \tag{8.4.28}$$

where f^{abc} are the structure constants of the isometry group.

One can proceed to a quantum description by studying (for example) Grassmann-algebra-valued wave functions of the form

$$\Psi\left(e^{\alpha}{}_{i}, \psi^{A}{}_{i}, v_{i}^{(a)}, \hat{\lambda}_{A}^{(a)}, n_{A}{}^{A'}\hat{\bar{\chi}}_{A'}^{J^*}, A^{I}, A^{J^*}\right).$$

The choice of $n_A{}^{A'}\hat{\bar{\chi}}_{A'}^{J^*}$ rather than $\hat{\chi}_A^J$ is designed so that the quantum constraint $\bar{S}_{A'}$ should be of first order in momenta. The momenta are represented by

$$\hat{p}_{\alpha}{}^{i} \rightarrow -i\hbar\delta/\delta e^{\alpha}{}_{i}, \tag{8.4.29}$$

$$\pi^{n(a)} \rightarrow -i\hbar\delta/\delta v_n^{(a)}, \tag{8.4.30}$$

$$\pi_I \rightarrow -i\hbar\delta/\delta A^I, \tag{8.4.31}$$

$$\pi_{I^*} \rightarrow -i\hbar\delta/\delta A^{I^*}, \tag{8.4.32}$$

$$\bar{\psi}^{A'}{}_{i} \rightarrow \frac{1}{\sqrt{2}}i\hbar D^{AA'}{}_{ji}\delta/\delta\psi^{A}{}_{j}, \tag{8.4.33}$$

$$\hat{\bar{\lambda}}^{(a)A'} \rightarrow -\sqrt{2}n^{AA'}\delta/\delta\hat{\lambda}^{(a)}, \tag{8.4.34}$$

$$\hat{\chi}^{IA} \rightarrow -\sqrt{2}n^{AA'}\delta^{IJ^*}\delta/\delta\hat{\bar{\chi}}^{J^*A'}. \tag{8.4.35}$$

Quantum-mechanically, one can order each term cubic in fermions in $\bar{S}_{A'}$ (using anticommutation) such that one 'momentum' fermionic variable is on the right, and two 'coordinate' fermionic variables are on the left. The ordering of the quantum constraint S_A is defined by taking the Hermitian adjoint with respect to the natural inner product (section 3.3). Then the terms in S_A cubic in fermions have two 'momenta' on the right and one 'coordinate' on the left.

Each term in $\bar{S}_{A'}$ contains no more than one functional derivative. Some terms, namely

$$-\sqrt{2}\epsilon^{ijk}e_{AA'i}{}^{3s}\tilde{\mathscr{D}}_{j}\psi^{A}{}_{k},$$

$$\left(2\sqrt{2}\right)^{-1}\epsilon^{ijk}e_{BA'k}\lambda^{(a)B}F_{ij}^{(a)},$$

$$\left(\sqrt{2}\right)^{-1}egD^{(a)}n^{A}{}_{A'}\lambda^{(a)}{}_{A}$$

and

$$-\sqrt{2}eg_{IJ^*}\left(\tilde{\mathscr{D}}_{i}A^{I}\right)\tilde{\chi}^{J^*}{}_{B'}n^{BB'}e_{BA'}{}^{i},$$

contain no derivative (the inhomogeneous terms), but are linear in a fermionic coordinate. Proceeding as in sections 4.3, 8.1, one can use the auxiliary fields to solve the classical $\tilde{S}_{A'} = 0$ constraint. Here one has to assume, as in section 4.3, that all fermion fields $\psi^{A}{}_{i}, \lambda^{(a)A}, \tilde{\chi}_{A'}^{J^*}$ are proportional to a single odd Grassmann number. Then the cubic terms in Eq. (8.4.22) drop out. This implies that the $\bar{S}_{A'}\Psi = 0$ constraint can be solved by taking Ψ to be the form $\Psi = \exp(-I_B/\hbar)$, where I_B is the bosonic Euclidean action of a four-dimensional classical solution of the theory with the fields $(e^{\alpha}{}_i, \psi^A{}_i, \ldots)$ as boundary data at a final surface, with only bosonic fields at an initial surface. Again, the terms in $\bar{S}_{A'}$ cubic in fermions do not contribute.

One now needs that $\Psi = \exp(-I_B/\hbar)$ should obey the $S_A\Psi = 0$ constraint, when auxiliary fields are included, with the above boundary conditions. This is straightforward in the case where the potential $P(A^I)$ vanishes, so that the last two explicit terms

$$-e \exp(K/2)[2Pn_A{}^{A'} e_{BA'}{}^i \psi^B{}_i - i(D_{I*}P)n_{AA'}\tilde{\chi}^{I*A'}]$$

corresponding to Eq. (8.4.22) are absent. This occurs, for example, when the Kähler manifold is compact [Witten & Bagger 1982]. Note that the remaining first-order constraints $Q_{(a)}\Psi = 0$ and the local rotation invariance are also satisfied automatically for $\Psi = \exp(-I/\hbar)$, because of the gauge invariance and local Lorentz invariance of the action I_B. Hence, $\Psi = \exp(-I_B/\hbar)$ obeys all the quantum constraints, and so defines a bosonic physical quantum state.

One then proceeds as in sections 4.4 and 8.1. A study of the Schrödinger (heat) equation shows that the path integral equals $\exp(-I_B/\hbar)$. The coefficients of all bosonic counterterms must be zero (except for one-loop topological terms, which give a overall constant prefactor in the amplitude). If all counterterms start with a bosonic term, followed by fermionic partners, then all fermionic amplitudes would be expected to be perturbatively finite. This indicates that $N = 1$ theories with zero potential $P(A^I)$ (as in the case of a compact Kähler manifold) might turn out to be finite theories of matter interacting with itself and with gravity, when one uses local fermionic boundary conditions. In principle one could test this result, by evaluating the one-loop divergences.

One might be surprised, on comparing this result with [van Nieuwenhuizen 1981], that there is no one-loop divergence. In an S-matrix formulation, using a spectral decomposition of the fermions into positive- and negative-frequency components at early and late times, there is, in the case of the scalar-spin-1/2 multiplet, a one-loop divergence proportional to $\int d^4x(T_{\mu\nu})^2$, i.e. to $\int d^4x(\partial_\mu\phi)^2(\partial_\nu\phi)^2$, in the bosonic sector (in the massless case), where ϕ indicates schematically a scalar field. This cannot be compared directly with the preceding paragraph, because the boundary conditions are different.

One can see how the constraint $\bar{S}_{A'}\Psi = 0$ leads immediately to infinities in the case of spectral boundary conditions. Taking, for example $\lambda_A^{(a)-}$ and $\bar{\lambda}_{A'}^{(a)-}$ as coordinates and $\lambda_A^{(a)+}, \bar{\lambda}_{A'}^{(a)+}$ as momenta, the cubic term $-\frac{1}{4}ie\,n^{BB'}\lambda_B^{(a)}\bar{\lambda}_{B'}^{(a)}K_{J*}\bar{\chi}_{A'}^{J*}$ in $\bar{S}_{A'}$ includes a contribution, due to the $\lambda\bar{\lambda}$ term, proportional to $\delta^3(0)$. Hence it is not possible to make sense of the quantum theory of supergravity coupled to supermatter, using spectral boundary conditions. The difference between local and spectral boundary conditions here is thus very striking.

9
Conclusion

There are many possibilities for further development of the ideas which have been presented here. These are concerned both with cosmological questions, such as that of the Hartle–Hawking state, and with quantum supergravity in the asymptotically flat context.

One question, which was raised already in section 2.4, concerns the effects of more complicated topology on the quantum amplitude. So far, the only topology to be studied in the asymptotically flat case has been $\mathbb{R}^3$, although the arguments of section 8.3 concerning quantum cosmology allow for different connected spatial topologies (so that there is expected to be a unique infilling classical gravitational solution) on the three-dimensional boundary. In the Hamiltonian definition of the Feynman path integral for the asymptotically flat case, one works with a fixed spatial topology, given by a connected three-surface Σ, so that the four-dimensional fields are defined on a spacetime manifold $[0, \tau] \times \Sigma$. In the asymptotically flat case, one would take Σ to be $\mathbb{R}^3$ minus a ball, outside some large sphere, but allow Σ to have more complicated topology inside the large sphere. The arguments which indicate that the bosonic amplitude is $\exp(-I_B/\hbar)$ and that the fermionic amplitude is given asymptotically by $(A + \hbar A_1 + \hbar^2 A_2 + \cdots)\exp(-I_B/\hbar)$, with finite loop terms $A, A_1, A_2, \ldots$ still proceed as before, subject to the provisos described in sections 8.1, 8.2. In the cosmological case of spectral fermionic boundary conditions with a compact bounding three-geometry, one must check that there are no zero modes of Eqs. (8.2.6),(8.2.8). If there are zero modes, one must find a way of including them in the quantum wave function.

One can also consider disconnected three-geometries, for the initial or final surface or both (or in the cosmological case), as discussed in section 2.4; see Fig. 9.1. One expects that there will be an infinite family of classical solutions joining the initial to the final gravitational data. Heuristically, these can be thought of as corresponding to the freedom of translating or rotating each connected portion of (e.g.) the initial data,

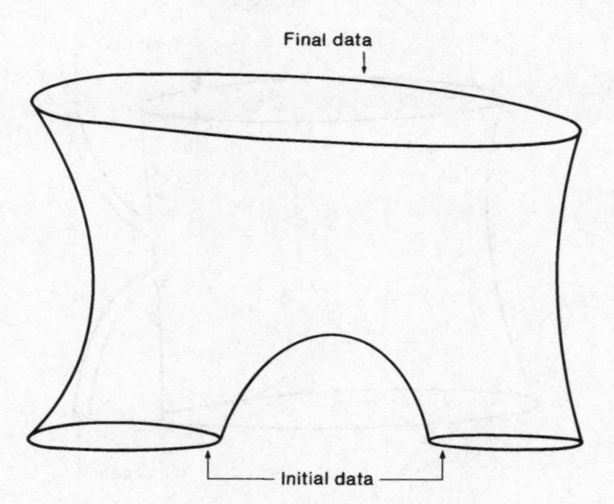

Fig. 9.1. The amplitude for an initial disconnected three-geometry to go to a final connected three-geometry.

Fig. 9.2. Wormholes in the quantum amplitude to go from initial to final data.

relative to the rest of the initial data. Hence the exact quantum amplitude, e.g. in the case of purely gravitational boundary data, will be given by a weighted average of semi-classical expressions $\exp(-I_B/\hbar)$, defined by the path integral. It would be very interesting to find this weighted average and understand its behaviour.

Topology can also be introduced into the path integral by allowing wormholes (chapter 6) into the four-geometries which are summed over (see Fig. 9.2). Suppose that there is just one wormhole. As in section 6.1, one can cut the wormhole at a three-surface Σ_0, and sum over all

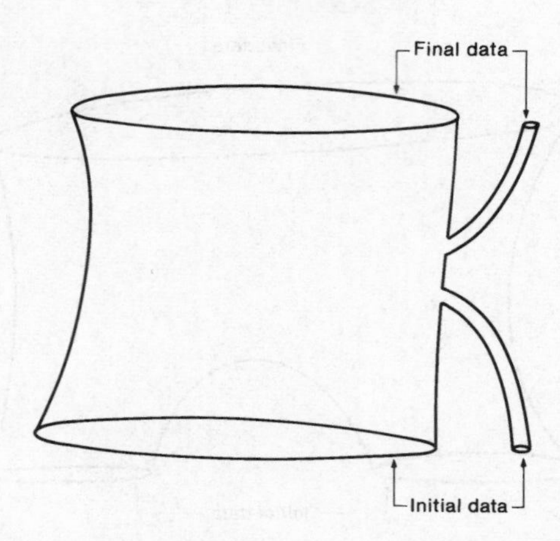

Fig. 9.3. The infilling field of Fig. 9.2, cut across the wormhole. One sums over all fields such as $e^{AA'}{}_{iW}$ on the resulting three-surface.

states on Σ_0, or equivalently (in the bosonic case) over all $e^{AA'}{}_i(x)$ on Σ_0. One can then assign one copy of Σ_0 to the initial surface, and one copy to the final surface (Fig. 9.3). Thus, the initial data consist of an asymptotically flat manifold, for example $\mathbb{R}^3$, with data $e^{AA'}{}_{iI}$, together with a disconnected compact manifold, for example S^3, with data $e^{AA'}{}_{iW}$, where W stands for wormhole. The final data consist of similar disconnected manifolds, with data $e^{AA'}{}_{iF}$ and $e^{AA'}{}_{iW}$. The path integral should be taken over connected four-geometries subject to these boundary data, and $e^{AA'}{}_{iW}$ should be summed over. As in the previous paragraph, for each wormhole geometry $e^{AA'}{}_{iW}$, the path integral will give a suitably weighted sum of terms $\exp(-I_B/\hbar)$. The contribution to I_B from the spherically symmetric part of $e^{AA'}{}_{iW}$ would be of the order of const. a^2, where a is the radius of a Friedmann universe (section 6.1), giving a wormhole quantum state (section 6.2). The contribution of inhomogeneous perturbations of the Friedmann universe to I_B is positive. Hence, when one integrates over $e^{AA'}{}_{iW}$, the dominant contribution will be from radii a of the order of the Planck length, giving a contribution to I_B of order $\hbar$. Hence, for the path integral with initial data $e^{AA'}{}_{iI}$, final data $e^{AA'}{}_{iF}$ and Euclidean time τ at spatial infinity, a wormhole has the effect of smearing out the amplitude $\exp(-I_B/\hbar)$ without wormholes, by an amount δI_B of order $\hbar$. Larger numbers of wormholes will smear the amplitude further.

Another question, of general interest, raised by this work concerns (as in section 4.4) the sign of the classical mass M and classical bosonic action I_B for the Euclidean classical boundary-value problem. One may conjecture, based on the discussion of section 4.4, that for the Euclidean gravitational boundary data $(e^{AA'}{}_{iF}\,;\, e^{AA'}{}_{iI}\,;\, \tau, \ldots)$, with the three-geometries h_{ijF} and h_{ijI} being asymptotically flat, the four-dimensional mass M of the classical infilling four-geometry $g_{\mu\nu}$ obeys $M \leq 0$. Here one expects $M = 0$ only for flat four-space. Of course, the three-geometries h_{ijF} and h_{ijI} may be freely chosen and can carry positive three-dimensional mass. However, one might expect that the extra term $i \int dS_i N_j \pi^{ij}$ of Eq. (4.4.2) in the Euclidean Hamiltonian would ensure that the complete mass M is non-positive.

Correspondingly, one may conjecture that the classical Euclidean action I_B for boundary data $(e^{AA'}{}_{iF}\,;\, e^{AA'}{}_{iI}\,;\, \tau, \ldots)$ obeys $I_B \geq 0$, with $I_B = 0$ only for flat four-space. If this were so, then the bosonic amplitude $\exp(-I_B/\hbar)$ would decay away from a maximum at flat space, as is already known in the case of weak gravitational fields.

As in Eq. (4.4.21), the mass $M(h_{ijF}\,;\, h_{ijI}\,;\, \tau, \ldots)$ and classical Euclidean action $I_B(h_{ijF}\,;\, h_{ijI}\,;\, \tau, \ldots)$ obey

$$\frac{\partial I_B}{\partial \tau} = M. \tag{9.1}$$

For $\tau \to 0_+$, one has [Eq. (4.4.20)]

$$M \sim -\tau^{-2}\, v(h_{ijF}\,; h_{ijI}\,; \ldots), \tag{9.2}$$

for some functional v, and correspondingly [Eq. (4.4.22)]

$$I_B \sim \tau^{-1}\, v(h_{ijF}\,; h_{ijI}\,; \ldots). \tag{9.3}$$

If the negative mass and positive action conjectures above are correct, then $v(h_{ijF}\,; h_{ijI}\,; \ldots) \geq 0$, with equality only for flat four-space. It is of interest to investigate the functional v; for example, in a more general context, the above discussion of topological effects on the wave function might be partially resolved by knowing the form of the amplitude for small τ. It may be possible to obtain understanding of the functional $v(h_{ijF}\,; h_{ijI}\,; \ldots)$ from the very detailed Hamilton–Jacobi investigations of rapidly-varying gravitational fields in [Salopek & Stewart 1992, Salopek *et al.* 1993].

Consider now the question of amplitudes which involve fermions. Following the treatment of section 4.1, one can perform explicit calculations [D'Eath & Wulf 1995] of one- and higher-loop factors $A, A_1, A_2, \ldots$ in the expansion [Eq. (4.1.13)] of the amplitude K to go from initial to final data:

$$K \sim (A + \hbar A_1 + \hbar^2 A_2 + \cdots)\exp(-I_{\text{class}}/\hbar). \tag{9.4}$$

This can be done with either local (section 8.1) or spectral (section 8.2) boundary conditions for the fermions. For simplicity, consider the local case. At one-loop order [Eqs. (4.1.17),(4.1.19)], the one-loop factor A

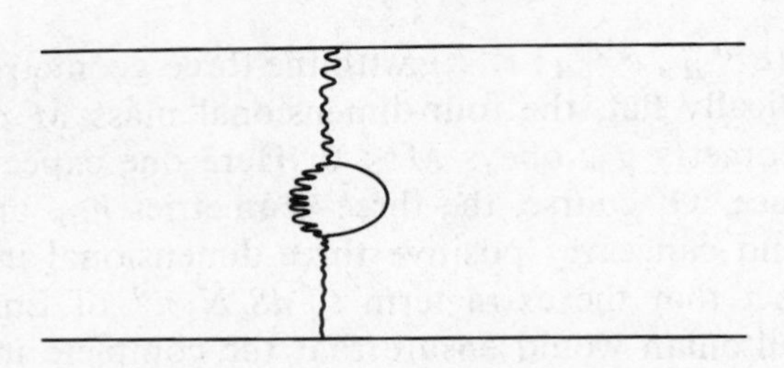

Fig. 9.4. Schematic representation of the simplest contribution to the one-loop boundary amplitude, quadratic in fermions.

is invariant under a primed supersymmetry transformation at the final surface:

$$\delta e^{AA'}{}_i = -i\kappa\tilde{\epsilon}^{A'}\psi^A{}_i, \quad \delta\psi^A{}_i = 0, \tag{9.5}$$

$$\delta A = 0. \tag{9.6}$$

Further [Eq. (4.1.20)] A transforms under an unprimed supersymmetry transformation [Eqs. (4.1.21),(4.1.22)] at the final surface:

$$\delta e^{AA'}{}_i = -i\kappa\epsilon^A\tilde{\psi}^{A'}{}_i \tag{9.7}$$

$$\delta\psi^A{}_i = 2\kappa^{-1}\,{}^{3s}D_i\epsilon^A - i\kappa\epsilon^B D^{AA'}{}_{ij}p_{BA'}{}^j + \tfrac{1}{2}i\kappa\epsilon^B\epsilon^{jkl}D^{AA'}{}_{ij}\psi_{Bk}\tilde{\psi}_{A'l}, \tag{9.8}$$

where $\tilde{\psi}^{A'}{}_i = \tilde{\psi}^{A'}{}_{iF}$ is the final primed gravitino field obtained by evolving the initial data $\tilde{\psi}^{A'}{}_{iI}$ classically to the final surface, and $D^{AA'}{}_{jk} = -2ih^{-1/2}e^{AB'}{}_k e_{BB'j}n^{BA'}$. Under the transformation (9.7), (9.8), A transforms [Eq. (4.1.20)] by

$$\delta A = \text{const.}\ \epsilon^A \frac{\delta}{\delta e^{AA'}{}_i}\left(D^{BA'}{}_{ji}\frac{\delta I_{\text{class}}}{\delta\psi_j^B}\right)A. \tag{9.9}$$

There are similar transformation rules for A under supersymmetry at the initial surface. In Eq. (9.9) the final expression $-iD^{BA'}{}_{ji}\delta I_{\text{class}}/\delta\psi_j^B$ is the final quantity $\tilde{\psi}^{A'}{}_i(x)$. Following the discussion of section 4.1, one can use these equations to build a series of finite expressions for the one-loop prefactor A near flat space, in powers of $\psi^A{}_{iI}, \psi^A{}_{iF}, (e^{AA'}{}_{iI} - \sigma^{AA'}{}_i)$ and $(e^{AA'}{}_{iF} - \sigma^{AA'}{}_i)$, where $\sigma^{AA'}{}_i$ is the flat-space spatial tetrad. These expressions may be represented diagrammatically; the first two are given in Figs. 9.4, 9.5. The finiteness at one loop, in the presence of boundaries, agrees with the results of section 4.2.

Further, one can use the same method to investigate (say) the two-loop factor A_1. This is also invariant under the primed supersymmetry transformation (9.5) at the final surface, and obeys an equation analogous

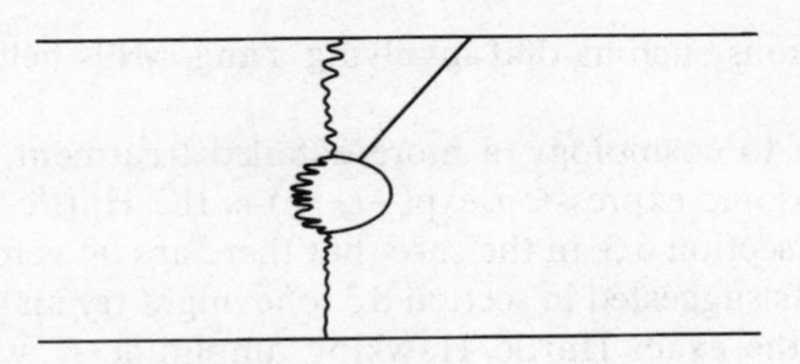

Fig. 9.5. Schematic representation of the one-loop boundary amplitude, at second order in fermions and first order in deviations from a flat geometry.

to (9.7), (9.8), (9.9) for unprimed supersymmetry at the final surface. By the two-loop counterterm results of section 4.2 [D'Eath 1986b], A_1 should be finite (even in the presence of boundaries), and the explicit calculation of A_1 will provide a good check on this. Similarly, one could in principle go on to higher-loop effects. The possibility of obtaining these explicit loop results from equations such as (9.5)–(9.9) shows that canonical methods may be used efficiently as a calculational tool, to compute quantities which are not so easily accessible by Feynman-diagram methods. As described above, these calculations will also provide a good check of whether or not the amplitude is finite.

These projected calculations are concerned with low orders in perturbation theory. It is also important to understand the summability of the complete asymptotic perturbation series (9.4), through understanding the behaviour of individual terms $\hbar^n A_n$(boundary data) as $n \to \infty$, with the boundary data fixed. In theories with a dimensionless coupling constant, Lipatov's method may be used [Lipatov 1977], and one can then investigate whether the perturbation series is Borel summable [Brézin *et al.* 1977a,b]. In supergravity, each power of $\hbar$ must be multiplied by two powers of momentum arising from the boundary data, and any investigation will need to take account of this.

Another question, already raised in section 8.4, is whether the higher-N gauged supergravity models [van Nieuwenhuizen 1981] may be finite. This can now be checked both for local and spectral boundary conditions, by seeing whether the quantum supersymmetry constraints have the correct form to admit purely bosonic quantum states of the form $\exp(-I_B/\hbar)$. Because gauged higher-N supergravity involves a (negative) cosmological term Λ, one expects to need to use the auxiliary fields corresponding to M in the $N = 1$ theory to cancel out, if necessary, the non-chiral contributions in the supersymmetry generators, as well as using the auxiliary fields corresponding to $b_{AA'}$ to help in satisfying the constraints. It would of course be extremely interesting if it turned out that higher-N gauged supergravity were finite, in part because this theory includes ordinary low-energy interactions such as that involving Yang–Mills fields and spin-1/2 fields.

Turning now to cosmology, a more detailed treatment is needed as to whether the bosonic expression $\exp(-I_B/\hbar)$ is the Hartle–Hawking state, as suggested in section 8.3, in the case that there are no zero modes of Eqs. (8.2.6),(8.2.8). As suggested in section 8.3, one might try an approach where one compares the exact Hartle–Hawking amplitude K with $\exp(-I_B/\hbar)$ for very small three-geometries. The classical four-geometry would in general be rapidly varying, somewhat analogously to the case $\tau \to 0_+$ of section 4.4, in which two different three-geometries are brought very close together. In the cosmological (compact) case one might use the small-geometry data as initial data for the evolution of the full amplitude K by means of the quantum constraint equations. In the case that there are fermionic zero modes, one would like to have a careful treatment of the Hartle–Hawking state. For example, one might ask whether the zero modes in general appear as factors in the amplitude, as does the zero mode ψ^A in the Friedmann geometry Hartle–Hawking state [Eq. (8.3.1)]

$$\Psi(a, \psi^A) = D\, \psi_A \psi^A \exp(3a^2/\hbar). \tag{9.10}$$

Note added in proof: As mentioned at the end of section 8.3, an argument involving large three-geometries shows that the Hartle–Hawking state in supergravity is of the form const. $\exp(-I_B/\hbar)$, provided that there are no zero modes.

A more thorough treatment of the Rarita–Schwinger harmonic equation [Eq. (8.2.4)]

$$\epsilon^{ijk} n_{AA'}\, {}^{4s}D_j \psi^A{}_k = -i\lambda \epsilon^{ijk} e_{AA'j} \psi^A{}_k, \tag{9.11}$$

and its conjugate, should yield more understanding of the conditions for a zero mode with $\lambda = 0$. The operator $\mathscr{L}$ defined by the left-hand side maps $\psi^A{}_k$ to a quantity $\phi_{A'}{}^i = \epsilon^{ijk} n_{AA'}\, {}^{4s}D_j \psi^A{}_k$. Equivalently, one can study the operator $\mathscr{L}^\dagger \mathscr{L}$, and ask for the condition that it has a zero eigenvalue.

Finally, there are applications of this work to certain amplitudes of interest in quantum gravity, concerning purely bosonic configurations, for which the amplitude in quantum supergravity is $\exp(-I_B/\hbar)$, with I_B the Euclidean bosonic action. For example, one can study the bosonic part of the ground state of quantum supergravity on $\mathbb{R}^3$, allowing for the inclusion of black-hole-like configurations. As in section 2.5, this is given by the Euclidean path integral with the initial three-geometry h_{ij}, and the condition that the four-geometry $g_{\mu\nu}$ tends to flatness as the Euclidean time at infinity τ tends to ∞. In supergravity, the path integral is $\exp(-I_B/\hbar)$, where I_B is the classical Euclidean action corresponding to these boundary data. Initial configurations without and with initial black-hole data are sketched in Figs. 9.6, 9.7. In Fig. 9.7, one can describe

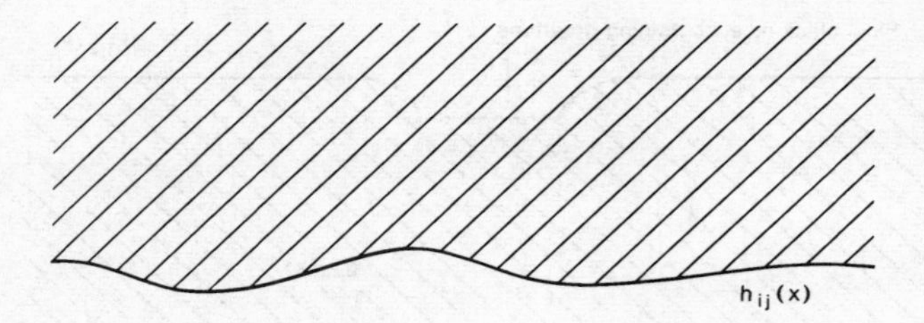

Fig. 9.6. A boundary configuration which does not correspond to a black hole.

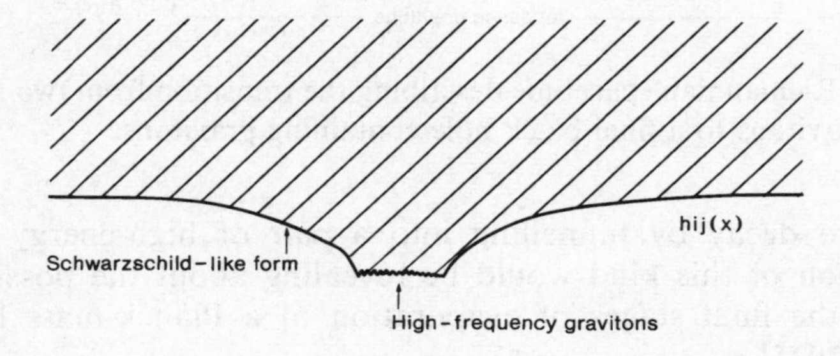

Fig. 9.7. A boundary configuration corresponding to a black hole.

the black hole by a Schwarzschild-like region at moderate distance from its centre, with (say) a region near the centre containing high-frequency gravitons, which in the four-geometry $g_{\mu\nu}$ will act as a source for the larger-scale Schwarzschild field. The high-frequency region and its source effect can be treated by the Isaacson approximation [Isaacson 1968a,b, Misner *et al.* 1973]. If one were working with a higher-N supergravity theory, then one could use lower-spin bosonic fields as the source, which should be interpreted as the matter which has already fallen into the black hole in the course of the black hole's formation.

More complicated configurations involve the Euclidean amplitude for black hole formation by (say) a pair of high-energy gravitons (Fig. 9.8). By a perturbative classical calculation, using the ideas of [D'Eath 1978, D'Eath & Payne 1992a,b,c, D'Eath 1996], it may be possible to find the action I_B and hence the wave function $\exp(-I_B/\hbar)$. The corresponding Lorentzian solution is the collision of two 'black holes' (as formed around each incoming graviton), which leads classically to formation of a larger black hole plus gravitational radiation. This process, with two gravitons colliding to form a black hole, is of interest in the study of quantum processes at Planckian energies, where gravity dominates over other forces ['t Hooft 1987]. (If one boosts a field of spin s by a large Lorentz factor γ, the largest effect is on spin 2.) Turning Fig. 9.8 upside down, one has the Euclidean amplitude for a black hole filled with (say) Planckian-energy

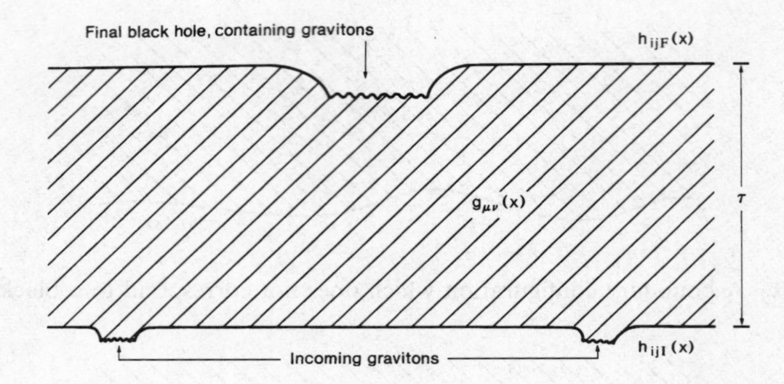

Fig. 9.8. A Riemannian spacetime describing the transition from two high-energy incoming gravitons to a final black hole containing gravitons.

gravitons to decay by tunnelling into a pair of high-energy gravitons. A calculation of this kind would be revealing about the possible mechanism for the final stages of evaporation of a Planck-mass black hole [Hawking 1975].

Black holes in thermal equilibrium are described by the partition function [Hawking 1979] $Z(\beta) = \exp(-I_B/\hbar)$, where I_B is the Euclidean action of the classical solution with period β in imaginary time τ. Here $Z(\beta)$ is given by a Euclidean path integral over all Riemannian geometries, subject to this periodicity requirement. The classical solution is the Schwarzschild solution with mass $M = \beta/8\pi$ and action $I_B = \beta^2/16\pi$. The resulting entropy is $4\pi M^2 = \frac{1}{4}A$, where A is the horizon area, without any quantum corrections.

References

Abramowitz, N. and Stegun, A.I. (1965). *Handbook of Mathematical Functions* (Dover, New York).

Allen, B. (1985). Vacuum states in de Sitter space. *Phys. Rev.* **D32**, 3136–49.

Allen, B. and Lütken, C.A. (1987). Two-point functions in de Sitter space. Tufts University preprint.

Alty, L.J., D'Eath, P.D. and Dowker, H.F. (1992). Quantum wormhole states and local supersymmetry. *Phys. Rev.* **D46**, 4402–12.

Arnowitt, R., Deser, S. and Misner, C.W. (1962). The dynamics of general relativity, in *Gravitation: An Introduction to Current Research*, ed. L. Witten (John Wiley and Sons, Inc., New York).

Asano, M., Tanimoto, M. and Yoshino, N. (1993). Supersymmetric Bianchi class A models. *Phys. Lett.* **B314**, 303–7.

Ashtekar, A. (1986). New variables for classical and quantum gravity. *Phys. Rev. Lett.* **57**, 2244–7.

Ashtekar, A. (1987a). New Hamiltonian form of general relativity. *Phys. Rev.* **D35**, 1587–1602.

Ashtekar, A. (1987b). *Asymptotic Quantization* (Bibliopolis, Naples).

Ashtekar, A. (1988). *New Perspectives in Canonical Gravity* (Bibliopolis, Naples).

Ashtekar, A. (1991). *Lectures on Non-Perturbative Canonical Gravity* (World Scientific, Singapore).

Ashtekar, A. and Pullin, J. (1990). *Ann. Israel Phys. Soc.* **9**, 66–79.

Atiyah, M.F. and Hitchin, N.J. (1985). Low energy scattering of non-abelian monopoles. *Phys. Lett.* **A107**, 21–5.

Atiyah, M.F., Patodi, V.K. and Singer, I.M. (1975). Spectral asymmetry and Riemannian geometry. I. *Math. Proc. Camb. Philos. Soc.* **77**, 43–70.

Balian, R. and Bloch, C. (1971). Asymptotic evaluation of the Green's function for large quantum numbers. *Ann. Phys. (N.Y.)* **63**, 592–606.

Balian, R. and Bloch, C. (1974). Solution of the Schrödinger equation in terms of classical paths. *Ann. Phys. (N.Y.)* **85**, 514–45.

Bao, D., Choquet-Bruhat, Y., Isenberg, J. and Yasskin, P.B. (1985). The well-posedness of ($N = 1$) classical supergravity. *J. Math. Phys.* **26**, 329–33.

Belinskii, V.A., Gibbons, G.W., Page, D.N. and Pope, C.N. (1978). Asymptotically Euclidean Bianchi-IX metrics in quantum gravity. *Phys. Lett.* **B76**, 433–5.

Berezin, F.A. (1966). *The Method of Second Quantization* (Academic Press, New York).

Berezin, F.A. and Marinov, M.S. (1977). Particle spin dynamics as the Grassmann variant of classical mechanics. *Ann. Phys. (N.Y.)* **104**, 336–62.

Birrell, N.D. and Davies, P.C.W. (1982). *Quantum Fields in Curved Space* (Cambridge University Press, Cambridge).

Brézin, E., Le Guillou, J.C. and Zinn-Justin, J. (1977a). Perturbation theory at large order. I. The ϕ^{2n} interaction. *Phys. Rev.* **D15**, 1544–57.

Brézin, E., Le Guillou, J.C. and Zinn-Justin, J. (1977b). Perturbation theory at large order. II. Role of the vacuum instability. *Phys. Rev.* **D15**, 1558–64.

Capovilla, R. and Guven, J. (1994). Super-minisuperspace and new variables. *Class. Quant. Grav.* **11**, 1961–70.

Capovilla, R. and Obregón, O. (1994). No quantum mini-superspace with $\Lambda \neq 0$. *Phys. Rev.* **D49**, 6562–5.

Carroll, S.M., Freedman, D.Z., Ortiz, M.E. and Page, D.N. (1994). Physical states in canonically quantized supergravity. *Nucl. Phys.* **B433**, 661–85.

Casalbuoni, R. (1976). On the quantization of systems with anticommuting variables. *Nuovo Cimento* **33A**, 115–25.

Cheng, A.D.Y. and D'Eath, P.D. (1995). Diagonal quantum Bianchi type IX models in $N = 1$ supergravity. Forthcoming.

Cheng, A.D.Y. and D'Eath, P.D. (1996). The Hartle–Hawking state in quantum supergravity. Forthcoming.

Cheng, A.D.Y., D'Eath, P.D. and Moniz, P. (1994). Quantization of the Bianchi-IX model in supergravity with a cosmological constant. *Phys. Rev.* **D49**, 5246–51.

Cheng, A.D.Y., D'Eath, P.D. and Moniz, P. (1995a). Quantization of a Friedmann-Robertson-Walker model in $N = 1$ supergravity with gauged supermatter. *Class. Quant. Grav.* **12**, 1343–53.

Cheng, A.D.Y., D'Eath, P.D. and Moniz, P. (1995b). Canonical quantization of $N = 1$ supergravity with supermatter: the general case and a Robertson-Walker model. *Gravitation and Cosmology* **1**, 1–11.

Choquet-Bruhat, Y. and Christodoulou, D. (1981). Elliptic systems in $H_{s,\delta}$ spaces on manifolds which are Euclidean at infinity. *Acta Mathematica* **146**, 129–50.

Choquet-Bruhat, Y. and Deser, S. (1973). On the stability of flat space. *Ann. Phys. (N.Y.)* **81**, 165–78.

Christodoulakis, T. and Papadopoulos, C.G. (1988). Quantization of Robertson-Walker geometry coupled to a spin-3/2 field. *Phys. Rev.* **D38**, 1063–8.

Christodoulakis, T. and Zanelli, J. (1984a). Quantization of Robertson-Walker geometry coupled to fermionic matter. *Phys. Rev.* **D29**, 2738–45.

Christodoulakis, T. and Zanelli, J. (1984b). Quantum mechanics of the Robertson-Walker geometry. *Phys. Lett.* **A102**, 2227–30.

Courant, R. and Hilbert, D. (1953). *Methods of Mathematical Physics*, vol.1 (Interscience, New York).

Csordás, A. and Graham, R. (1995). Nontrivial fermion states in supersymmetric minisuperspace. Forthcoming.

Das, A., Fischler, M. and Roček, M. (1977). Massive, self-interacting scalar multiplet coupled to supergravity. *Phys. Lett.* **B69**, 186–8.

D'Eath, P.D. (1976). On the existence of perturbed Robertson-Walker universes. *Ann. Phys. (N.Y.)* **98**, 237–63.

D'Eath, P.D. (1978). High-speed black hole encounters and gravitational radiation. *Phys. Rev.* **D18**, 990–1019.

D'Eath, P.D. (1981). Perturbation methods in quantum gravity. The multiple-scattering expansion. *Phys. Rev.* **D24**, 811–28.

D'Eath, P.D. (1984). Canonical quantization of supergravity. *Phys. Rev.* **D29**, 2199–2219.

D'Eath, P.D. (1986a). Boundary counterterms in supergravity. *Nucl. Phys.* **B269**, 665–90.

D'Eath, P.D. (1986b). Surface counterterms in supergravity, in *Supersymmetry and its Applications*, ed. G.W. Gibbons, S.W. Hawking, and P.K. Townsend (Cambridge University Press, London).

D'Eath, P.D. (1993) Quantization of the Bianchi-IX model in supergravity. *Phys. Rev.* **D48**, 713–18.

D'Eath, P.D. (1994). Quantization of the supersymmmetric Bianchi-I model with a cosmological constant. *Phys. Lett.* **B320**, 12–15.

D'Eath, P.D. (1995). Existence of solutions of the classical supersymmetry constraints. Forthcoming.

D'Eath, P.D. (1996). *Black Holes: Gravitational Interactions* (Oxford University Press, Oxford).

D'Eath, P.D., Dowker, H.F. and Hughes, D.I. (1991). Supersymmetric quantum wormhole states, in *Quantum Gravity*, ed. M.A. Markov, V.A. Berezin and V.P.Frolov (World Scientific, Singapore).

D'Eath, P.D. and Halliwell, J.J. (1987). Fermions in quantum cosmology. *Phys. Rev.* **D35**, 1100–23.

D'Eath, P. D., Hawking, S.W. and Obregón, O. (1993). Supersymmetric Bianchi models and the square root of the Wheeler-DeWitt equation. *Phys. Lett.* **B300**, 44–8.

D'Eath, P.D. and Hughes, D.I. (1988). Supersymmetric mini-superspace. *Phys. Lett.* **B214**, 498–502.

D'Eath, P.D. and Hughes, D.I. (1992). Mini-superspace with local supersymmetry. *Nucl. Phys.* **B378**, 381–409.

D'Eath, P.D. and Payne, P.N. (1992a). Gravitational radiation in black hole collisions at the speed of light: I. Perturbation treatments of the axisymmetric collision. *Phys. Rev.* **D46**, 658–74.

D'Eath, P.D. and Payne, P.N. (1992b). Gravitational radiation in black hole collisions at the speed of light: II. Reduction in two independent variables and calculation of the second-order news function. *Phys. Rev.* **D46**, 675–93.

D'Eath, P.D. and Payne, P.N. (1992c). Gravitational radiation in black hole collisions at the speed of light: III. Results and conclusions. *Phys. Rev.* **D46**, 694–701.

D'Eath, P.D. and Wulf, M. (1995). Explicit loop calculations in N=1 supergravity. Forthcoming.

Deser, S., Kay, J.H. and Stelle, K.S. (1977a). Hamiltonian formulation of supergravity. *Phys. Rev.* **16**, 2448–55.

Deser, S., Kay, J.H. and Stelle, K.S. (1977b). Renormalizability properties of supergravity. *Phys. Lett.* **B38**, 527–30.

Deser, S. and Zumino, B. (1976). Consistent supergravity. *Phys. Lett.* **B62**, 335–8.

DeWitt, B.S. (1967). Quantum theory of gravity 1. The canonical theory. *Phys. Rev.* **160**, 1113–48.

DeWitt, B.S. (1984a). The spacetime approach to quantum field theory, in *Relativity, Groups, and Topology II*, ed. B.S. DeWitt and R. Stora (North-Holland, Amsterdam).

DeWitt, B.S. (1984b). *Supermanifolds* (Cambridge University Press, Cambridge).

DeWitt-Morette, C.M., Maheshwari, A. and Nelson, B. (1979). Path integration in non-relativistic quantum mechanics. *Phys. Rep.* **50C**, 255–372.

Dieudonné, J. (1960). *Foundations of Modern Analysis* (Academic Press, New York).

Dirac, P.A.M. (1933). The Lagrangian in quantum mechanics. *Physikalische Zeitschrift der Sowjetunion,* Band 3, Heft 1, 64–72.

Dirac, P.A.M. (1950). Generalized Hamiltonian dynamics. *Canad. J. Math.* **2**, 129–48.

Dirac, P.A.M. (1958a). Generalized Hamiltonian dynamics. *Proc. Roy. Soc. London* **A246**, 326–32.

Dirac, P.A.M. (1958b). The theory of gravitation in Hamiltonian form. *Proc. Roy. Soc. London* **A246**, 333–43.

Dirac, P.A.M. (1959). Fixation of coordinates in the Hamiltonian theory of gravitation. *Phys. Rev.* **114**, 924–30.

Dirac, P.A.M. (1965). *Lectures on Quantum Mechanics* (Academic Press, New York).

Duff, M. (1982). Ultraviolet divergences in extended supergravity, in *Supergravity '81*, ed. S. Ferrara and J.G. Taylor (Cambridge University Press, Cambridge).

Eguchi, T., Gilkey, P.B. and Hanson, A.J. (1980). Gravitation, gauge theories and differential geometry. *Phys. Rep.* **66**, 214–393.

Erdélyi, A., Magnus, W., Oberhettinger, F. and Tricomi, F.G. (1954). *Integral Transforms* (McGraw-Hill, New York).

Esposito, G. *et al.* (1995). One-loop amplitudes in Euclidean quantum gravity. Forthcoming.

Faddeev, L.D. (1976). Introduction to functional methods, in *Methods in Field Theory*, ed. R. Balian and J. Zinn-Justin (North-Holland, Amsterdam).

Faddeev, L.D. and Slavnov, A.A. (1980). *Gauge Fields: Introduction to Quantum Theory* (Benjamin Cummings, Reading, MA).

Feynman, R.P. and Hibbs, A.R. (1965). *Quantum Mechanics and Path Integrals* (McGraw-Hill, New York).

Fradkin, E.S. and Vasiliev, M.A. (1977). Hamiltonian formalism, quantization and S matrix for supergravity. *Phys. Lett.* **B72**, 70–4.

Fradkin, E.S. and Vilkovisky, G. (1977). Preprint TH.2332-CERN (unpublished).

Garabedian, P.R. (1964). *Partial Differential Equations* (Wiley, New York).

Gerlach, U.H. and Sengupta, U.K. (1978). Homogeneous collapsing star: Tensor and vector harmonics for matter and field asymmetries. *Phys. Rev.* **D18**, 1773–97.

Gibbons, G.W. (1979). Quantum field theory in curved spacetime, in *General Relativity*, ed. S.W. Hawking and W. Israel (Cambridge University Press, Cambridge).

Gibbons, G.W. and Hawking, S.W. (1977). Cosmological event horizons, thermodynamics and particle creation. *Phys. Rev.* **D15**, 2738–51.

Gibbons, G.W., Hawking, S.W. and Perry, M.J. (1978). Path integrals and the indefiniteness of the gravitational action. *Nucl. Phys.* **B138**, 141–50.

Gibbons, G.W. and Pope, C.N. (1979). The positive action conjecture and asymptotically Euclidean metrics in quantum gravity. *Commun. Math. Phys.* **66**, 267–90.

Giddings, S. and Strominger, A. (1989). Axion-induced topology change in quantum gravity and string theory. *Nucl. Phys.* **B30**, 890–907.

Goroff, M.H. and Sagnotti, A. (1985). Quantum gravity at two loops. *Phys. Lett.* **B160**, 81–5.

Graham, R. and Luckock, H. (1994). The Hartle-Hawking state for the Bianchi-type IX model in supergravity. *Phys. Rev.* **D49**, R4981–4.

Green, M.B., Schwarz, J.H. and Witten, E. (1987). *Superstring Theory*, vol.1 (Cambridge University Press, Cambridge).

Grishchuk, L.P. (1974). Amplification of gravitational waves in an isotropic universe. *Zh. Eksp. Teor. Fiz.* **67**, 825–38.

Grishchuk, L.P. (1977). Graviton creation in the early universe. *Ann. N.Y. Acad. Sci.* **302**, 439–44.

Halliwell, J.J. and Hawking, S.W. (1985). The origin of structure in the universe. *Phys. Rev.* **D31**, 1777–91.

Halliwell, J.J. and Louko, J. (1989a). Steepest-descent contours in the path-integral approach to quantum cosmology. I. The de Sitter minisuperspace model. *Phys. Rev.* **D39**, 2206–15.

Halliwell, J.J. and Louko, J. (1989b). Steepest-descent contours in the path-integral approach to quantum cosmology. II. Microsuperspace. *Phys. Rev.* **D40**, 1868–75.

Hanson, A., Regge, T. and Teitelboim, C. (1976). *Constrained Hamiltonian Systems* (Accademia Nazionale dei Lincei, Rome).

Hartle, J.B. (1986). Quantum cosmology, in *High Energy Physics 1985*, vol. 2, ed. M.J. Bowick and F. Gürsey (World Scientific, Singapore).

Hartle, J.B. and Hawking, S.W. (1983). Wave function of the universe. *Phys. Rev.* **D28**, 2960–75.

Hawking, S.W. (1975). Particle creation by black holes. *Commun. Math. Phys.* **43**, 199–220.

Hawking, S.W. (1977). Zeta function regularisation of path integrals in curved space-time. *Commun. Math. Phys.* **56**, 133–48.

Hawking, S.W. (1979). The path-integral approach to quantum gravity, in *General Relativity. An Einstein Centenary Survey*, ed. S.W. Hawking and W. Israel (Cambridge University Press, London).

Hawking, S.W. (1982). In *Astrophysical Cosmology*, ed. H.A. Bruck *et al.* (Pontificia Academiae Scientarium, Vatican City) **48**, 563–80.

Hawking, S.W. (1984). The quantum state of the universe. *Nucl. Phys.* **B239**, 257–76.

Hawking, S.W. (1987). Quantum cosmology, in *Three Hundred Years of Gravitation*, ed. S.W. Hawking and W. Israel (Cambridge University Press, Cambridge), 631–51.

Hawking, S.W. (1988). Wormholes in space-time. *Phys. Rev.* **D37**, 904–10.

Hawking, S.W. (1995). *Nature of space and time.* (Princeton University Press, Princeton).

Hawking, S.W. and Ellis, G.F.R. (1973). *The Large Scale Structure of Space-Time* (Cambridge University Press, Cambridge).

Hawking, S.W. and Page, D.N. (1986). Operator ordering and the flatness of the universe. *Nucl. Phys.* **B264**, 185–96.

Hawking, S.W. and Page, D.N. (1990). Spectrum of wormholes. *Phys. Rev.* **D42**, 2655–63.

Hawking, S.W. and Wu, Z.-C. (1985). Numerical calculations of minisuperspace cosmological models. *Phys. Lett.* **B151**, 15–20.

Henneaux, M. (1983). Poisson brackets of the constraints in the Hamiltonian formulation of tetrad gravity. *Phys. Rev.***D27**, 986–9.

Henneaux, M. and Teitelboim, C. (1982). Relativistic quantum mechanics of supersymmetric particles. *Ann. Phys. (N.Y.)* **143**, 127–59.

Higgs, P.W. (1958). Integration of secondary constraints in quantized general relativity. *Phys. Rev. Lett.* **1**, 373–4.

Higgs, P.W. (1959). Integration of secondary constraints in quantized general relativity. Errata. *Phys. Rev. Lett.* **3**, 66–7.

Hosoya, A. and Ogura, W. (1989). Wormhole instanton solution in the Einstein-Yang-Mills system. *Phys. Lett.* **B225**, 117–20.

Hughes, D.I. (1986). Bosonic states in supersymmetric Bianchi models. Unpublished paper.

Hughes, D.I. (1990a). Supersymmetric quantum cosmology. Ph.D. thesis, University of Cambridge. Unpublished.

Hughes, D.I. (1990b). Symbolic computation with fermions. *J. Symbolic Computation* **10**, 657–64.

Isaacson, R.A. (1968a). Gravitational radiation in the limit of high frequency, I: The linear approximation and geometrical optics. *Phys. Rev.* **166**, 1263–71.

Isaacson, R.A. (1968b). Gravitational radiation in the limit of high frequency, II: Non linear terms and the effective stress tensor. *Phys Rev.* **166**, 1272–80.

Isham, C.J. and Nelson, J.E. (1974). Quantization of a coupled Fermi field and Robertson-Walker metric. *Phys. Rev.* **10**, 3226–34.

Itzykson, C. and Zuber, J.-B. (1980). *Quantum Field Theory* (McGraw Hill, New York).

Jacobson, T. (1988). New variables for canonical supergravity. *Class. Quantum Grav.* **5**, 923–35.

Jacobson, T. and Smolin, L. (1988). Nonperturbative quantum geometries. *Nucl. Phys.* **B299**, 295-345.

Kasner, E. (1921). Geometrical theorems on Einstein's cosmological equations. *Am. J. Math.* **43**, 217–21.

Kodama, H. (1990). Holomorphic wave function of the universe. *Phys. Rev.* **D42**, 2548–65.

Kuchař, K. (1970). Ground state functional of the linearized gravitational field. *J. Math. Phys.* **11**, 3322–34.

Kuchař, K. (1981). Canonical methods of quantization, in *Quantum Gravity 2*, ed. C.J. Isham, R. Penrose, and D.W. Sciama (Oxford University Press, Oxford).

Ladyzhenskaya, O.A. and Uraltseva, N.N. (1968). *Linear and Quasi-linear Equations of Elliptic Type* (Academic Press, New York).

Lifschitz, E.M. and Khalatnikov, I.M. (1963). Investigations in relativistic cosmology. *Adv. Phys.* **12**, 185–249.

Linde, A. (1987). Inflation and quantum cosmology. *Three Hundred Years of Gravitation*, eds. S.W. Hawking and W. Israel (Cambridge University Press, Cambridge).

Lipatov, L.N. (1977). Divergence of the perturbation-theory series and the quasi-classical theory. *JETP* **45**, 216–23.

Louko, J. (1995). Chern–Simons functional and the no-boundary problem in Bianchi-IX quantum cosmology. *Phys. Rev.* **D51**, 586–90.

Lyons, A. (1989). Fermions in wormholes. *Nucl. Phys.* **B324**, 253–75.

Lyons, G. (1992). Complex solutions for the scalar field model of the universe. *Phys. Rev.* **D46**, 1546–50.

Macías, A., Obregón, O. and Ryan, M.P. (1987). Quantum cosmology: the supersymmetric square root. *Class. Quant. Grav.* **4**, 1477–86.

Matschull, H.J. (1994). About loop states in supergravity. *Class. Quant. Grav.* **11**, 2395–2410.

Misner, C.W. (1972). Minisuperspace, in *Magic Without Magic*, ed. Klauder, J.R. (Freeman, San Francisco).

Misner, C.W., Thorne, K.S. and Wheeler, J.A. (1973). *Gravitation* (Freeman, San Francisco).

Moncrief, V. and Ryan, M.P. (1991). Amplitude-real-phase exact solutions for quantum mixmaster universes. *Phys. Rev.* **D44**, 2375–79.

Moss, I.G. and Poletti, S.J. (1994). Conformal anomalies on Einstein spaces with boundary. *Phys. Lett.* **B333**, 326–30.

Nelson, J.E. and Teitelboim, C. (1978). Hamiltonian formulation of the theory of interacting gravitational and electron fields. *Ann. Phys. (N.Y.)* **116**, 86–104.

Obregón, O., Pullin, J. and Ryan, M.P. (1993). Bianchi cosmologies: new variables and a hidden supersymmetry. *Phys. Rev.* **D48**, 5642–7.

Olver, F.W.J. (1974). *Asymptotics and Special Functions* (Academic, New York).

Penrose, R. and Rindler, W. (1984). *Spinors and Space-Time*, vol. 1 (Cambridge University Press, Cambridge).

Pilati, M. (1978). The canonical formulation of supergravity. *Nucl. Phys.* **B132**, 138–54.

Regge, T. and Teitelboim, C. (1974). Role of surface integrals in the Hamiltonian formulation of general relativity. *Ann. Phys. (N.Y.)* **88**, 285–318.

Reula, O. (1982). Existence theorem for solutions of Witten's equation and nonnegativity of total mass. *J. Math. Phys.* **23**, 810–14.

Reula, O. (1987). A configuration space for quantum gravity and solutions to the Euclidean Einstein equations in a slab region. Max-Planck-Institut für Astrophysik, preprint **MPA** 275.

Rey, S.J. (1990). Space-time wormholes with Yang-Mills fields. *Nucl. Phys.* **B336**, 146–56.

Roček, M. (1981). An introduction to superspace and supergravity, in *Superspace and Supergravity*, ed. S.W. Hawking and M. Roček (Cambridge University Press, Cambridge).

Rubakov, V.A., Sazhin, M.V. and Veryaskin, A.V. (1982). Graviton creation in the inflationary universe and the grand unification scale. *Phys. Lett.* **B115**, 189–92.

Ryan, M.P. (1972). *Hamiltonian Cosmology* (Springer, Heidelberg).

Ryan, M.P. and Shepley, L.C. (1975). *Homogeneous Relativistic Cosmologies* (Princeton University Press, Princeton).

Salopek, D.S. and Stewart, J.M. (1992). Hamilton-Jacobi theory for general relativity with matter fields. *Class. Quant. Grav.* **9**, 1943–67.

Salopek, D., Stewart, J.M. and Parry, J. (1993). The semi-classical Wheeler-DeWitt Equation: solutions for long-wavelength fields. *Phys. Rev.* **D48**, 719–27.

Sano, T. (1992). The Ashtekar formalism and WKB wave functions of $N = 1, 2$ supergravities. University of Tokyo preprint **UT-621**.

Sano, T. and Shiraishi, J. (1993). The non-perturbative canonical quantization of the N=1 supergravity. *Nucl. Phys.* **B410**, 423–47.

Schiff, L.I. (1968). *Quantum Mechanics* (McGraw-Hill, New York).

Schoen, R. and Yau, S.-T. (1979). Proof of the positive-action conjecture in quantum relativity. *Phys. Rev. Lett.* **42**, 547–8.

Simon, B. (1974). *The $P(\phi)_2$ Euclidean (Quantum) Field Theory* (Princeton University Press, Princeton).

Smoot, G.F. *et al.* (1992). A measurement of the cosmic microwave background temperature at 7.5 GHz. *Astrophys. J. Lett.* **396**, L3–9.

Starobinsky, A. (1980). A new type of isotropic cosmological model without singularity. *Phys. Lett.* **B91**, 99–102.

Teitelboim, C. (1977a). Supergravity and square roots of constraints. *Phys. Rev. Lett.* **38**, 1106–10.

Teitelboim, C. (1977b). Surface integrals as symmetry generators in supergravity theory. *Phys. Lett.* **B69**, 240–4.

Teitelboim, C. (1980). The Hamiltonian structure of space-time, in *General Relativity and Gravitation*, vol. 1, ed. A. Held (Plenum, New York).

't Hooft, G. (1987). Graviton dominance in ultra-high-energy scattering. *Phys. Lett.* **B198**, 61–3.

Townsend, P.K. (1977). Cosmological constant in supergravity. *Phys. Rev.* **D15**, 2802–4.

van Nieuwenhuizen, P. (1981). Supergravity. *Phys. Rep.* **68**, 189–398.

Verbin, Y. and Davidson, A. (1989). Quantized non-abelian wormholes. *Phys. Lett.* **B229**, 364–7.

Verlinde, H. and Verlinde, E. (1992). Scattering at planckian energies. *Nucl. Phys.* **B371**, 246–68.

Vilkovisky, G. (1984). The gospel according to DeWitt, in *Quantum Theory of Gravity*, ed. S.M. Christensen (Adam Hilger, Bristol).

Wald, R.M. (1984). *General Relativity* (University of Chicago Press, Chicago).

Wess, J. and Bagger, J. (1992). *Supersymmetry and Supergravity* (Princeton University Press, Princeton).

Wheeler, J.A. (1964). Geometrodynamics and the issue of the final state, in *Relativity Groups and Topology 1963*, ed. C.M. DeWitt and B.S. DeWitt (Gordon and Breach, New York).

Wheeler, J.A. (1968). Superspace and the nature of quantum geometrodynamics, in *Battelle Rencontres 1967*, ed. C.M. DeWitt and J.A. Wheeler (Benjamin, New York).

White, M., Scott, D. and Silk, J. (1994). Anisotropies in the cosmic microwave background. *Ann. Rev. Astron. Astrophys.* **32**,319–70.

Witten, E. (1981). A new proof of the positive energy theorem. *Commun. Math. Phys.* **80**, 381–402.

Witten, E. and Bagger, J. (1982). Quantization of Newton's constant in certain supergravity theories. *Phys. Lett.* **B115**, 202–6.

Wulf, M. (1995). Investigations concerning the algebra of constraints in $N = 1$ supergravity. Unpublished diploma thesis, University of Hamburg.

Index

adiabatic approximation 61, 81–3
Ashtekar variables 190–209
  algebra of constraints (general relativity) 196–7
  Hamiltonian form (general relativity) 193–5
  Lagrangian form (general relativity) 192, 193
  quantum constraints (general relativity) 197
  reality conditions (general relativity) 195, 196
  supergravity 198–209: algebra of constraints 202; Chern–Simons state 204–8; cosmological constant 203; Hamiltonian form 198–200; Lagrangian form 198; quantum constraints 202, 203; reality conditions 200–202

Borel summation 237
boundary-value problem, well-posed 28–30

classical constraints 25, 114
Codazzi's equation 24
conformally invariant wave equation in four dimensions 46
conformal transformation 43, 44, 46

de Sitter spacetime 53–5, 58, 83, 84, 171
  temperature 63, 85
DeWitt metric 39, 40

Einstein–Dirac theory
  action 65, 66, 74, 75, 81
  creation and annihilation operators 77, 78
  Dirac brackets 68, 70, 71
  Dirac equation 75
  Hamiltonian form 67–79
  particle creation 83
  particle detector 84–5
  spectral boundary conditions 72–85
Euclidean path integral 11–14, 41–85

Feynman path integral, non-relativistic quantum mechanics 31–7
  Euclidean action 37
  Euclidean path integral 37

Gauss' equation 23
general relativity
  Euclidean action 10, 41–5, 49–58, 79–83
  Euclidean amplitude 41–4
  Hamiltonian form 9–30, 38–85
  Lorentzian action 27

Hamilton–Jacobi equation 168, 169
Hartle–Hawking state 11–14, 42–85, 142, 143, 154–6, 169–71
  semi-classical expansion 50–62

inner product 50

mass 97, 109, 133–7, 211, 234, 235

mini-superspace models 11–14, 45–62, 141–71
  Bianchi models 143, 162–70
  Friedmann model 141–62, 170, 171: quantization 152–62; reduction of four-dimensional supergravity 144–52; with Λ-term 170–1; with scalar-spin-1/2 matter 156–62
momentum constraints 10, 39–85

parametrized particle dynamics 14–17
perturbations of a Friedmann universe
  bosonic 13, 58–62
  spin-1/2 13, 14, 62–85

quantum wormhole state 142, 143, 156, 173–89
  ground state 178, 180, 186–9
  supergravity–supermatter model 181–9

Schrödinger equation 35, 36, 39, 60, 78, 79
second fundamental form 20–8, 41–6, 66, 69
spinors, two-component 63–5
supergravity
  action 90, 107–14, 118–21, 137, 210–14, 218, 219, 221–3, 230, 235–40
  algebra of constraints 96, 106
  auxiliary fields 131–3, 210–31
  boundary terms at infinity 108–10
  black-hole evaporation 239, 240
  black-hole initial data 238, 239
  black hole in thermal equilibrium 240
  canonical momenta 92–4: quantum representation 97–101
  canonical quantization 87, 88, 97–107
  classical boundary-value problem 113–18
  classical constraints 94–6
  in cosmology, with spectral boundary conditions 220–3: exactly semi-classical Hartle–Hawking state 223, 238
  Dirac brackets 93–4
  disconnected bounding three-geometries 232
  finiteness 140, 203, 210–31, 236–7: higher-N supergravity 237, 238; two-loop order with boundaries 121–31
  Hamiltonian form 86–111
  Hartle–Hawking state 220–3, 232, 238
  heat equation 133–5, 211, 212
  infinitesimal coordinate transformation 91
  inner product 99–103
  local boundary conditions 210–14, 223–41
  local Lorentz transformation 91, 104
  local supersymmetry transformation 91, 105, 110, 111
  multiple-scattering expansion 213
  path integral 107–11, 112: semi-classical expansion 118, 119, 235–7
  quantum amplitude 112–40, 210–14, 218, 219, 229–30: exact semi-classical form 212, 213, 218, 219, 222, 223
  spectral boundary conditions 214–19: time evolution 133–40
  with supermatter 223–31
  quantum constraints 103–7, 110–11, 119, 141, 143, 153, 154, 160, 161, 165, 167, 181– 5, 212, 213, 218, 219, 224
  torsion 89, 90

Wheeler–DeWitt equation 9, 10, 12, 13, 39–85, 107, 141, 143, 153, 154, 162, 164, 165, 178, 179, 183
wormholes 172–7, 233–4
  effective particle interactions 176, 177, 189